Exploring Natural Hazards

A Case Study Approach

Exploring Natural Hazards

A Case Study Approach

Edited by
Darius Bartlett
Ramesh P. Singh

CRC Press
Taylor & Francis Group
Boca Raton London New York

CRC Press is an imprint of the
Taylor & Francis Group, an **informa** business

A CHAPMAN & HALL BOOK

CRC Press
Taylor & Francis Group
6000 Broken Sound Parkway NW, Suite 300
Boca Raton, FL 33487-2742

First issued in paperback 2020

© 2018 by Taylor & Francis Group, LLC
CRC Press is an imprint of Taylor & Francis Group, an Informa business

No claim to original U.S. Government works

ISBN-13: 978-0-367-57192-4 (pbk)
ISBN-13: 978-1-138-05442-4 (hbk)

Library of Congress Cataloging-in-Publication Data

Names: Bartlett, Darius J., 1955- author. | Singh, R. P. (Ramesh P.), author.
Title: Exploring natural hazards : a case study approach / Darius Bartlett and Ramesh Singh.
Description: Boca Raton, FL : CRC Press, 2018.
Identifiers: LCCN 2017045766 | ISBN 9781138054424 (hardback : alk. paper)
Subjects: LCSH: Natural disasters. | Hazard mitigation.
Classification: LCC GB5014 .B376 2018 | DDC 363.34--dc23
LC record available at https://lccn.loc.gov/2017045766

Visit the Taylor & Francis Web site at
http://www.taylorandfrancis.com

and the CRC Press Web site at
http://www.crcpress.com

Contents

Foreword

The Sendai Framework for Disaster Risk Reduction 2015–2030 has identified four priority areas for Disaster Risk Reduction: understanding disaster risk; strengthening disaster risk governance to manage disaster risk; investing in disaster risk reduction for resilience and enhancing disaster preparedness for effective response; and to "Build Back Better" in recovery, rehabilitation and reconstruction. This excellent book, *Exploring Natural Hazards: A Case Study Approach*, which was edited by Darius Bartlett and Ramesh P. Singh, directly contributes to the knowledge that is needed as a basis for these actions.

This book provides a case study approach across 13 chapters addressing issues related to tropical cyclones and typhoons, desertification, floods, lightning as a hazard and the need for alert systems, and ends with an analysis of natural hazards in Poland, as an example. These weather–climate related disastrous events have a dominant impact on global societies. These cases provide the reader with a broad range of issues and approaches which are important for providing the improved understanding to enable the prediction of the occurrence of these hazardous events. These predictions need to be both for the near term, to alert populations who can then take actions to reduce their exposure, and for the longer term, to enable broader strategies for reducing risks. The chapter 'Alert Systems in Natural Hazards: Best Practices' addresses these issues.

The international Integrated Research on Disaster Risk (IRDR) Programme, co-sponsored by the International Council for Science, is guided by research objectives including the characterisation of hazards, vulnerability and risk, addressing the gaps in knowledge, methodologies and information that are preventing the effective application of science to averting disasters and reducing risk; and reducing risk and curbing losses through knowledge-based actions. The legacy of the IRDR programme 'would be an enhanced capacity around the world to address hazards and make informed decisions on actions to reduce their impacts'. This book provides the information for moving ahead on the research program and directly contributes to the Sendai Framework and hence on reducing risks and impacts on planet Earth.

Gordon McBean, C.M., O.Ont, PhD, FRSC, FAGU, FIUGG
Professor and President, International Council for Science
Institute for Catastrophic Loss Reduction
Professor Emeritus, Department of Geography, Western University
London, Canada

Preface

The ancient Greeks believed that the world was made up of four types of matter or 'elements': earth, air, fire and water. Similar philosophies were advanced in India, Babylonia and elsewhere. While modern science has greatly expanded our knowledge of how the world is constructed, and has discarded this fourfold division in favour of a much richer and more complex understanding of the nature of matter, echoes of these four traditional classical elements may be found in the present-day concepts of the geosphere, the atmosphere, the hydrosphere and the biosphere, the study of which lies at the heart of Earth system science. To these, we may also add the anthroposphere (the part of the environment that is made or modified by humans). This book is about the workings of these 'spheres' and the interactions between them. Specifically, it is about the risks to human society that arise from living on a sometimes very mobile and dynamic planet. These interactions and their potentially adverse consequences come under the heading of natural hazards.

Although tremendous progress has been made in recent decades in understanding the workings of the Earth system and, in particular, its impacts on and responses to human actions, there remains a continuing and pressing need for knowledge that will allow society to simultaneously reduce exposure to global environmental hazards, while also meeting economic development goals. In 2009, the International Council for Science (ICSU) and the International Social Science Council (ISSC) identified five 'Grand Challenges' that they proposed should guide future research in the Earth sciences, and to help reduce human vulnerability to hazard. These were to improve the usefulness of forecasts of future environmental conditions and their consequences for people; develop, enhance and integrate observation systems to manage global and regional environmental change; determine how to anticipate, avoid and manage disruptive global environmental change; determine institutional, economic and behavioural changes to enable effective steps toward global sustainability; and encourage innovation (and mechanisms for evaluation) in technological, policy and social responses to achieve global sustainability.

The chapters in this book, and in our second book, *Natural Hazards: Earthquakes, Volcanoes, and Landslides* (Singh and Bartlett, 2018), address many of the issues raised by these Grand Challenges. In *Natural Hazards: Earthquakes, Volcanoes, and Landslides*, the primary focus is on the processes and dynamics of the geosphere. The present volume is mainly concerned with hazards arising from processes within the hydrosphere and atmosphere, triggered or exacerbated by inputs to and transfers of energy between environmental components. It examines the causes of these phenomena, but also ways in which improved policy making, sometimes coupled with the application of appropriate modern technologies, can help to reduce people's exposure to harm.

We hope this book will have value for practitioners and professionals alike, for researchers, and for students and others who work at the intersection between environmental hazards, sustainable development, and social justice. It could not have come to completion without the assistance and support of many individuals and institutions. In particular, both editors would like to thank Irma Shaglia-Britton and her team at CRC Press/Taylor & Francis, for encouraging us to take on the project, and for continued guidance and support at all stages thereafter. It also goes without saying that we recognize and appreciate with gratitude the work of all the authors who made inputs to the chapters that follow. Without them, there would be no book.

The editors would also like to thank the following referees, for helping to review and select the chapters for inclusion in each of the two books: A.M. Foyle (USA); A.M. Martinez-Granna (Spain); Adrian Grozavu (Romania); Anabel Alejandra Lamaro (Argentina); Ajay Srivastava (India); B.K. Rastogi (India); B.R. Arora (India); C. Wauthier (Belgium); Carolina Garcia (Italy); Devendra Singh (India); E. Sansosti (Italy); Ezatollah Karami (Iran); Fuqiong Huang (China); Gerassimos Papadopoulos (Greece); Gilda Currenti (Italy); Guido Pasquariello (Italy); Hans von Suchodoletz (Germany); James Hower (USA); Jan Altman (Czech Republic); Kenneth Hewitt (Canada); Laxmansingh S. Chamyal (India); Leah Courtland (USA); M. Hayakawa (Japan); Masahiro Chigira (Japan); Matthew Blackett (UK); Minakshi Devi (India); Mukesh Gupta (USA and Spain); Namrata Batra (USA); N.C. Mishra (USA); Nicolas D'Oreye (Luxembourg); Partha Bhattacharjee (USA); Paula Dunbar (USA); Qi Feng (China); Rachel G. Mauk (USA); Rasik Ravindran (India); Ritesh Gautam (USA); S.K. Jain (India); Simona Colombelli (Italy); Sudipta Sarkar (USA); T.A. Przylibski (Poland); T.N. Singh (India); U.S. Panu (Canada); Vikrant Jain (India); and Yoav Me-Bar (Israel).

At an individual level, Darius Bartlett would like to acknowledge the encouragement and support of his former colleagues in the Department of Geography at University College Cork, both prior to his retirement and subsequently; and of course, as always, his wife Mary-Anne. Ramesh Singh acknowledges the help of his wife, Alka Singh, to sacrifice her time to support him during preparation of this book. Ramesh Singh dedicates this book to his late father Rama N. Singh on the tenth anniversary of his death to remember his love, affection and motivation for his efforts in using satellite data to explore Earth systems and natural hazards.

Darius Bartlett
Ramesh P. Singh

REFERENCE

Singh, R. P. and Bartlett, D., editors. (2018). *Natural Hazards: Earthquakes, Volcanoes, and Landslides*. CRC Press/Taylor & Francis: Boca Raton, Florida.

Editors

Darius Bartlett is an earth scientist and geomorphologist by training, with particular specialism in geographic information systems and the application of geo-information technologies to coastal zone science. Now retired from active teaching, he is currently researching sites of possible historical tsunami impacts on the coast of Ireland. He is the co-editor of *Marine and Coastal Geographical Information Systems* (with Dawn Wright, 2000), *GIS for Coastal Zone Management* (with Jennifer Smith, 2005) and *Geoinformatics for Marine and Coastal Management* (with Louis Celliers, 2017), all published with CRC Press/Taylor & Francis.

Ramesh P. Singh has been a professor in the School of Life and Environmental Sciences, Chapman University, Orange, California, since 2009. He graduated from Banaras Hindu University, Varanasi, India. He was a postdoctoral and Alberta Oil Sands Technology and Research Authority (AOSTRA) fellow with the Department of Physics, University of Alberta, Canada (1981–1986). He joined the Department of Civil Engineering, Indian Institute of Technology, Kanpur, India, as a faculty member in 1986 and remained there until 2007.

Dr. Singh joined George Mason University as a distinguished visiting professor from 2003 to 2005 and later as a full professor from 2007 to 2009. He then moved to Chapman University, where he has been since 2009.

Dr. Singh has published more than 200 research papers, and supervised several MTech and PhD students. He was chief editor of the *Indian Journal of Remote Sensing* from 1999 to 2007; chief editor of *Geomatics, Natural Hazards and Risk*; and associate editor of the *International Journal of Remote Sensing*, the latter two of which are published by Taylor & Francis. He has edited books and several special issues of journals.

Dr. Singh is a recipient of the Alexander von Humboldt Fellowship and a JSPS Fellowship. He is the recipient of Indian National Remote Sensing, Indian National Mineral and Hari Om Ashram Prerit awards, as well as a fellow of the Indian Remote Sensing Society and the Indian Geophysical Union. He is a member of the editorial board of Aerosol Air Quality Research.

Dr. Singh has been a member of the International Union of Geodesy and Geophysics Commission on GeoRisk since 2004 and has served as vice president. He is currently (2017–2018) president of the American Geophysical Union Natural Hazards Focus Group. He is also a member of the International Union of Geodesy and Geophysics–Electromagnetic Studies of Earthquakes and Volcanoes Bureau.

Contributors

Jan Franklin Adamowski is an associate professor in the Department of Bioresource Engineering in the Faculty of Agricultural and Environmental Sciences at McGill University. He is also the director of McGill's Integrated Water Resources Management Program and the interim director of McGill's Brace Centre for Water Resources Management. Adamowski came to McGill University in 2009 after working as a post-doctoral associate at the Massachusetts Institute of Technology. He completed his graduate studies in the United States and Europe at several universities (Cambridge, MIT, London Business School, HEC Paris, NHH Bergen and Warsaw Technical University) and his undergraduate degree in Canada at the Royal Military College of Canada. Adamowski's research explores critical engineering, as well as social-economic and management problems in the field of water resources with the aim of developing highly innovative and useful approaches to help reduce the vulnerability, as well as enhancing the resilience, adaptive capacity and sustainability of water resources systems around the world in the face of increasing challenges, uncertainty and climate change variability. In recognition of his contributions to research in hydrology and water resources engineering and management, Adamowski was recently elected as a member of the College of New Scholars, Artists and Scientists of the Royal Society of Canada.

Anteneh Belayneh is a PhD candidate at the School of Public Policy and Administration (SPPA), where he started in 2014. His research interests relate to water resource allocation and management. His areas of expertise are bioresource management and hydrology. His recent work involves studying the economic impact of mining on indigenous communities in the Canadian subarctic and drought forecasting in sub-Saharan Africa. Belayneh holds a master's degree in bioresource engineering and a BSc (honours) in hydrology from Montreal's McGill University. He has presented his work at conferences of the Canadian Society of Civil Engineers, the North American Society of Bioresource Engineering, and Resources and Sustainable Development in the Arctic. In 2012, he interned with the United Nations Convention on Biodiversity, working on economic planning in developing countries. Belayneh is also a member of the Canadian Water Network's Student and Young Professionals Committee.

Sekhar Chandra Dutta is presently serving as a professor in the Department of Civil Engineering at IIT (ISM Dhanbad). A Fulbright-Nehru senior research fellow at the University of California–Davis, his broad areas of research are earthquake engineering, soil–structure interaction, and so on. Professor Dutta has 162 publications to his credit with h-index 16 and i-10 index 24. He has supervised eight PhD scholars and carried out a number of research projects including those under BRNS, CSIR, DST and UGC. He was recognized as a resource person since 2006 under the National Programme for Capacity Building of Engineers in Earthquake Risk Management launched by the government of India, and he has been involved in dissemination of knowledge related to earthquake-resistant design.

T.I. Eldho is an institute chair professor in the Department of Civil Engineering, IIT Bombay. He has 25 years of experience in the area of water resources and environmental engineering as a scientist, teacher and consultant. He works in the areas of coastal hydrodynamics and climate change impact on water resources, CFD, groundwater flow and contaminant transport. He has guided 17 PhDs and 35 masters theses, and has published more than 380 research papers in various international journals and conferences. He is editor/associate editor/editorial board member of five journals. He is also a reviewer for more than 30 international journals.

Montserrat Ferrer-Julià is a senior researcher in hydrological data acquisition through GIS and remote sensing analysis, and has worked in different public organizations and private companies. Currently, she is with the University of León and University Isabel I in Spain, teaching about spatial analysis with GIS and remote sensing in biological sciences and environmental sciences degrees, as well as flood risk analysis in the MSc of natural hazards. Her main research lines are related to hydrological resources, water footprints, natural hazards and geomorphological cartography. She has participated in several international and national projects and has written more than 15 articles and book chapters on these topics.

Lindsey M. Harriman holds a BS in geographic science from James Madison University and an MS in environmental sciences and policy from Johns Hopkins University. She has worked in both the public and private sectors, using a wide range of remote sensing and GIS data for a variety of environmental applications. She is currently employed by Stinger Ghaffarian Technologies as a contractor to the U.S. Geological Survey and serves as the science integration lead for NASA's Land Processes Distributed Active Archive Center.

Paul Hirst has been a UK police officer for 30 years, spending much of his career involved in the police use of firearms, public order and latterly in emergency communications. He was a qualified tactical level ('Silver') incident commander and was also a nationally qualified hostage and critical incident negotiator. He holds a masters in police studies from the University of Exeter and a diploma in management from the Institute of Management. Since his retirement at the rank of superintendent, he has worked for the British APCO on a number of EU- and UK-funded emergency communications projects, representing the needs and views of the end user in this field.

S. Taku Ide is the chief executive officer of Koveva Ltd., which has a focus on understanding methane flow through subsurface coal seams and associated consequences, including its effects on underground coal fires. From 2000 to 2011, he attended Stanford University, earning degrees in chemical engineering (BS) and in petroleum engineering (MS, PhD). He founded Koveva in 2011, based on his dissertation, 'Anatomy of Subsurface Coal Fires: A Case Study of a Coal Fire on the Southern Ute Indian Reservation'. While with Koveva, he has helped characterize and extinguish coal fires around the world. His coal fire research has appeared in a number of peer-reviewed journals, and he is a co-author of U.S. Patent US8770306 B2, to control underground coal fires using inert gas injection.

Amal Kar was attached with Central Arid Zone Research Institute, Jodhpur, from 1974 to 2012. He has been involved in research on desert geomorphology, desertification, natural resources assessment, and applications of geomatics for more than 40 years. Apart from holding research and administrative positions at CAZRI, he was a visiting professor at the University of Tokyo (2001) and contributed to the UN system on desertification. He was an editor of the journals *Annals of Arid Zone* (Jodhpur) and *Sustainability Science* (Springer). Dr. Kar now lives in Kolkata, is associated with some learned societies, and works as an independent researcher.

Rita Kovordanyi is an associate professor at the Department of Computer and Information Science at Linköping University. Her previous research includes artificial neural network models of human visual attention and visual perception, human risk perception and behaviour adaptation to risk, complex event processing and decision support systems. Her present research focuses on automatic image segmentation and image annotation using biologically inspired recurrent artificial neural networks.

Claudia Kuenzer earned her PhD in remote sensing from Vienna University of Technology in 2005 and is the head of the Department of 'Land Surface' at the German Earth Observation Centre (EOC), German Aerospace Centre (DLR). The currently over 50 scientists of this department are involved in Earth observation data analyses, monitoring dynamic processes of urban environments, forest eco-systems, agro-ecosystems and the cryosphere, amongst others. Kuenzer's personal interest over the past years is on the potential of Earth observation for the quantification of land surface dynamics – especially based on time series analyses. Kuenzer has coordinated numerous large research consortia, and frequently lectures for national and international universities and research organizations. She has authored and co-authored >100 Science Citation Index (SCI) journal papers, >35 book chapters and >150 conference contributions, and published three books with Springer. Kuenzer is an associate editor of the *International Journal of Remote Sensing* and a member of the editorial board of the journal *Remote Sensing*, and she acts as a reviewer for 15 SCI journals from the field of remote sensing, environmental sciences and geophysics.

Anand T. Kulkarni holds a PhD in water resources engineering from IIT Bombay. Prior to his PhD, he had more than 4 years of experience in the water regulation and hydro power sector. After his PhD he has been working with one of the world's leading modelling companies as a flood modeler. He has published more than 20 research papers in various international journals and conferences.

Zbigniew W. Kundzewicz is a full professor of earth sciences, a corresponding member of the Polish Academy of Sciences and a member of the Academia Europaea. He holds a doctorate and habilitation in hydrology from the Institute of Geophysics, Polish Academy of Sciences, Warsaw, Poland. His main scientific contributions have been in the areas of extreme hydrological events (and floods in particular) and climate change impacts on water resources. He has led the Climate and Water Department in the Institute for Agricultural and Forest Environment, Polish Academy of Sciences in Poznan, Poland (since 1990). He was a senior scientist in the Potsdam Institute for Climate Impact Research (PIK) in Potsdam, Germany. He was an Alexander von Humboldt-Foundation research fellow at the University of Karlsruhe and a scientific officer at the World Meteorological Organization in Geneva. He was an editor and a co-editor of the *Hydrological Sciences Journal*. He has received several distinctions, including a prestigious International Hydrology Prize of IAHS-UNESCO-WMO and a Dooge Medal, aimed to recognize fundamental contributions to the science of hydrology.

Miguel Mendes holds a civil protection engineering degree, is experienced in emergency management planning as well as R&D activities for the development of emergency management solutions at the European level. He has been involved in the development of multiple emergency management plans at the municipal as well as the regional level in Spain, and has participated as a researcher in several European-funded R&D projects. A non-exhaustive list of these projects includes Alert4All (a European alert system), PHAROS (platform to respond to multihazard events) and HEIMDALL (operative platform to support efficient response planning and the building of multidisciplinary scenarios).

Parthasarathi Mukhopadhyay is presently serving as a professor in the Department of Architecture, Town and Regional Planning at IIEST Shbipur. A fellow of the Indian Institute of Architects, his areas of research include disaster-resistant architecture, bamboo as a design element and so on. Professor Mukhopadhyay has been recognized as a resource person under the National Programme for Capacity Building of Architects in Earthquake Risk Management launched by the government of India in 2006. He is a co-innovator of 'low bamboo-energy bamboo concrete flooring', which was one of the Top 7 entries in Lafarge Invention Awards in 2011. Since 2016, he has been an alternate member of the Panel for Timber and Bamboo under NBC, India.

Franklin M. Orr, Jr. is the Keleen and Carlton Beal professor emeritus in the Department of Energy Resources Engineering, Stanford University. He served as the dean of the School of Earth Sciences at Stanford from 1994 to 2002, as the director of the Stanford Global Climate and Energy Project from 2002 to 2008 and as the director of the Stanford Precourt Institute for Energy from 2009 to 2013. He was the undersecretary for Science and Energy at the U.S. Department of Energy, Washington, DC, from 2014 to 2017. Prior to joining Stanford in 1985, he was employed by the U.S. Environmental Protection Agency in Washington, DC (1970–1972), Shell Development Company in Houston, TX (1976–1978) and the New Mexico Institute of Mining and Technology in Socorro, NM (1978–1985). He holds a PhD from the University of Minnesota and a BS from Stanford University, both in chemical engineering. He is a member of the National Academy of Engineering, the board of directors of the Monterey Bay Aquarium Research Institute and the board of directors of the ClimateWorks Foundation.

Natalia Perevalova is a leading researcher in the Institute of Solar-Terrestrial Physics of Siberian Branch of the Russian Academy of Sciences (ISTP SB RAS), Irkutsk, Russia. She received her PhD in physics of atmosphere and hydrosphere in 1996. Her primary interests include use of the Global Navigation Satellite Systems (GPS, GLONASS) for studying the ionosphere; the ionospheric response to geomagnetic storms, solar flares, solar eclipses, tropical cyclones and rocket launches; and lithosphere–atmosphere–ionosphere interactions. The results of her studies have been reported in more than 130 papers in different journals, and in 4 books.

Joanna Pociask-Karteczka is a full professor of earth sciences at the Department of Hydrology, Institute of Geography and Spatial Management, Jagiellonian University in Kraków, Poland. Her research is focused on high mountain hydrology, climate groundwater–surface water interactions, hydrological significance of the North Atlantic Oscillation, the human impact on water circulation, limits of human activity in legally protected areas, protection of springs in southern Poland and tourism in mountain regions.

Colin Price is a full professor in the School of Geosciences at Tel Aviv University, and the head of the Porter School for Environmental Studies. He is an atmospheric physicist, specializing in the Earth's weather and climate, with a focus on thunderstorms, climate change and natural hazards. He was born in Johannesburg, South Africa, in 1962. After starting his university studies in South Africa, he transferred to Tel Aviv University in 1982 where he completed his BSc and MSc in geophysics and atmospheric sciences. He received his PhD at Columbia University, New York, in 1993, while working at NASA's Goddard Institute for Space Studies in New York. After completing a postdoctoral fellowship at the Lawrence Livermore National Laboratory in California, he returned to Israel and joined the faculty of Tel Aviv University in 1995. He has published more than 125 scientific papers.

Atta-ur Rahman is an associate professor in the Department of Geography, University of Peshawar, Pakistan. Dr. Rahman holds a bachelor of sciences in bio-geo-chemistry, a master of science in geography with distinction, a master of science in urban and regional planning with Gold Medal, an MPhil in disaster risk management, a post-graduate diploma in GIS and remote sensing, a doctor of philosophy in environmental geography and a post-doctorate in flood risk modelling and management from Kyoto University, Japan. He started his commissioned service career in 1998 from Education Department Government of Khyber Pakhtunkhwa (1998–2002) and then joined the University of Peshawar-Pakistan in February 2002, where he is still serving. In addition, he is also a visiting faculty in the Centre for Disaster Preparedness and Management, University of Peshawar, Pakistan. He has specialization in HRD, disaster risk reduction and environmental impact assessment. He is also working with international organizations on various aspects of HRD and DRR, and is supervising research students in the field of population studies and disaster risk management. Currently, he is a member of the editorial boards of reputed journals and has authored books and numerous research articles in prestigious journals.

Agnieszka Rajwa-Kuligiewicz is an assistant in the Department of Hydrology at the Institute of Geography and Spatial Management, Jagiellonian University in Kraków (Poland). She received her MSc in physical geography from the Jagiellonian University and PhD in geophysics from the Institute of Geophysics, Polish Academy of Sciences, Warsaw, Poland. Her research interests focus on fluvial hydraulics, water quality and methods of spatio-temporal analysis of environmental data.

Joaquín Ramírez is the founder and principal consultant at Technosylva, providing sophisticated fire behaviour analysis and management systems, for natural hazards and wildland fire in particular. Ramírez is well respected as a leading emergencies technologist and software architect in Europe and North America, and is the chief designer of the Wildfire Analyst™, Tactical Analyst™ and fiResponse™ software products. As a professor at the University of Leon, he teaches the first class on geotechnologies in emergencies at the first European MSc in forest fires (www.masterfuegoforestal.es). Since 2014, he has been based at the Qualcomm Institute at the University of California, San Diego.

Chandan Roy is an associate professor at the Department of Geography and Environmental Studies at Rajshahi University, Bangladesh. His previous research includes mapping short- and long-term changes in land use and land cover and monitoring severe weather phenomena, such as cyclone on the basis of satellite image. His current research takes in developing weather forecasting techniques using biologically inspired recurrent artificial neural networks.

Robert Twardosz is a member of the research and teaching staff of the Department of Climatology, Institute of Geography and Spatial Management, Jagiellonian University, Kraków, Poland. He holds doctorate and habilitation in meteorology and climatology. In his research, he mainly focuses on human bioclimatology, climate change, probabilistic modelling of maximum precipitation, atmospheric circulation and precipitation and thermal anomalies. He has taken part in several research projects, including on climate variability in the Baltic Sea basin and on temporal variability, as well as on the form and type of precipitation in Kraków in the context of atmospheric circulation. He has published, including joint publications, more than 100 original research contributions, including an atlas of atmospheric precipitation in the Polish Carpathian Mountains and a monograph *Thermal Anomalies in Europe (1951–2010)* [in Polish *Anomalie termiczne w Europie (1951–2010)*].

Xidong Wang obtained his PhD in physical oceanography at the South China Sea Institute of Oceanology, Chinese Academy of Sciences in 2011. Between 2006 and 2016, he was an assistant research fellow/associate research fellow/research fellow at the National Marine Data and Information Service, China, with a year spent in 2013 as a visiting scientist at the University of Miami/Atlantic Oceanographic and Meteorological Laboratory in Florida. Since 2016, he has been a full professor at the College of Oceanography, Hohai University in China. His research interests include multiscale interactions between tropical cyclones and the ocean, ocean circulation and climate and ocean data assimilation and climate prediction.

P.E. Zope holds a PhD in water resources engineering from IIT Bombay. He has more than 25 years of experience as an engineer in Mumbai Municipal Corporation, India, in various departments. Presently, he is working as an independent consultant in the area of water resources. He has published more than 10 research papers in various international journals and conferences.

1 Tropical Cyclone and Track Forecasting

Chandan Roy and Rita Kovordanyi

CONTENTS

1.1   INTRODUCTION

Tropical cyclones (TCs), also known as hurricanes or typhoons, are among the most remarkable and deadliest meteorological phenomena. TCs are unique among all meteorological disasters because of their capacity to create both strong up-shore winds and high storm surges. In addition to these characteristics, TCs are well known for their ability to cause substantial property damage and human casualty. If we rank historical meteorological disasters (disasters that occurred during the period 1900–2011) based on inflicted human casualty and property damage, the historically largest TCs occupy positions 1 through 10 in both of these aspects. These 10 most lethal and most destructive meteorological disasters killed nearly 1 million people and caused $281,800 million in property damage.

The techniques used to predict cyclone track are not easily accessible to the layman. As cyclones are governed by thermodynamic and fluid-dynamic processes, the combination of factors underlying cyclone formation and movement is extremely complex and difficult to fully understand even for expert meteorologists. Moreover, as the formation period and movement pattern of TCs are ocean-basin-specific, this makes track forecasting even more difficult for meteorologists. Though several TC formation theories have been formulated based on the factors and processes involved in TC formation, these theories are still controversial. In contrast, TC track forecasting techniques are relatively well established in the meteorologist community. Factors governing TC movement can be of external and internal origin. A complex interaction between these factors and the TC vortex initiates horizontal motion to the low-pressure system. Principal factors responsible for TC motion include large-scale environmental circulations, Earth's vorticity, upper wind-driven air circulations, barotropic instability, Rossby waves, sea–air interaction, relative humidity and boundary layer influences (influences of the Earth's surface on the lower part of the troposphere). TC track forecasting techniques either directly model these factors' influence on TC motion or consider the historical and/or recent past motion characteristics of TC, without actually modelling the underlying factors and forces, to predict the statistically most likely future movement direction of a cyclone.

When TCs are accompanied by high surges, they can create huge property damage and human casualties (Figure 1.1), as here the force of cyclonic wind is coupled with the impact of flooding.

Prediction accuracy of track forecasting techniques, both for shorter time periods (up to 24 hours) and longer time periods (>24 hours) depend mainly on the ability of the various techniques to capture the factors that influence TC motion. Prediction accuracy of statistically based techniques is additionally affected by the availability of suitable historical predictors (datasets from which the forecasts are produced). The main purpose of this chapter is to provide a straightforward exposition of the processes and factors governing TC formation and motion so far as they are understood today. This chapter also provides a detailed description of various TC track

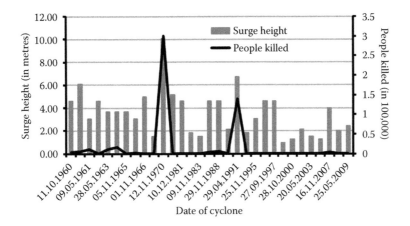

FIGURE 1.1 Heights of tropical cyclone-induced surges along the Bay of Bengal coast and number of human casualties caused by those cyclones. Left vertical axis of the graph presents the surge level, right vertical axis presents the number of casualties and the horizontal axis presents dates on which the tropical cyclones made landfall.

forecasting techniques where the focus is on explaining the predictors, clarifying the equations used to produce forecasts, comparing the performance of different forecasting techniques and describing each technique's computational resource requirements.

1.2 TROPICAL CYCLONE (TC) FORMATION

Oceans in the tropical and subtropical regions receive immense amounts of energy from the sun. This energy is subsequently released into the atmosphere through an evaporation process. The energy released from the warm waters through evaporation is the main driving force of a tropical cyclone. This released energy warms up the air at the lower atmospheric level near the surface of the ocean. When this large air mass over the sea becomes convectively unstable and moist compared to the surrounding environment, an upward movement of air is initiated and a low-pressure zone is formed beneath. The upward movement of air creates a temporary empty space that draws the surrounding air toward the low-pressure centre. This movement of air toward the low-pressure centre sets off a cyclonic air circulation. Subsequently, the combined effect of the Earth's rotational force and the centrifugal force of the existing cyclonic air circulation counteract and delay the movement of air toward the low-pressure centre. The delayed movement of air causes further pressure fall at the centre of the low-pressure system. This cumulative process makes the low-pressure system grow progressively stronger until it intensifies into a cyclone. To summarize, cyclone formation can be assumed to go through two successive stages (Zehr, 1992):

- *Stage 1*: Convective activity of the kind described above drives the development of a mesoscale vortex that can be 100 to 200 km in diameter. Formation of these mesoscale vortices is quite common in the TC formation basins, that is, in ocean basins located near the equator.

- *Stage* 2: Under favourable environmental and internal conditions, these mesoscale vortices can intensify into a full-blown TC: In essence, if the central air pressure continues to decrease and the speed of spiralling wind flow continues to increase, the mesoscale vortex gradually takes the form of a TC.

Simpson et al. (1997) further divided stage 2 into two stages: a preformation and an initial development stage.

1.2.1 FACTORS AFFECTING TC FORMATION

The formation of low-pressure zones and the development of these into cyclones can be affected by several factors. Although low-pressure zones frequently form over tropical oceans, only a fraction of these reach the TC intensity (Anthes, 1982).

1.2.1.1 TC Formation

Factors affecting TC formation can be either environmental or internal (Jeffries and Millar, 1993). Both types of factors are equally important for TC formation. Environmental factors that influence the formation of low-pressure zones and the development of these low-pressure zones into TCs include (Syōno, 1953; Yanai, 1964; Gray, 1968, 1979, 1998; Perrone and Lowe, 1986; Zehr, 1992; Chan and Kepert, 2010):

- *The Earth's vorticity*: TCs have not been observed to form within about 5° of the equator (Gray, 1968). Though the Earth's vorticity, the Coriolis force, increases toward the poles, this does not necessarily increase the likelihood of TC formation (Jeffries and Miller, 1993). In addition to the Coriolis force, other factors also have to be present for a TC to be formed.
- *Low-level relative vorticity*: As atmospheric disturbances are necessary for TC formation, cloud clusters that develop into tropical depressions tend to be located in regions where relative vorticity is low (Gray, 1968). Such low-level relative vorticity regions include the Intertropical Convergence Zone (ITCZ) (Figure 1.2), the monsoon trough and the near equatorial trough (Perrone and Lowe, 1986).
- *Vertical wind shear*: Latent heat release during convection plays a key role in the forming of a TC vortex by working as the primary supplier of energy for TC formation. If the heat generated through convection is removed from the TC vortex by a strong vertical wind shear, the low-pressure system will not have sufficient energy to grow into a mature cyclone. Hence, for a TC to be formed, the speed of vertically moving air should be less than 40 km/h. Any larger values will result in removal of heat from the cumulus column and will prevent TC formation (Gray, 1968). Vertical wind shear also changes the thermal structure near the storm centre and these changes have a significant impact on TC intensity (DeMaria, 1996).
- *Thermal energy content of the ocean*: The ocean water mass works as the main energy source for TCs with an energy transfer taking place at the ocean surface. As cyclonic wind can cause surface/internal water circulation, the ocean water needs to be sufficiently warm to a certain depth. For

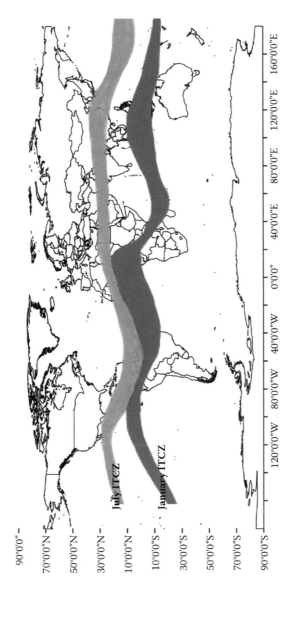

FIGURE 1.2 Location of ITCZ during the months of July and January over the globe.

TC formation, the temperature of ocean water must be ≥26.5°C to 50 m depth. If the upwelled water is cooler than 26.5°C, the ocean will no longer be able to drive the TC development process.

- *Relative humidity at the mid-tropospheric level*: When a low-pressure system forms over the ocean surface, the air moves upward, sucking the surrounding air toward the rising vortex. For a TC to develop, convection must be forceful, which in turn requires that the rising air parcel is saturated with moisture. To keep the air parcel sufficiently saturated during its ascent, relative humidity of the mid-tropospheric level needs to be at least 50–60%.
- *Potentially unstable troposphere*: This is a measure of the difference of equivalent potential temperature between air at the sea surface level and air at a 500 mbar pressure level. This difference must be less than or equal to 10° K for convection to occur. Potentially unstable tropospheric conditions like this are often observed over tropical oceans, and these observations confirm the critical 10° K threshold (Gray, 1968; Perrone and Lowe, 1986).

The above six factors are key ingredients of TC formation, and can also explain seasonal variations in TC frequency for all cyclone formation basins (Gray, 1998). The derivatives of the first three factors specify the dynamic potential of a TC and the derivatives of the last three factors yield the TC's thermal potential. A seasonal tropical cyclogenesis parameter can be obtained through multiplying the dynamic and thermal potentials; that provides a good estimate of the seasonal frequency of TC formation (Gray, 1998).

Once the low-pressure system that was formed through convection has become more structured, the importance of external influences diminishes, and instead internal factors become more important. As tropical cyclone formation is not confined to particular locations within the cyclone formation basins, the conditions for individual cyclone development can differ in nature and degree from the previously mentioned standard conditions of cyclone development (Palmen, 1948; Chan and Kepert, 2010).

1.2.2   Theories of Cyclone Formation

The problem of TC formation is one of the most difficult in meteorology. Diverse views concerning this problem resulted in the formulation of various theories of TC formation. However, none of the suggested TC formation theories are universally accepted. In the following, we will describe some of the most widely accepted theories of TC formation.

1.2.2.1   Conditional Instability Theory

Tropical cyclones can be considered as large-scale instabilities associated with the tropical atmosphere which is conditionally unstable (Montgomery and Farrell, 1993). If the air is saturated it becomes unstable under a smaller temperature change compared to the dry air. Situations that cause instability in the moist and saturated air but not in the dry air are called conditionally unstable situations. Conditionally unstable atmospheric situations are considered to be significant during the initial stage of TC formation. Early theoretical work focused on defining the conditional instabilities of a calm atmosphere in a rapidly rotating environment. Through these early efforts, meteorologists managed to properly define small-scale cumulus instabilities, but failed to capture the instabilities that can occur in a large-scale rotating air mass like a TC (Haque, 1952; Lilly, 1960). A new instability theory was then proposed to properly explain the growth of tropical depressions Charney and Eliassen (1964). Unlike earlier theories, the new theory explains the instability of the atmosphere as Conditional Instability of the Second Kind (CISK). Reciprocal feedback between small-scale convection and large-scale atmospheric circulation constitutes the basis of CISK. According to CISK, tropical low-pressure systems intensify using the potentially available convective energy that arises through this reciprocal feedback (Montgomery and Farrell, 1993). Though CISK can better explain the growth of tropical depressions compared to the earlier theories, it also has some drawbacks. The major drawback of CISK is the suggested exponential growth rate's sensitivity to vertical heat distribution within the TC vortex. As vertical heat distribution within the storm vortex can be influenced by various internal and external factors, exponential growth rate calculated by CISK can diverge from the real growth rate (Pedersen and Rasmussen, 1985; Pedersen, 1991).

1.2.2.2 Air–Sea Interaction Theory

An important prerequisite for CISK is a widespread conditional instability in the atmosphere; at the same time, the tropical atmosphere has been observed to be nearly stable. So, to better describe the formation of TC, *the cyclogenesis process*, Emanuel (1986) and Rotunno and Emanuel (1987) proposed an air–sea interaction theory as an alternative to CISK. According to the air–sea interaction theory, TCs are initially formed and further develop entirely by self-induced irregular fluctuations of the humidity enthalpy from the sea surface, which suggests that atmospheric instabilities do not seem to be necessary for TC formation (Emanuel, 1986). Instead, wind-induced sea surface heat transfer (sea to air energy transfer) is considered to be the primary source of energy here. Though air–sea interaction theory does not require any pre-existing atmospheric instabilities for TC formation and development, sufficiently developed axisymmetric disturbances are considered necessary to initiate the air–sea energy transfer process (Montgomery and Farrell, 1993).

1.2.2.3 Eddy Angular Momentum Theory

The significance of the eddy angular momentum fluxes in TC intensity change was proposed long ago by Pfeffer (1958). Later, Challa and Pfeffer (1980) suggested that eddy angular momentum fluxes not only influence TC intensity at later stages but also partake in the initial TC development process. This theory focuses on angular momentum fluxes that are associated with tropical disturbances, and proposes that these are responsible for TC development (Challa and Pfeffer, 1980; Molinari et al., 2004; Yu and Kwon, 2005). Eddy angular momentum fluxes work on the cumulus convection as an external force, which contrasts this theory with competing ideas that TC development is driven only by internal forces like cumulus convection and frictional convergence in the boundary layer (Challa and Pfeffer, 1980).

1.2.2.4   Latent Heat Release Theory

Like CISK, latent heat release theory was also proposed as a reaction to the previously proposed conditional instability theories which could not properly describe the formation of large rotating air masses, such as cyclones (Kuo, 1965). Charney and Eliassen (1964) and Kasahara (1961) among others proposed various solutions to this problem. A different approach was taken by Riehl and Malkus (1961), who developed a theory focusing on the central role of the cumulus vortex in releasing latent heat. Based on Riehl and Malkus' work, Kuo (1965) proposed a latent heat release theory of cyclone formation and intensification. This theory assumes that the latent heat release of condensation is accomplished by the vertical motion associated with the cumulus convection, and not through the vertical motion of the wind field. By doing this, latent heat release theory properly includes the heating effect of cumulus convection on a synoptic scale motion. This latent heat release (heat release during condensation of water vapour) works as one of the main driving forces of TC development and intensification.

1.2.3   Cyclone Formation Basins and Their Characteristics

Tropical oceans where cyclones form on a more or less regular basis are called cyclone formation basins. TCs are observed to form over all tropical oceans (between 30° N and 30° S latitudes) except the South Atlantic and the eastern South Pacific (Figure 1.3). Due to insufficient earth vorticity, TCs do not form near the equator, between 5° N and 5° S latitudes. Of all the global TCs, 87% have their origin within 20° N and 20° S of the equator (Gray, 1968). The tropical oceans are divided into seven basins where TCs have been observed to form on a regular basis (Figure 1.3):

- Northwest Pacific basin
- Northeast Pacific basin
- North Atlantic basin
- North Indian Ocean basin (Bay of Bengal and Arabian Sea)
- Southwest Pacific basin
- Southwest Indian Ocean basin
- Southeast Indian Ocean basin

Due to dissimilar topography, geography, oceanography and large-scale air flow patterns (like the monsoon and easterly winds), the causes, number and period (season) of TC formation may vary between basins. Also, TC formation basins are not equally active every year, and even within the cyclone season, the TC formation basins could have shorter active (cyclone formation frequency is high) and inactive (cyclone formation frequency is low) periods (Harr and Elsberry, 1991). In general, TCs are mostly formed during the seasonal transition periods, when the atmosphere remains unstable over most of the cyclone formation basins. For the North Atlantic basin this scenario is different, as in this basin TC formation is mainly associated with easterly winds from Africa (Elsner and Kara, 1999; Arnault and Roux, 2011). In the following paragraphs we describe the causes and periods of TC formation in the different basins.

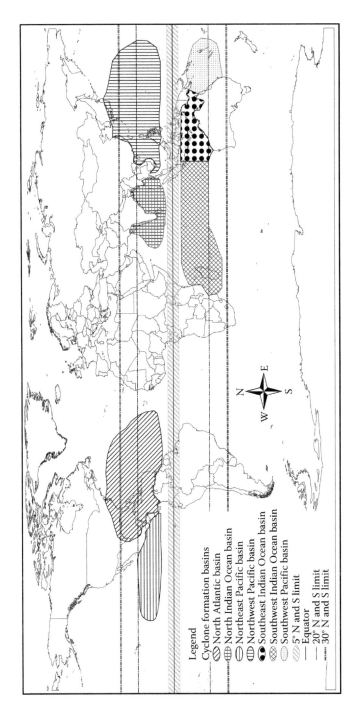

FIGURE 1.3　Locations of tropical cyclone formation basins. Different basins have been presented with different shades. Equator, 20° N and S limit, and 30° N and S limit are shown with continuous, single discontinuous and double discontinuous lines.

The *Northwest Pacific basin* is the most active among all cyclone formation basins. About one-third of all global TCs are formed in this basin (Figure 1.4). Low values of vertical wind shear (Gray, 1968), and very warm ocean temperatures over a broad area are the main reasons for this basin being the most active. TCs are observed to form here during any month of the year, although the formation frequency is highest during July through November. The monsoon trough over Southern China and the Philippine Seas is where most of the TCs (about 80%) in this basin originate. Though it is well known that the monsoon trough is where most TCs in the Northwest Pacific basin are formed, the formation mechanism is not yet properly understood. Except for the monsoon trough, shifting of the ITCZ and an easterly wave-like disturbance in the ITCZ may also cause cyclogenesis here (Fett, 1968; Heta, 1990). Variations in sea surface temperature and locational variation of the monsoon trough causes interannual variation in TC formation in this basin (Chen et al., 1998). Cyclones formed in the Northwest Pacific basin affect the Philippines and Southeast Asia including China, Taiwan and Japan.

The *Northeast Pacific basin* is the second most active in the world (Figure 1.4). This basin remains most active from mid-May to the end of November. Almost all the cyclones in this basin are observed to form in the ITCZ. Like for the Northwest Pacific basin, also for this basin there are some disagreements regarding the development of TC. Avila (1990) argues that easterly winds from the Caribbean Sea have a significant effect on cyclone formation in this basin. TCs formed in this basin affect Western parts of Mexico and Hawaii.

The *North Atlantic basin* is the most widely studied among all TC formation basins. Many of the factors responsible for TC formation are unique for this basin. Here the TC formation season extends from the beginning of June to the end of November. In this basin, the location of TC formation varies widely throughout the season. During the beginning of the season, TCs usually form in the southwest Caribbean and in the Gulf of Mexico. With the progress of the cyclone season, TC formation location gradually shifts to the central Atlantic. During the end of September, the western

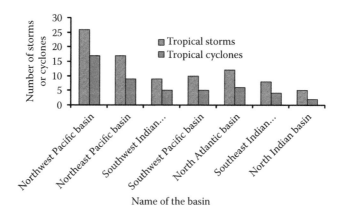

FIGURE 1.4 Annually averaged tropical storms and cyclone formation records by basin based on observations from 1981–2010 for the Northern Hemisphere and from 1981–1982 to 2010–2011 for the Southern Hemisphere.

Caribbean becomes the most active location for TC formation once again. Movement of the equatorial trough, variation in vertical wind shear and the strength of the easterly winds from Africa are largely responsible for these shifts in TC formation location (Gray, 1968). Around half of the TCs that form in this basin originate from the easterly winds from Africa (Avila, 1990; DeMaria et al., 2001). In the remaining cases, TC formation is driven by a mid-latitude frontal system and upper-level tropospheric trough (Jeffries and Miller, 1993). Regions that are affected by cyclones formed in this basin include the Mexican gulf coast in the United States and Mexico, parts of Canada, the Caribbean islands, and Central America.

The *North Indian Ocean basin* is divided into two cyclone-formation areas: the Arabian Sea and the Bay of Bengal. Of the two, the Bay of Bengal is more active, with about five to six times more TCs being formed in the Bay of Bengal compared to the Arabian Sea (Sampson et al., 1995). TCs usually form in this basin before and after the monsoon period, that is, during the premonsoon (in May) and postmonsoon (in November). In contrast, TCs are normally not formed during the postmonsoon (winter) period in the Arabian Sea. During the monsoon period itself, TC formation is suppressed, as the ITCZ shifts northward and as vertical wind shear becomes stronger. Countries mainly affected by cyclones formed in the North Indian Ocean include Bangladesh, India and Myanmar.

In the *Southwest Pacific basin*, TCs form west of 160° W longitude between 5° S and 20° S latitudes. During El Niño years, TC formation location shifts north and eastward in this basin. In this basin, TCs are mainly formed within the South Pacific Convergence Zone (SPCZ) (Sampson et al., 1995). Though SPCZ is distinct from the ITCZ, they may merge together to form a single zone of convergence. Of the total low-pressure zones that form along the zone of convergence, only 4–5% develop into TCs. Cyclones formed in this basin affect Northeastern parts of Australia.

The *Southeast Indian Ocean basin*, also known as the Australian basin, is located to the west of the Southwest Pacific basin. However, the boundary between these two basins is difficult to draw. The Australian basin can be divided into three areas: northwest Australia, Gulf of Carpentaria and northeast Australia based on TC formation and motion. January to March is the period when cyclones form in the northwest area but in the other two areas remarkable cyclone activity is observed during the months of December to April. Though low-pressure zones frequently form in the Gulf of Carpentaria, very few of them reach the intensity of TC due to its restrictive size. More than half of the TCs formed in the northwest area are associated with the monsoon trough over land (McBride and Keenan, 1982). Regeneration of TCs after crossing the land is a unique characteristic of this basin. Previous statistics shows that about 12% of the annual cyclones are regenerated after crossing over the Australian continent (Sampson et al., 1995). Cyclones formed in this basin most often hit the Northwestern part of Australia.

The *Southwest Indian Ocean basin* is the most poorly studied among all cyclone formation basins. Starting from October through April is the period when TCs form in this basin and the highest frequency is observed during the month of January. TC formation in this basin is associated with ITCZ; as a result, the formation location changes with the shifting of the ITCZ. Cyclones formed in this basin affect Madagascar and parts of southern Africa.

1.2.4 LIFE CYCLE OF A TROPICAL CYCLONE

After formation over warm sea waters, tropical depressions accumulate energy and intensify into TCs under favourable environmental and internal conditions (Jeffries and Miller, 1993). At the same time, it also travels several hundred kilometres over sea and finally decays over land or sea. The time period starting from genesis up to decay is known as the cyclone life cycle. The life cycle of a tropical cyclone can be divided into four stages (Abdullah, 1954; Kuo, 1965; Jeffries and Miller, 1993):

- *Formation or genesis stage*: TC genesis is associated with pre-existing large-scale disturbances (Ooyama, 1982). This stage is characterized by minor air circulation with no closed isobars surrounding the low-pressure zone (Figure 1.5). The genesis stage ends with the formation of closed isobars and clearly defined air circulation.

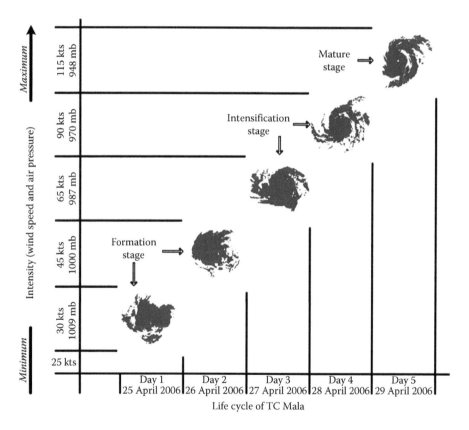

FIGURE 1.5 Life cycle of Tropical Cyclone Mala. After formation, a tropical cyclone usually goes through several stages under favourable environmental conditions before decaying over land or sea. In the figure, wind speed and air pressure are presented on the vertical and the days Mala passed to reach the mature stage after formation are presented on the horizontal axis. Satellite images of the corresponding dates have been given on the right upward axis.

- *Intensification or deepening stage*: In this stage, central air pressure of the low-pressure system falls and velocity of the surface wind increases (Figure 1.5). As a result, the low-pressure system gradually takes the form of a cyclone. If the TC continues to grow, an eye develops at the centre of the associated cloud mass.
- *Mature stage*: For a TC, the mature stage is the period when it reaches the maximum intensity. Lowest central pressure and highest surface wind speed are two main characteristic features of this stage (Figure 1.5). At this stage, the diameter of the storm vortex reaches its maximum.
- *Decay stage*: TCs do not remain stationary at their formation location. When a cyclone changes its location, the environmental conditions surrounding it also change. These changes may in some circumstances cause a weakening of convective activity. For example, if the TC moves over land or over cool water, the main source of energy (energy transfer from the warm ocean water) will be cut off and the TC would decay gradually. Hence, in the absence of favourable conditions, a TC may not go through all development stages before decaying over land or sea.

1.3 TROPICAL CYCLONE STRUCTURE

The structure of a fully developed TC can be defined by three components (Wang and Wu, 2004): storm-scale structure, inner-core structure and spiral cloud bands. The storm-scale structure of a TC is determined by wind and pressure in the lower tropospheric level, air temperature and moisture near the sea surface, vertical thermal stratification, winds in the middle and upper troposphere and vertical motion fields. The inner-core structure includes the cyclone eye, eyewall and convective adjustments in the eyewall. The cloud band structure comprises different types of cloud bands and their influences on TCs. Among all the structural components, the cyclone eye, the eyewall and the spiral cloud bands are considered to be the most important as they effectively determine intensification of the TC (Emanuel, 1997; May and Holland, 1999). The formation and characteristics of the cyclone eye, eyewall and spiral cloud bands and how they influence TC intensification are discussed in the following sections.

1.3.1 CYCLONE EYE AND EYEWALL

The cyclone eye is a characteristic feature of a fully developed TC (i.e. a TC with maximum sustained wind speed above 119 km/h), while a cyclone eye cannot be seen in a premature cyclone. The eye of a cyclone is surrounded by circular and vertically rising clouds, which can be up to 15 km high. This thick wall consisting of cloud masses is known as the eyewall. The eyewall is considered to be the most destructive region within a cyclonic system. Here rainfall is heaviest and wind speeds can reach up to 305 km/h (based on 1-minute sustained wind speed measured in the strongest TC in recorded history). Within a cyclone, barometric pressure usually increases if we travel outward from the eye. This pressure variation is more rapid near the centre compared to the peripheral regions. Rapid pressure variation near the centre creates large pressure gradients (wind speed acceleration due to pressure difference), which causes the strongest winds to occur at the eyewall. The conservation of angular momentum could also be another reason for the strongest wind at the eyewall. Any structural change in cyclone eye and eyewall can create changes in the wind speed of a TC.

In satellite images of a mature TC, the eye appears as a roughly circular hole in the centre of the TC's cloud masses. The TC eye is the area of the TC where barometric pressure is the lowest, and it is characterized by calm winds, little or no precipitation and often a cloud-free sky (Figure 1.6). Usually, wind speeds within the cyclone eye do not exceed 24 km/h. The typical diameter of an eye ranges between 32 and 64 km. Though these characteristics of the cyclone eye can easily be recognized in meteorological data, the mechanisms by which a cyclone eye is formed are still not properly understood. According to the current state of research, two processes are considered to be involved in cyclone eye formation: convection and subsidence (Smith, 1980; Vigh, 2010). While convection is necessary for the development of the cyclone as such, subsidence at the centre of the storm vortex particularly forms the eye.

Another hypothesis suggests that a combination of the conservation of angular momentum and centrifugal force could have an impact on TC eye formation. The conservation of angular momentum is a property of a spinning object, where the spinning rate of an object increases if it moves towards the centre of circulation. In the case of TC, as air moves in a spiral toward the storm centre, the spinning speed of the air increases. At the same time, the spinning of air also creates an outward-directed force called the centrifugal force which is proportional to the spinning speed of the air. As a result, the faster spinning air near the cyclone centre will produce a greater centrifugal force compared to the slower spinning air in the peripheral regions of the cyclone. The faster rotation of air near the cyclone centre pushes air toward the eyewall, creating an empty space at the centre and causing an eyewall to be formed.

FIGURE 1.6   Satellite image of a mature cyclone where the cyclone eye, eyewall and spiral cloud bands are clearly visible.

It also makes the air rise up from the centre indirectly causing an outflow of air at the top of the centre, preventing cloud formation and creating an area of clear sky at the centre. This area of clear sky is known as the cyclone eye.

From the previous paragraph it is clear that various structural changes are associated with cyclone eye formation. Several mechanisms have also been offered to explain why and how these structural changes occur within a cyclone. A combination of some or all of the following mechanisms can partake in the eye formation process (Smith, 1980; Emanuel, 1997; Vigh, 2010):

- Kinematic and thermodynamic aspects of the vortex which create subsidence at the centre
- Frictional processes at lower atmospheric levels which lead to a rising of air
- Changes in the cyclone's convective morphology, such as the surrounding cloud bands or spinning air mass, with significant influence on the formation of central subsidence
- Distribution of vorticity within the cyclone, which works by isolating the central air mass from the surroundings

1.3.2 SPIRAL CLOUD BANDS

Spiral cloud bands are one of the most prominent features of a TC (Figure 1.6). These cloud bands can extend up to 1000 km from the cyclone centre and spiral inward into the eyewall. Based on their location, cloud bands can be classified into two types (Guinn and Schubert, 1993): inner and outer cloud bands. Cloud bands that are located more than 500 km away from the centre are usually called outer cloud bands. Inner cloud bands in contrast are located within 500 km of the vortex centre. Due to the dense cirrus overcast, these cloud bands are usually not visible in the visible band satellite images, but they are evident on radar images.

Each strand of spiral cloud band is usually 20–40 km in width, leaving a cloud-free zone within 40 km of the cyclone centre (Atkinson, 1971). Cloud bands are usually classified on two different bases (Chow et al., 2002): motion and location. Based on motion, spiral cloud bands can be classified into two types (Willoughby et al., 1984): stationary and dynamic cloud bands. Dynamic cloud bands rotate around the cyclone centre at a speed close to the maximum sustained wind speed of the TC. Stationary cloud bands in contrast remain at a fixed location relative to the vortex and are present in almost all cyclones regardless of their intensity. They occur nearer the centre of a weak or premature cyclone and more toward the periphery of an intense cyclone. Stationary cloud bands can be characterized according to their principal band, connecting band and secondary band (Willoughby et al., 1984).

1.3.2.1 Theories of Cloud Band Formation

Several theories of cloud band formation have been proposed, but at this point there is no consensus of how cloud bands arise. Current theories can be divided into two groups. The first group of theories considers inertia-gravity waves (gravity waves in a rotating system) to be responsible for cloud band formation (Dierks and Anthes,

1976; Kurihara, 1976; Willoughby, 1978). In contrast, the second group of theories postulates that spiral bands are generated by the vortex Rossby waves (Guinn and Schubert, 1993; Montgomery and Kallenbach, 1997; Chen and Yau, 2001). More generally, various processes seem to be involved in outer and inner cloud band formation. Nonlinear effects during the breakdown of ITCZ through barotropic instability results in the formation of the outer cloud bands (Guinn and Schubert, 1993; Ferreira and Schubert, 1997).

The formation process of inner cloud bands is different from the outer bands. With the intensification of a TC, the potential vorticity (PV) field becomes nearly circular with the highest values of PV observed at the TC centre. The nonlinear breakdown of vortex Rossby waves creates an irreversible change to the high-PV core and this change finally results in the formation of inner cloud bands (Guinn and Schubert, 1993). According to Willoughby et al. (1984), nonlinear breakdown of the ITCZ cannot properly explain the formation process of the outer moving spiral bands, since these outer bands are moving at a speed that is larger than the speed of local mean wind. He suggests that moving bands are inertia-gravity waves which are formed by the oscillatory motions of the vortex centre.

Spiral cloud bands do not have any direct effect on TC intensity. However, changes in the kinematic energy and angular momentum correlate with the presence of cloud bands and these budgets have a significant impact on TC intensity (Diercks and Anthes, 1976). Correlation with kinematic energy and angular momentum are observed to occur in a number of ways (Barnes et al., 1983; Powell, 1990a,b). The mass convergence into the cloud bands may decrease the mass convergence in the eyewall, which will affect the convection process and finally result in TC intensity reduction (Wang, 2009). In addition, subsidence resulting from the convective activities in the spiral cloud bands introduces dry air (air with low humidity) in the mid and low tropospheric levels. This dry air can effectively reduce the eyewall convection and limit TC intensity (Shapiro and Willoughby, 1982; Willoughby et al., 1982). As the cloud bands spiral into the eyewall, they are capable of affecting the PV field of the centre or they can bring a decaying of the eyewall. This decaying of the eyewall will weaken the TC (Wang, 2002). Cyclonic PV irregularities generated in the spiral cloud bands of the mid-tropospheric level could be a considerable PV source to the TC core and thus could increase the intensity of a TC (May and Holland, 1999).

1.4 TROPICAL CYCLONE MOTION

The dynamics responsible for TC motion are complex, with the result that a cyclone's track can be straight, curved or even looping. TC motion arises from a wide variety of external and internal forces and interactions between the storm vortex and the surrounding environment. In general, external influences are more important compared to internal influences in the determination of future TC track in all the TC formation basins (Namias, 1955; Gray, 1968; Jeffries and Miller, 1993). In addition to the influences on cyclone formation and track, these circulations create inter-annual variability in TC characteristics (Shapiro, 1982; Gray, 1984; Nicholls, 1985). The external/environmental influences on TC track include large-scale environmental air currents and boundary layer conditions, such as air-surface friction and air fluxes that originate from the ocean surface. In contrast, forces associated with the low-pressure system itself constitute the internal influences, which include irregularities in convection process, the beta effect (variation of Coriolis force with latitude), circulations within the storm vortex, and instability of the outflow layer (Sampson et al., 1995; Bin et al., 1999).

Without the environmental influences, the circulations associated with a TC would be axially symmetric, and the TC would not move. However, the symmetric circulation is disturbed as the TC vortex interacts with the surrounding environment since its formation. An asymmetric circulation may also result from interactions between the TC and the beta effect. Though the influence of the Earth's vorticity gradient on a TC is small, the resulting asymmetric circulation has a significant effect on TC motion (Bin et al., 1999). In the absence of any external influences, the beta effect causes TCs to move to the northwest at a speed of several hundred kilometres per day in the Northern Hemisphere (Rossby, 1948, 1949; Holland, 1983; Fiorino and Elsberry, 1989; Smith, 1993; Li and Wang, 1994; Jones, 1995). In the Northern Hemisphere, if a TC forms due to easterly winds, the beta effect causes it to move faster and slightly to the right of the steering flow. The beta effect also causes the northwestward moving TCs to move faster compared to the northeastward moving cyclones in the Northern Hemisphere (Jeffries and Miller, 1993).

Though TCs may form at any location within a basin, some locations appear to be more productive compared to others (Gray, 1984). Examination of historical TC track data suggests that some of these formation locations are associated with straight moving TCs, while others are more likely to produce TCs with recurving tracks. If similar TC track patterns occur during a period, the controlling climatological circulations usually have a large spatial coverage. Identification of these controlling circulations could be used to produce extended range forecasts of TCs' likely track patterns (Harr and Elsberry, 1991).

TC motion is only influenced by the Earth's topography if it passes an island. Though an island's topography has a large influence on TC motion, details of these interactions are not properly understood. Island topography is assumed to affect the TC track by modifying the basic wind speed and direction associated with the storm (Bender et al., 1985, 1987). When a TC comes closer to an island, the topography dissipates and disturbs the outer wind flow, and as a result the TC is deflected toward the island (Bin et al., 1999). TCs are also observed to decay faster over land and will have a changed rainfall pattern when they interact with mountainous topography (Hamuro et al., 1969; Brand and Blelloch, 1973).

During the last few decades considerable progress has been made in understanding how various factors affect TC formation, as well as in the understanding of the dynamics of TC motion (Bin et al., 1999). Better understanding of these effects on TC motion enabled us to take an integrative view on these influences in the application of TC track forecasting techniques. Almost all the currently used TC track forecasting techniques use wind information from a single or multiple pressure levels as predictors. Prior to the integration of environmental steering factors into track forecasting techniques, only climatological data (historical cyclone track data) and recent past movement information of the current cyclone could be used as predictors.

Since the beginning of the 1960s, when the first weather satellite TIROS-1 was launched, considerable progress has also been made in satellite imaging technology. Now it is possible to get detailed information on sea surface temperature, ocean heat content, rainfall, wind and pressure information directly from the satellites. Better spatial coverage and frequent availability of new data (satellite data can be updated several times a day) has contributed to the recognition of satellite-derived information as one of the most useful predictors in TC track forecasting (Roy and Kovordányi, 2012).

1.5 TC TRACK FORECASTING TECHNIQUES

TC track forecasting techniques are usually described in terms of which predictors are used, how the predictors are combined, that is, which equations are used to translate the predictors into a track forecast, how accurate the described technique is and what computational resources are required to produce a forecast. We will follow the same structure in the following sections.

1.5.1 PREDICTOR SELECTION AND USE

Selection and use of predictors is an important issue in the case of TC track forecasting because it strongly influences the performance of the forecasting technique. Based on predictor selection and use, TC track forecasting techniques can be classified into four categories (Figure 1.7).

1.5.1.1 Techniques Using Recent Past Track and Climatological Data (Historical Cyclone Track Data) as Predictors

Examples of techniques or a group of techniques using the same predictors that belong to this first category include *extrapolation* (XTRP), *climatology* (CLIM), *analogue*, and *climatology and persistence*.

XTRP was developed by the Joint Typhoon Warning Center (JTWC), Hawaii. This technique is used to forecast the future motion of a TC through extrapolating and averaging the cyclone's current and recent past movement (usually 6 to 12 hours back in time). As XTRP does not consider any other predictors except the recent past movement, it is considered to be the simplest among all forecasting techniques (Jeffries et al., 1993).

CLIM, also developed by the JTWC, considers motion characteristics of historical cyclones in the same basin as predictors. Instead of using all historical data, CLIM chooses predictor datasets on the basis of two criteria: historical cyclones that originated within $6° \times 6°$ latitude and longitude of the current cyclone's formation location and was formed during the same time of the year as the current cyclone. If none of the historical cyclone tracks satisfy both criteria, cyclones that originated within $6° \times 6°$ latitude and longitude of the current cyclone's formation location are used as predictors (Jeffries et al., 1993). In the Northwest Pacific basin, subsets of the historical storms that occurred between 1945 and 1981 are used as predictors. In contrast, in the North Indian basin, subsets of the cyclones that occurred between 1900 and 1981 are used as predictors.

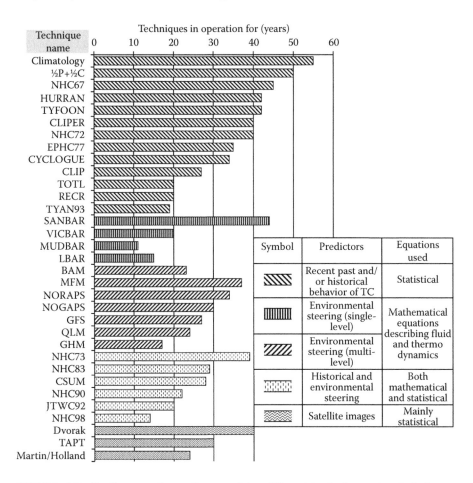

FIGURE 1.7 Predictors and equations used by different track forecasting techniques. Techniques using similar predictors and equations are presented with identically shaded bars. Lengths of the bars present how many years they have been in operation starting from the year 2012 at different meteorological offices.

Like CLIM, analogue track forecasting techniques also consider historical TC motion characteristics as predictors. The only difference between analogue forecasting techniques and CLIM is that analogue forecasting techniques employ some additional criteria in selecting predictors from the historical datasets. These criteria include speed and movement direction. Analogue techniques consider only those historical cyclones that have the same speed and movement direction as the current cyclone. Examples of analogue cyclone track forecasting techniques include *HURRAN* (Hope and Neumann, 1970), *TYFOON* (Jarrell and Somervell, 1970), *CYCLOGUE* (Annette, 1978), *TOTL* (Mautner and Guard, 1992) and *TYAN93* (Jeffries et al., 1993).

Climatology and persistence techniques use both the recent past movement characteristics of the current cyclone (persistence) and the movement characteristics of historical cyclones as predictors (climatology). The first step toward the development of climatology and persistence techniques was taken at the Royal Observatory in Hong Kong. The first version of climatology and persistence technique assigned equal weights to both predictor sets when producing a forecast. Hence, it was named ½P+ ½C (Bell, 1962). Examples of climatology and persistence techniques include CLIPER (CLImatology and PERsistence), East Pacific CLIPER (EPCLIPER) and Northwest Pacific Climatology and Persistence (WPCLPR or CLIP). CLIPER was developed by Neumann (1972) as an alternative to HURRAN for the North Atlantic basin and uses the following eight predictors to produce forecasts for cyclones:

1. Latitude of the current cyclone
2. Longitude of the current cyclone
3. Current cyclone's u-component of motion (represents the east-west component of the wind motion)
4. 12-hour-old u-component of motion
5. Current cyclone's v-component of motion (represents the north-south component of the wind motion)
6. 12-hour-old v-component of motion
7. Day number (Julian date)
8. Maximum sustained wind speed (period during which a temporarily maximum wind speed is observed within a cyclonic system, measured in minutes)

For short-term prediction (24 hours), CLIPER uses the eight above-mentioned predictors but for long-term prediction (48 to 72 hours) cross-products from the historical dataset are also used (Neumann, 1979). EPCLPR uses all these predictors except number eight to forecast cyclone track in the East Pacific basin (Neumann, 1979). WPCLPR, which was developed and is used by JTWC, also considers eight predictors: (1) climatological information, (2) present cyclone position, (3) present cyclone intensity, (4) past 12-hour cyclone position, (5) past 12-hour cyclone intensity (6–7) 24-hour cyclone position and intensity, and (8) date to forecast TC tracks in the Northwest Pacific basin (Mautner and Guard, 1992; Jeffries et al., 1993).

1.5.1.2 Techniques Using Steering Information as Predictors

Height over sea level and air pressure is inversely related. The gradual change in air pressure with increasing height causes remarkable changes in the composition and characteristics of the atmosphere (Chin, 1976). Techniques of the second category take these changes into account by including wind and other atmospheric information from single or multiple air pressure/steering levels as predictors.

Among the factors affecting TC motion, the influence of large-scale environmental steering is the most prominent. Large-scale environmental steering determines 70–90% of the movement characteristics of a TC (Neumann, 1992). Observations indicate that the mean of different steering levels (deep layer mean) has a better correlation with the storm motion than any single steering level (Dong and Neumann, 1983). As intense TCs extend higher in the troposphere, their steering levels tend to lie higher up compared to weak TCs (Dong and Neumann, 1986). Though deep layer mean steering can better explain the TC motion compared to a single level steering, there are situations when forecasting of TC motion must be based on a single level steering. These situations include (Jeffries and Miller, 1993):

- When deep layer mean information on the TC is not available. In such situations using wind information from the 500 mbar level is the best for predicting future TC motion.
- In the case of weak TCs, it is better to use steering wind information from a single level (usually from 700 mbar level).

Synoptic-climatological techniques were the first collection of techniques that considered atmospheric information from selected pressure levels (synoptic information) as one of the predictors in forecasting TC track (Neumann, 1979). In addition to this atmospheric information, these techniques also consider climatological data in their predictor list.

Among the synoptic-climatological forecasting techniques, the techniques developed by Veigas et al. (1959), Chen et al. (1999), and Tse (1970) have been in use for a long time and are well known for their high performance. The technique of Veigas et al. (1959) considers not less than 447 historical cyclones and augments this with dynamically chosen subsets of 91 sea level pressure readings in the Atlantic basin as predictors (Chin, 1965; Miller and Chase, 1966). The Chen et al. technique uses the same eight predictors as WPCLPR, which constitute this technique's climatology part, and in addition two predefined synoptic patterns in the Northwest Pacific (Carr and Elsberry, 1995), which constitute the synoptic part of its predictor set. In contrast to this, the TC track forecasting technique developed by Tse (1970) does not take climatological data into account and relies exclusively on synoptic patterns at the 700 mbar level to produce TC forecasts. Kumar and Prasad (1973) attempted to increase the average prediction accuracy of Tse's technique by introducing previous-movement information as an additional predictor (Kumar and Prasad, 1973). However, using historical track data as an additional predictor did not increase the prediction accuracy of Tse's technique. Instead, Kumar and Prasad (1973) encountered an increase in error when forecasting TCs with recurving track, where the TC suddenly changes direction (Chin, 1976). This result could be because recurving introduces an element of unexpectedness, which can be especially difficult to forecast looking at historical data.

As previously mentioned, environmental steering can explain up to 70–90% of TC motion. This observation led to the development of steering air flow determination techniques (Keenan, 1982). The basic difference between steering air flow determination and synoptic-climatological techniques is that steering techniques do not use climatological or persistence predictors like the synoptic techniques did. Instead, steering techniques are aimed at determining which air pressure levels (steering levels) best describe the recently observed motion of a cyclone, and uses the airflow information of these layers as predictors to forecast TC track. Which layer or layers are best to use as steering layers is still a point of debate among the meteorological community. Different steering techniques consider different air pressure levels as steering levels (Riehl and Shafer, 1944; Miller and Moore, 1960; Renard et al., 1973). Though there is some disagreement between steering forecasting techniques in the identification of suitable pressure levels, it is broadly accepted that airflows in the mid-tropospheric levels (between 500 and 700 mbar) are the most suitable for forecasting TC track (George and Gray, 1976; Chan and Gray, 1982). In special cases, the steering level may be set outside the normal range, either higher or lower depending on the height of the cyclone vortex.

Dynamical-numerical techniques simulate the physical behaviour of the atmosphere through numerical approximation of mathematical equations which describe the dynamics of weather systems and the physical forces behind TC formation and development (Jeffries et al., 1993; Elsberry, 1995). These techniques produce forecast for TCs through modelling the basic dynamical and physical mechanisms of tropical circulations, so they are not dependent on the availability of the climatological data for the particular basin. Any force influencing the motion of a TC can be considered as a predictor in dynamical-numerical techniques. Moreover, dynamical-numerical techniques may consider these forces in a vertically averaged atmosphere (barotropic forecasting techniques) or in a vertically arranged atmosphere (baroclinic forecasting techniques) (Elsberry, 1995).

Examples of barotropic forecasting techniques include SANders BARotropic (SANBAR) (Sanders and Burpee, 1968), Vic Ooyama BARotropic (VICBAR) (Demaria et al., 1992), Limited Area Sine Transform Barotropic (LBAR) (Horsfall et al., 1997) and Multigrid Barotropic (MUDBAR) (Fulton, 2001). SANBAR considers averaged wind direction and speed from 10 pressure levels (1000 to 100 mbar) and averaged pressure-height data as a predictor. VICBAR also relies on vertically averaged winds and pressure heights to produce forecasts. This wind and pressure height information is obtained from pressure levels between 200 and 850 mbar. Layers below 850 mbar and above 200 mbar are influenced by boundary-layer processes and storm outflow, respectively, so VICBAR does not take any information from these layers into account (Demaria et al., 1992; Horsfall et al., 1997). Once the wind and height information are available, they are applied to three horizontal domains (the synoptic-scale domain, the storm environment domain and the vortex domain) to produce the final forecast. The purpose of considering three domains by VICBAR is to focus on an increasingly smaller area (from an area larger than the cyclone to the cyclone vortex) for better representation of the surrounding environment's influence on TC motion. Both LBAR and MUDBAR use the averaged wind direction and speed information from the same pressure levels as VICBAR (Vigh et al., 2003). LBAR, in contrast to VICBAR, uses wind and pressure information also from the boundary layer, as it is considered important for forecasting TC motion.

Atmospheric situations change sharply with the change in barometric pressure and taking these changes into account could considerably enhance the prediction ability of TC track forecasting techniques. Baroclinic forecasting techniques have been developed to account for these atmospheric changes and use them for TC motion forecasting. The movable fine mesh (MFM) (Hovermale and Livezey, 1977), the Navy Operational Regional Atmospheric Prediction System (NORAPS) (Bayler and Lewit, 1992), the Navy Operational Global Atmospheric Prediction System (NOGAPS) (Bayler and Lewit, 1992), the global forecast system (GFS) (Kanamitsu, 1989), the Quasi-Lagrangian Model (QLM) (Mathur, 1988, 1991), the Beta and Advection Model (BAM) (Marks, 1992), and the Geophysical Fluid Dynamics

Laboratory Multiply-Nested Moveable Mesh Hurricane Model (GHM) (Kurihara et al., 1995; Kurihara et al., 1998) are examples of baroclinic TC track forecasting techniques.

MFM was originally developed for improving forecasts of precipitation associated with TCs making landfall. It considers the physical processes (moisture dynamics, cumulus parameterization and aerodynamics in the boundary layer), and wind information at different pressure levels (from 10 vertically arranged layers while developed) as predictors (Hovermale and Livezey, 1977; Fiorino et al., 1982).

NORAPS and NOGAPS are the regional and global versions of the navy operational atmospheric prediction system. Both models were developed for general purpose atmospheric prediction but can also be used for forecasting TC motion (Baler and Lewit, 1992). NORAPS is characterized by 21 vertical levels and uses u and v wind components, temperature, geopotential heights, mixing ratio (water vapour mass per kilogram of dry air), sea surface temperature and surface air pressure as predictors (Hodur, 1982). NOGAPS in contrast uses information from 42 vertical levels (an early version had 18 vertical levels) and produces forecasts making use of predictors such as vorticity, divergence, virtual potential temperature, specific humidity, surface pressure, ground temperature, ground wetness and cloud patterns (Bayler and Lewit, 1992; Chan and Kepert, 2010).

GFS was developed with a view to produce weather forecasts for aviation worldwide (Sela, 1980), but the model's current use also includes cyclone motion and intensity forecast. GFS has gone through several modifications and improvements since it became operational in 1985 (Bonner et al., 1986; Bonner, 1989). Air temperature, east and northward wind velocity, rainfall rate, relative humidity, mean sea level pressure and net downward long-wave and short-wave radiation fluxes are used by GFS as predictors (Kanamitsu, 1989). Presently GFS integrates 64 vertical layers and produces both short- and long-term TC track forecasts.

QLM is characterized by 16 vertical layers and use both internal and external processes affecting the motion of TC as predictors. These predictors include surface pressure, relative vorticity, divergence, virtual temperature, mixing ratio, surface friction effects, sea–air exchange of heat, convective release of heat, horizontal diffusion and isobaric condition of water vapour (Prasad and Rama Rao, 2003).

BAM, like other barotropic techniques, is using averaged wind information from different pressure levels. However, unlike other barotropic techniques, BAM does this averaging in three different versions: BAM shallow (considers wind information from 850 to 700 mbar), BAM medium (considers wind information from 850 to 400 mbar) and BAM deep (considers wind information from 850 to 200 mbar). BAM also includes a correction term in that it considers the Coriolis force's influence on TC motion at different latitudes (Holland, 1983; Marks, 1992).

1.5.1.3   Techniques Using Both Historical and Steering Information as Predictors

Techniques of the third category possess characteristics both from statistical and dynamical forecasting techniques and are known as statistical-dynamical forecasting techniques. As these techniques aim to combine the strengths of both statistical and dynamical forecasting techniques, they use historical track data as well as wind information from single or multiple pressure levels associated with the current cyclone as predictors. Examples of statistical-dynamical forecasting techniques include the National Hurricane Center (NHC) statistical-dynamical techniques, the Colorado State University Model (Matsumoto, 1984) and the Joint Typhoon Warning Center 92 (JTWC92) (Englebretsn, 1992).

1.5.1.4 Techniques Using Satellite-Derived Information as Predictors

Any satellite image or product of the image that holds useful information about TC track can in principle be considered as a predictor (Kokhanovsky and von Hoyningen-Huene, 2004). Though most of today's operational TC track forecasting techniques use satellite-derived information as predictors to enhance their forecast accuracy, in the fourth category of techniques we considered only those techniques that use satellite images as primary predictors. These satellite-derived predictors include the visible channel image, the infrared channel image, the ocean heat content image, the sea surface temperature image, the water vapour image, the precipitable water image, the surface wind image and the mean sea level pressure image.

Satellite-image-based cyclone forecasting techniques interpret these images to produce track forecast for TCs (Leslie et al., 1998; Tomassini et al., 1998; Velden et al., 1998; Isaksen and Stoffelen, 2000; Xiao et al., 2002; Wang et al., 2006; Goerss, 2009; Langland et al., 2009; Ma and Tan, 2009). Examples of satellite-image-based TC track forecasting techniques include the Lajoie–Nicholls technique (Lajoie and Nicholls, 1974), the U.S. Air Force Defense Meteorological Satellite Program (DMSP) (Fett and Brand, 1975), and techniques using artificial neural networks (ANN). The selection of satellite images as predictors varies largely between these techniques. The Lajoie–Nicholls technique uses only visible channel satellite images as predictors. This technique was primarily aimed for and tested in the north Australian region. Forecasts were based on 68 satellite images of 16 TCs (Lajoie, 1976). The DMSP technique uses both visible and infrared channel satellite images as predictors and was tested on 31 separate datasets of 7 cyclones in the Northwest Pacific basin (Fett and Brand, 1975). Unlike the Lajoie–Nicholls and the DMSP techniques, ANN-based techniques allow for more freedom to choose the predictors, as statistical patterns detected in any dataset can be easily combined into a coherent forecast. For example, the nonlinear neural network developed by Pickle (1991) uses wind speed, wind direction and geopotential heights of five different pressure levels (700, 500, 400, 300 and 200 mbar) as predictors (Pickle, 1991). The biologically based hierarchical ANN technique developed by Kovordányi and Roy (2009) uses only visible channel NOAA-AVHRR satellite images as predictors, but this technique is open to further inclusion of additional satellite images as predictors. Other ANN-based techniques that use visible or infrared channel images as predictors include the use of Hybrid Radial Basis Function (HRBF) networks (Lee and Liu, 2000), SEMO-MAMO (Feng and Liu, 2004), and an ANN-based technique developed by Ali and co-workers (Ali et al., 2007).

1.5.2 EQUATION-SYSTEM USE

An alternative way to categorize TC track forecasting techniques is on the basis of their equation use. This categorization runs orthogonally to the previously described grouping of forecasting techniques based on their predictor selection. While the previously described groups were based on their selection of input data, in the sections below we describe how these data are processed and combined. With equation use in mind, TC track forecasting techniques can be classified broadly into the following four categories (Figure 1.7).

1.5.2.1 Techniques Using Statistical Equations

Techniques that belong to the first category use regression equations (Miller and Chase, 1966), along with other statistical equations like averaging, extrapolation and bivariate normal distribution as forecast equations. These techniques are considered to be the simplest among all TC track forecasting techniques (Chin, 1976). As previously mentioned, TC motion characteristics are related to TC formation location and time. This entails that the motion characteristics of historical TCs could provide valuable information about the motion of the current TC in the same basin. These types of observations led to the development of several TC track forecasting techniques. Extrapolation, climatology and analogue forecasting techniques, for example, make use of the motion characteristics of the current and historical TCs in the same basin to produce track forecasts for the current cyclone.

XTRP is the simplest among all TC track forecasting techniques that use *extrapolation* as the prediction equation. This technique extrapolates the cyclone's current and recent past movement (6 to 12 h back in time) to produce a forecast for the next 12 to 24 h (Jeffries et al., 1993). CLIM produces 24-, 48- and 72-hour forecasts through an unweighted averaging of the motion of historical cyclones in the same basin. TYAN93, in contrast, produces forecasts through a weighted averaging of the historical datasets (Jeffries et al., 1993). Rather than direct weighted/unweighted averaging of historical TC tracks, HURRAN preprocesses the historical analogues before using them for forecast. This preprocessing includes relocation of the historical analogues' origin to a common origin and rotation of the historical analogues' heading to a common heading (Neumann, 1979). Once the preprocessing is done, all historical analogues' positions after 12, 24, 36, 48 and 72 h are fitted to a bivariate normal distribution and the locations of the centroids of these distributions are taken as the most likely forecasted track of the current cyclone. HURRAN also produces probability ellipses for these time periods, so that any location within these ellipses is considered to be a less likely forecasted location of the current cyclone (Hope and Neumann, 1970).

CLIPER and EPCLPR both accomplish the forecasting task applying the least squares method to fit discrete data to continuous polynomials. This process actually relates future zonal and meridional displacements of a TC to a set of predictors, and on the basis of this relationship a forecast is produced (Neumann, 1972, 1979). CLIP was developed for the Northwest Pacific basin. Also, this technique uses polynomial regression equations to produce forecasts for cyclones (Xu and Neumann, 1985). CLIP applies 12 regression equations, 6 of which are used to predict zonal speeds and the remaining 6 are used to predict meridional speeds in 12-h increments. Finally, the resulting predictions are combined to forecast TC positions for the next 24, 48 and 72 h (Jeffries et al., 1993).

Statistical synoptic techniques produce track forecasts through correlating geopotential height or a combination of geopotential height data with zonal and meridional motion of a TC (Neumann, 1979). As part of the forecast, the heights of the pressure surfaces measured from the mean sea level (the geopotential heights) are usually shown on a storm-centred grid of latitudes and longitudes. Correlation coefficient fields between current geopotential heights and the TC's future (12 h) zonal and meridional motion are also calculated and presented on the same grid. A stepwise screening regression method automatically selects the grid with maximum correlation. A TC is supposed to move in a direction where the correlation is maximal for the next 12 h. As heights of the pressure surfaces do not remain the same for longer time periods, for long-term forecast (72 h) climatological predictors are used along with geopotential height data (Chin, 1976).

TC motion is mainly controlled by large-scale flow surrounding it. These observations led to the development of the steering airflow technique of TC movement (Chan and Gray, 1982). Steering airflow techniques consider the cyclone as a point vortex embedded in the surrounding environment in such a way that the direction and speed of the vortex can be approximated by the direction and speed of surrounding winds (George and Gray, 1976). To produce motion forecasts using the steering airflow determination technique, speed and direction of the surrounding wind at different pressure levels are plotted on a cyclone centred grid of suitable coordinate system, which could be cylindrical or geographical. Wind information of the TC vortex is also plotted on a separate grid using the same coordinate system. This grid containing wind information of the surrounding environment is then rotated to match the direction of the current TC's motion so that the cyclone motion can always be kept at a 0° or 360° heading. The difference between the movement direction of the surrounding wind and the movement of the TC vortex is calculated for different pressure levels. Finally, the level/levels where the least amount of difference is found is considered as steering level/levels and used to produce a forecast (George and Gray, 1976; Dong and Neumann, 1986).

1.5.2.2 Techniques Using Fluid Dynamics and Thermodynamics Equations

A combination of equations describing the fluid dynamics and thermodynamics of the atmosphere constitutes the computational basis of the techniques of the second category (Elsberry, 1995). These equations include (Gates et al., 1955; Hubert, 1957; DeMaria, 1985; Radford, 1994):

- *Mathematical equations* describing the large-scale horizontal movement of the atmosphere.
- *Shallow water equations* describing the hydrodynamics of shallow water. This is motivated by the fact that the dynamics of the atmosphere under the influence of the Coriolis force correspond to the hydrodynamics of shallow water. These observations led to the development of the shallow water equation of TC track forecast.
- *Primitive equations* are a set of nonlinear differential equations (mainly describing the conservation of momentum, the thermal energy and the conservation of mass). These equations approximate the dynamic, thermodynamic and the static state of the atmosphere.

- *Barotropic equations* assume that measurements of atmospheric variables affecting TC motion do not change with height above sea level. Therefore, these equations consider predictors in a vertically averaged atmosphere for producing TC track forecasts.
- *Baroclinic equations*, as opposed to barotropic techniques, assume that measurements of atmospheric variables change with height above sea level. For this reason, baroclinic models consider atmospheric variations at different vertically arranged pressure levels, and can therefore better represent the dynamics within a TC compared to barotropic techniques.

Though the above-mentioned equations constitute the computational basis for the dynamical and numerical TC forecasting techniques, the performance of these techniques depends on a number of auxiliary elements. These elements include (Vanderman, 1962; Haltiner and Williams, 1980; Leslie and Holland, 1995; Wang, 1998) the following.

- *A grid of points to hold the predictors.* All dynamical and numerical techniques start with atmospheric fields on a grid array. This grid array holds all the predictors required for the techniques to run. Though the grids can be of any shape, most often they are square in shape. Commonly used grids include various types of map projections, such as the Mercator, Lambert or polar stereographic projections. The grid resolution is selected based both on the data resolution and the region of interest.
- *Suitable means of representing the vertical structure of the atmosphere.* All multilevel models use some form of vertical coordinate system to properly describe the vertical structure of the atmosphere. Vertically arranged pressure levels are most commonly used as the base of this vertical stratification.
- *Analysis and assimilation processes to obtain the initial fields for running the technique.* Data used as predictors by the forecasting techniques are obtained from a wide variety of sources and instruments. Therefore, these data need to be converted into a suitable form so that TC track prediction techniques can use them easily.
- *Means of bogussing to provide additional information in the data-scarce regions.* Forecasts from the dynamical models can be inaccurate if the observational values for any region of the model grid are not available. In such data-scarce regions, it is necessary to include false observations derived from human interpretation or empirical relationships of the predictors. Two approaches of bogussing are commonly used for TC track prediction: moisture bogussing and TC vortex bogussing.
- *Means of handling boundary conditions.* Efficient handling of boundary conditions is very important for all dynamical and numerical techniques (Errico and Baumhefner, 1987). As atmospheric disturbances generated at the boundary can rapidly propagate throughout the model domain, it is necessary to exclude or at least minimize these disturbances to avoid unexpected errors in the forecast. Therefore, dynamical techniques are designed to handle two types of boundary conditions: the real boundaries

at the Earth's surface and top of the atmosphere, and the pseudo horizontal boundaries that divide the whole model area into smaller squares or triangles. This is required because of the computational resource limitations.

- *Some form of physical parameterization.* Dynamical and numerical techniques cannot adequately handle small-scale variations in the atmosphere, such as the interchange of energy between the surface and the ocean. Small-scale processes can have remarkable impact on the TC track, and are included in the technique by parameterizing their effects on a larger scale.
- A *method for carrying the model equations forward in time.* After initialization, dynamical and numerical techniques determine the tendencies of change that various predictors possess. This allows the forecasting techniques to extrapolate the TC motion over a short time interval and to generate a new set of initial data. This whole process is repeated until the forecasts for the desired time frame are obtained (Hubert, 1957).

1.5.2.3 Techniques Using a Combination of Mathematical and Statistical Equations

Techniques of the third category use both statistical and dynamical equations to produce TC track forecast. Fluid-dynamics and thermodynamics equations can effectively capture various internal and external factors affecting the current TC's motion. Statistical techniques in contrast are good at considering the climatological characteristics of TCs. Many of the currently used statistical-dynamical forecasting techniques, such as those developed and used at the NHC, and JTWC92 do not approximate the physical behaviour of the atmosphere or calculate the historical TC movement pattern themselves. Instead, these techniques integrate outputs from statistical and dynamical techniques to produce a more balanced forecast for TCs (Neumann and Lawrence, 1975; Neuman, 1988; Newman and McAdie, 1991).

1.5.2.4 Techniques Using Other Types of Equations

In addition to these statistical and dynamical equation-based techniques, there are alternative techniques that employ some form of statistical equations to produce forecasts for TCs. As these statistical equations are not widely implemented for TC forecast, they are discussed separately in a fourth category of equations used for TC forecasting. Equations used in this category include the statistical equations widely employed in ANNs and empirical and hybrid forecasting techniques.

Application of ANN as a TC track forecasting technique is fairly recent compared to other forecasting techniques. As these techniques are still in the phase of development and require a huge effort to build and train, they are not as widely adopted as other forecasting techniques. Almost all ANN-based TC forecasting techniques use satellite images as predictors. Extraction of useful information from the satellite images that could be translated into a TC track forecast often involves processes such as identification of variations in very low contrast cloud features, identification of spectral and spatial patterns in the images and integration of information collected through various satellite sensors (Villman et al., 2003). ANN-based techniques can efficiently be used to process these data and produce forecasts for TCs.

ANNs consist of many artificial units, grouped into layers. Units in adjacent layers are connected with each other, as a rule in an all-to-all fashion (Figure 1.8). The number of input and output layers depends on the number of predictors used in the forecast and the number of variables to be predicted (track, intensity, etc.) (Kasiri et al., 2008). Various types of satellite images are fed into the corresponding input layers for processing. A weighted sum of all these inputs constitutes the information forwarded to the hidden and output layers. As part of the transfer of signals to the hidden layers, an activation function is usually used to determine the activation pattern of nodes in the hidden layer (Villmann et al., 2003). The hidden layers forward the processed information to the output layers where it undergoes another transformation, and finally the outputs are generated (Møller, 1993; Mnih and Hinton, 2010).

ANN-based TC track forecasting techniques that use supervised learning include the nonlinear ANN technique developed by Pickle (1991), the Hybrid Radial Basis Function (HRBF) network technique developed by Lee and Liu (2000), the SEMO-MAMO technique developed by Feng and Liu (2004) and the genetic learning artificial neural network (GLANN) developed by Yang and Wang (2005). In contrast to these techniques, the biologically based hierarchical ANN technique developed by Kovordanyi and Roy (2009) uses a more advanced learning algorithm where supervised learning is intertwined with unsupervised learning, in this way allowing for an effective development of feature detectors, while also optimizing overall network performance (O'Reilly, 1998; O'Reilly and Munakata, 2000; Aisa et al., 2008).

Though different ANN-based TC track forecasting techniques can have different architectures and can use different learning algorithms, all ANN-based techniques work through two phases: training and testing. During the training phase, the network

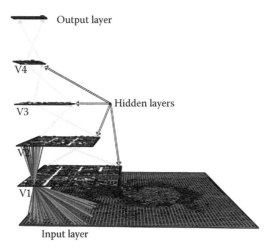

FIGURE 1.8 A neural network with six hierarchically arranged layers. Layers at the bottom and at the top represent the input and output layers, accordingly. Four layers between the input and output layers are called hidden layers. All the layers except input are bidirectionally connected. This is because the input layer only sends information to the hidden layer and the other layers both send and receive information from the next hierarchical layer.

learns to detect statistically indicative information from the predictor datasets. In the testing phase the network interprets new sets of data and produces forecasts using knowledge of reoccurring patterns that the network extracted during the training phase.

An important subcategory of forecasting techniques comprise 'manual' techniques where forecasting to a large extent relies on human expertise. These empirical forecasting techniques were developed based on the assumption that the current TC will move in the same direction and with the same speed as most of the past storms which occurred on the same location and at the same time of the year (Chin, 1976). The prediction accuracy of these techniques depends on the human forecasters' skill in recognizing the TC movement pattern. Though empirical techniques are not in operation at meteorological offices today, forecasters' skill in recognizing patterns can eliminate forecasting errors associated with recurving, rapid-moving and looping TCs (Jeffries et al., 1993). Examples of empirical forecasting techniques include the Dvorak technique (Dvorak, 1984), the Martin–Holland technique (Mautner and Guard, 1992) and the typhoon acceleration prediction technique (TAPT) (Weir, 1982; Mautner and Guard, 1992).

Hybrid forecasting techniques combine the output of two or more forecasting techniques using statistical methods. All TC track forecasting techniques are not equally skilful in handling various predictors. Hybrid forecasting techniques were developed with the aim to accumulate the strengths of different forecasting techniques into a single technique. When two or more forecasting techniques are combined, the resultant technique appears to produce more accurate forecasts and can handle a broader set of predictors compared to the constituent individual techniques. The major weakness of hybrid forecasting techniques is that the process of combining the output of these techniques causes the error levels of the parent techniques to accumulate (Jeffries et al., 1993).

Among the hybrid techniques, the half persistence and climatology (HPAC) combines the outputs from XTRP and CLIM with equal weights to produce TC track forecast (Mautner and Guard, 1992). The combined confidence weighted forecast (CCWF) combines the outputs of the one-way influence tropical cyclone model (OTCM), the Colorado State University Model (CSUM) and the HPAC to produce TC track forecast. The inverse of the covariance matrices computed from historical and real-time cross-track and along-track errors defines the weighting function for CCWF during combination (Jeffries et al., 1993). The blended forecasting technique (BLND) uses the average of outputs of six forecasting techniques (OTCM, CSUM, FBMA, JTWC92, CLIP and HPAC) to produce TC track forecast (Dillon and Andrews, 1997). The weighted forecast technique (WGTD) uses outputs from the same six techniques as BLND, but here the weights are adapted based on the component technique's previous performance. The proportion of weights currently used by WGTD are: OTCM—29%, CSUM—22%, FBMA—14%, JT92—14%, HPAC—14% and CLIP—7% (Dillon and Andrews, 1997).

1.5.3 Prediction Length and Performance

Prediction accuracy is considered to be the most important property of TC track forecasting techniques. Due to the variations in geographical and climatological conditions and the use of predictors and prediction equations, different forecasting techniques can perform differently in the same basin. Conversely, similar forecasting techniques can perform differently in different basins. The statistics we provide here on how various TC track forecasting techniques perform are based on the annual tropical cyclone reports published by the Joint Typhoon Warning Center, and the forecast verification reports published by the National Hurricane Center.

The performance of individual forecasting techniques cannot be directly derived from which predictors and which forecast equations are used in a particular technique. As a result, it is difficult to categorize individual forecasting techniques based on their performance. For this reason, we compared groups and subgroups of TC track forecasting techniques according to the grouping proposed by Roy and Kovordanyi (2012). In addition, if the forecasting techniques within a group or subgroup are operational in different basins, their performance is presented on different vertical axes of the same graph.

Forecasting techniques that apply some form of averaging across cyclone occurrences, also known as extrapolation techniques, can produce forecasts with acceptable accuracy for shorter time periods (12 to 24 h). As extrapolation techniques are not reliable for longer-time predictions (>48 h), meteorological offices around the world do not use these techniques as operational forecasting techniques. To give an example of the typical performance of these techniques, during the period 1970–1990 the average TC track prediction error produced by XTRP in the Northwest Pacific basin was around 200 km for 24 h and 500 km for 48 h (Figure 1.9).

Cyclone mean motion detection techniques, also known as climatology techniques, are capable of producing forecasts for longer time periods compared to extrapolation techniques. For 24-h forecasts, CLIM performs similarly both in the Northwest Pacific and North Indian basins. For 48- and 72-h forecasts, performance is better in the Northwest Pacific basin (Figure 1.10). This could be due to the better availability of suitable predictors in the historical datasets, as the Northwest Pacific is the most active among all TC formation basins.

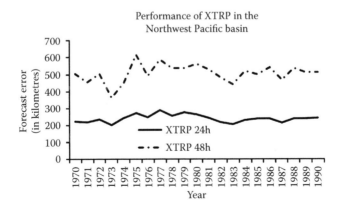

FIGURE 1.9 Performance of extrapolation (XTRP) technique in the Northwest Pacific basin during the period 1970–1990.

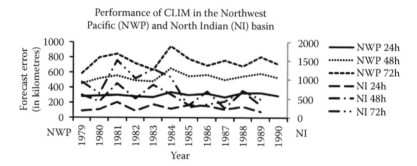

FIGURE 1.10   Performance of CLIM in the Northwest Pacific and North Indian basins for 24, 48 and 72 hours during the period 1979–1990. Performance in the Northwest Pacific is presented on the left and performance in the North Indian basin is presented on the right vertical axis.

For a 24-h forecast, the performance of HURRAN in the North Atlantic basin is similar to the performance of TYFOON in the Northwest Pacific basin (Figure 1.11). For longer time periods (48 and 72 h), TYFOON performs better compared to HURRAN in the respective basins (Figure 1.11). Though longer-term forecasts are generally associated with larger forecast error, TYFOON produces smaller forecast errors for 72-h compared to 48-h forecasts in the Northwest Pacific basin.

For 24-h forecasts, climatology and persistence techniques perform similarly both in the Northwest Pacific and in the North Atlantic basin. In the case of longer-range forecasts (48 and 72 h), performance is better in the Northwest Pacific basin

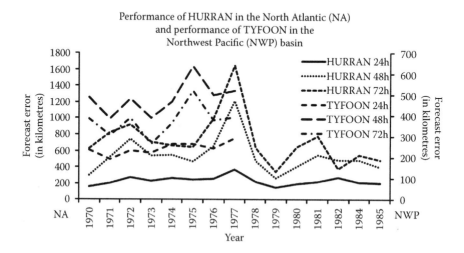

FIGURE 1.11   Performance of HURRAN in the North Atlantic basin for 24-, 48- and 72-h forecasts during the period 1970–1985 (presented on the left vertical axis) and performance of TYFOON in the Northwest Pacific for 24-, 48- and 72-h forecasts during the period 1970–1977 (presented on the right vertical axis).

than in the North Atlantic basin (Figure 1.12). The relatively poor performance of climatology and persistence techniques in the North Atlantic basin could be attributed to the irregular pattern of cyclone formation and the strong influence of external forces on the subsequent TC track.

In the North Atlantic basin, statistical synoptic technique National Hurricane Center 67 (NHC67) performs similar to NHC72 for short-term forecast (12 h). For 24-, 48- and 72-h forecasts, NHC72 performs better than NHC67 (Figure 1.13). As geopotential height data is usually available only for the last 24 h, statistical synoptic techniques perform better for shorter-term forecasts (up to 24 h). Forecast

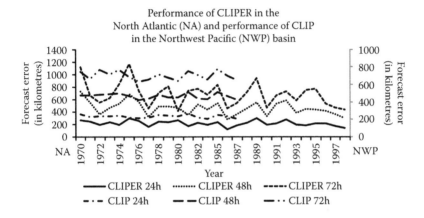

FIGURE 1.12   Performance of CLIPER in the North Atlantic for 24-, 48- and 72-h forecasts during the period 1970–1998 and performance of CLIP in the Northwest Pacific for 24-, 48- and 72-h forecasts during the period 1970 to 1987. CLIPER is presented on the left and CLIP is presented on the right vertical axis.

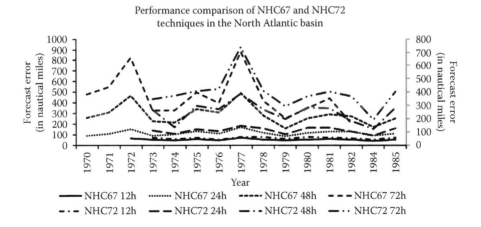

FIGURE 1.13   Performance of NHC67 (presented on the left vertical axis) and NHC72 (presented on right vertical axis) for 12-, 24-, 48- and 72-h forecasts in the North Atlantic basin.

performance varies across years for longer-term predictions (>24 h), as climatological predictors are used along with geopotential height data (Figure 1.13).

Barotropic forecasting techniques use vertically averaged wind information in their predictor datasets. Averaged wind direction and speed yield forecasts with acceptable accuracy for shorter time periods (up to 24 h). For longer time (>24 h) the averaged wind field method is not enough to properly describe the environmental effects on TC motion, which means that forecast errors become larger (Figure 1.14).

Three different versions of BAM, namely BAM Shallow, BAM Medium and BAM Deep, are operational in the North Atlantic basin. The 12-h and 24-h performance of BAMD and BAMS are very similar. However, for 48-h and 72-h forecasts, BAMS outperforms BAMD (Figure 1.15a). Likewise, the performance of BAMD and BAMM are similar for shorter-term 12-, 24- and 48-h forecasts; for longer-term 72-h forecasts, BAMM performs slightly better than BAMD (Figure 1.15b). Fnmoc Beta and Advection Model (FBAM) uses a weighted average of the wind direction and speed lying between 1000 mbar and 100 mbar pressure levels as the steering level. This technique is operational in the Northwest Pacific basin. As FBAM uses wind information from identical pressure levels like BAMD, these two techniques show nearly equal performance both for short- and long-term forecasts in the respective basins (Figure 1.16).

Combining wind information from different pressure levels allows baroclinic techniques to produce forecasts with improved accuracy compared to barotropic techniques. Though the early baroclinic techniques perform at the same level as barotropic forecasting techniques, newer baroclinic techniques outperform barotropic

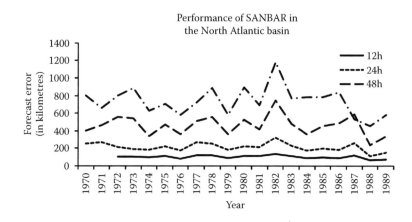

FIGURE 1.14   SANBAR performance in the North Atlantic basin for 12-, 24-, 48- and 72-h forecasts during the period 1970–1989.

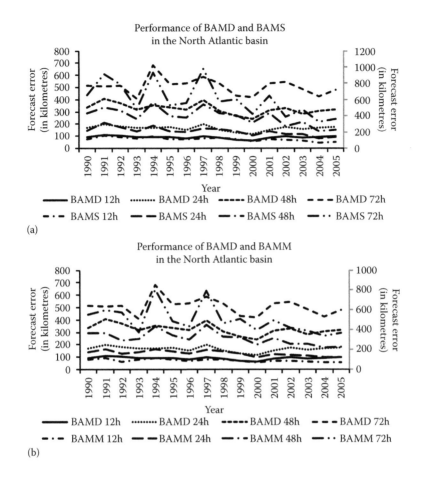

FIGURE 1.15 Performance of BAMD, BAMM and BAMS in the North Atlantic basin. (a) For 12-, 24- and 48-h BAMD and BAMS perform almost similar but for 72-h BAMS performs better than BAMD. (b) If the performance of BAMM is compared to BAND we will get a similar scenario. For 12-, 24- and 48-h forecasts, both of the techniques perform almost the same, but for 72-h BAMM performs better compared to BAMD.

forecasting techniques in the same basin (Cooper and Falvey, 2008). For example, in the North Atlantic basin, the MFM technique performs almost at level with SANBAR for 12- and 24-h forecasts, but outperforms SUNBAR for 48-h forecasts (Figure 1.17a). Then again, the forecast error of the GHM, the NOGAPS or NGPS and the QLM are significantly lower for 12-, 24-, and 48-h predictions compared to MFM (Figure 1.17b–d).

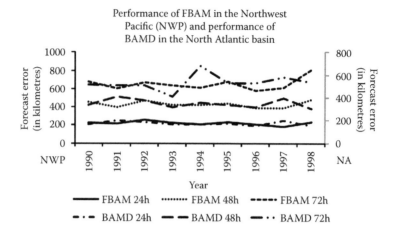

FIGURE 1.16   Comparison between the performances of FBAM (presented on the left vertical axis) and BAMD (presented on the right vertical axis). Both of the techniques perform
similarly for 12-, 24- and 48-h forecasts in the respective basins.

The OTCM is another baroclinic technique that produces forecasts for cyclones
in the Northwest Pacific basin. OTCM and NGPS produce similar forecast errors in
this basin (Figure 1.18).

The NHC statistical-dynamical techniques represent a series of four techniques
(NHC73, NHC83, NHC90 and NHC98). Each of the successive techniques in this
series has been developed through extensions and adjustments of the preceding technique, to achieve better performance for TC track forecasts in the North Atlantic
basin (Figure 1.19a,b).

The CSUM is also a statistical-dynamical forecasting technique and is operational in the Northeast Pacific and North Indian Ocean basins. This technique is
especially good at producing forecasts for slow moving and erratic moving or looping cyclones. CSUM performs at the same level of accuracy as CLIP for 24- and
48-h forecasts. For 72-h forecasts, the CSUM performs slightly better than CLIP
(Figure 1.20).

1.5.4   COMPUTATIONAL RESOURCE REQUIREMENTS

The computational resources required by various forecasting techniques depend
on the predictors used for the forecast, forecast length and on the complexity of
the equations used to produce the forecast. As it is difficult to categorize the TC
forecasting techniques based on computational resource requirements, we present this information for individual groups and subgroups without any comparison
(Table 1.1).

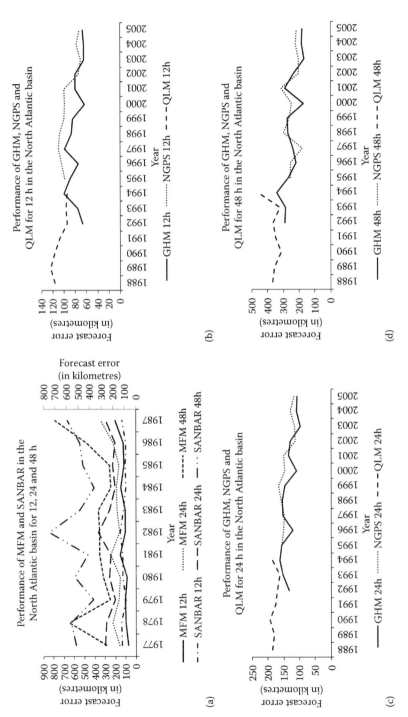

FIGURE 1.17 (a) Performance of MFM and SANBAR in the North Atlantic basin for 12-, 24- and 48-h forecasts. (b) GHM, NGPS and QLM performance for 12 h in the North Atlantic basin. (c) GHM, NGPS and QLM performance for 24 h in the North Atlantic basin. (d) GHM, NGPS and QLM performance for 48 h in the North Atlantic basin. For 12-, 24- and 48-h forecasts, GHM, NGPS and QLM perform almost the same.

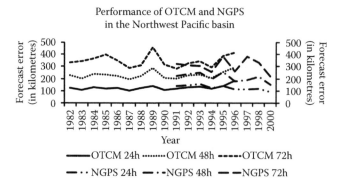

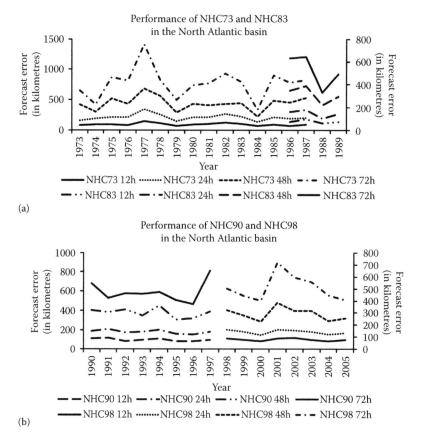

FIGURE 1.18   Performance of baroclinic forecasting techniques OTCM (presented on the left vertical axis) and NGPS (presented on the right vertical axis) in the Northwest Pacific basin.

FIGURE 1.19   (a) Performance of NHC73 (on the left vertical axis) and HNC83 (on the right vertical axis) in the North Atlantic basin. (b) Performance of HNC90 (on the left vertical axis) and HNC98 (on the right vertical axis) in the same basin.

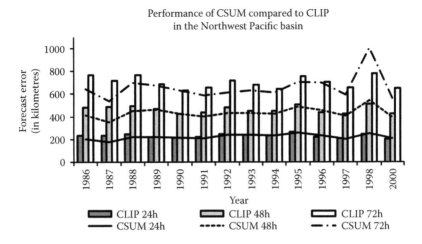

FIGURE 1.20 Performance of CSUM and CLIP in the Northwest Pacific basin for 24-, 48- and 72-h forecasts.

TABLE 1.1

Computational Resource Requirements of the Groups and Subgroups of TC Track Forecasting Techniques

Group of Forecasting Technique	Subgroup	Data Requirements	Computational Resource Requirements
Averaging across occurrences	–	Cyclone's current and recent-past positions	Can be run on PCs
Statistical	Climatology	Past cyclone tracks in the same basin	Can be run on PCs
	Climatology and persistence	Current cyclone's position, wind direction and past tracks	Can be run on PCs
	Synoptic	Past cyclone track data and atmospheric information from different pressure levels	Can be run on PCs
	Steering airflow	Wind information from different pressure levels surrounding the cyclone eye	Can be run on PCs
	Satellite image-based	Satellite images	Can be run on PCs
	Neural network	Satellite images of previous and current cyclones. Other data can also be used	Can be run on PCs
Dynamical and numerical	–	Atmospheric information from different vertically arranged pressure levels	Parallel processing required
Statistical-dynamical	–	Combines forecasts from different statistical and numerical models	Parallel processing required
Hybrid	–	Combines forecasts from different techniques	Depends on which techniques are being combined

1.6   CONCLUSIONS

TCs are highly varying and extremely complex weather phenomena, whose formation and motion characteristics seem to depend on a complex interaction between
external and intrinsic forces. What is more, these forces seem to act differently in
different ocean basins. Consequently, although it would be important to be able to
forecast if and where a TC will hit land, today there is no consensus on which factors are more influential in tropical cyclone formation and motion, and today it is
difficult to pinpoint a single forecasting technique as the generally best performing
for cyclone track forecasting. In particular, continued research is needed in order to
understand the mechanisms that drive an irregular motion pattern, such as a sudden
recurving or looping of a TC track, and in connection with this we also need to better understand how small disturbances, such as island topography, affect TC motion.

In general, it seems that future TC forecasting techniques need to be more dynamically adaptable to changing circumstances, such as those driving sudden changes
in TC track. One avenue for further research could be the development of large-
scale ANN techniques that can be automatically adapted through online (i.e. during
operation) learning as the ANN is exposed for irregular data in everyday operation.

REFERENCES

Abdullah, A. J. (1954). A proposed mechanism for the development of the eye of a hurricane, *J. Atmos. Sci.*, 11, 189–195.

Aisa, B., Mingus, B. and O'Reilly, R. (2008). The emergent neural modeling system, *Neural Networks*, 21(8), 1146–1152. doi: 10.1016/j.neunet.2008.06.016. Epub 2008 Jul 1.

Ali, M. M., Kishtawal, C. M. and Jain, S. (2007). Predicting cyclone tracks in the north Indian Ocean: An artificial neural network approach, *Geophys. Res. Lett.*, 34(4), L04603.

Annette, P. (1978). Cyclogue: Analogic prediction of the course of tropical cyclones by national meteorological analysis center, Australia Bureau of Meteorology, Melbourne, Australia, Tech. Rep. No. 28.

Anthes, R. A. (1982). Tropical cyclones: Their evolution, structure and effects, *Meteorological Monographs,* 19(41). American Meteorological Society.

Arnault, J. and Roux, F. (2011). Characteristics of African easterly waves associated with tropical cyclogenesis in the Cape Verde Islands region in July-August-September of 2004–2008, *Atmospheric Research*, 100(1), 61–82.

Atkinson, G. D. (1971). Forecasters' guide to tropical meteorology, US Air Force, Scott AFB, IL, AWS-TR 240.

Avila, L. A. (1990). Atlantic tropical systems of 1989, *Monthly Weather Review*, 118, 1112–1125.

Barnes, G. M., Zipser, E. J., Jorgensen, D. et al. (1983). Mesoscale and convective structure of a hurricane rainband, *J. Atmos. Sci.*, 40(9), 2125–2137.

Bayler, G. and Lewit, H. (1992). The Navy operational global and regional atmospheric pre-diction systems at the fleet numerical oceanography center, *Weather Forecast*, 7(2), 273–279.

Bell, G. J. (1962). Predicting the movement of tropical cyclones in the region of the China Sea, *Proc. Interregional Seminar on Tropical Cyclones*, Tokyo, Japan, 1962, pp. 18–31.

Bender, M. A., Tuleya, R. E. and Kurihara, Y. (1985). A numerical study of the effect of a mountain range on a landfalling tropical cyclone, *Monthly Weather Review*, 113(4), 567–582.

Bender, M. A., Tuleya, R. and Kurihara, Y. (1987). A numerical study of the effect of island terrain on tropical cyclones, *Monthly Weather Review*, 115(1), 130–155.

Bin, W., Elsberry, R. L., Yuqing, W. et al. (1999). Dynamics in tropical cyclone motion: A review, *Chinese J. Atmos. Sci.*, 22(4), 416–434.

Bonner, W. (1989). NMC overview – Recent progress and future plans, *Weather Forecast*, 4, 275–285.

Bonner, W. D., White, G. H., Tracton, M. S. et al. (1986). Global analysis and prediction at NMC Washington, Wellington, New Zealand, pp. 1–5.

Brand, S. and Blelloch, J. W. (1973). Changes in the characteristics of typhoons crossing the Philippines, *J. Appl. Meteorol.*, 12, 104–109.

Carr III, L. E. and Elsberry, R. L. (1995). Systematic and Integrated Approach to Tropical Cyclone Track Forecasting Part 1. Approach Overview and Description of Meteorological Basis, Naval Postgraduate School, Monterey, CA, Tech. Rep. No. NPS-MR-94-002.

Challa, M. and Pfeffer, R. L. (1980). Effects of eddy fluxes of angular momentum on model hurricane development, *J. Atmos. Sci.*, 37, 1603–1618.

Chan, J. C. L. and Gray, W. M. (1982). Tropical cyclone movement and surrounding flow relationships, *Monthly Weather Review*, 110(10), 1354–1374.

Chan, J. C. L. and Kepert, J. D. Eds. (2010). *Global Perspectives on Tropical Cyclones: From Science to Mitigation.* World Scientific.

Charney, J. G. and Eliassen, A. (1964). On the growth of the hurricane depression, *J. Atmos. Sci.*, 21(1), 68–75.

Chen, J. M., Elsberry, R. L., Boothe, M. A. et al. (1999). A simple statistical-synoptic track prediction technique for western North Pacific tropical cyclones, *Monthly Weather Review*, 127(1), 89–102.

Chen, T. C., Weng, S. P., Yamazaki, N. et al. (1998). Interannual variation in the tropical cyclone formation over the western North Pacific, *Monthly Weather Review*, 126(4), 1080–1090.

Chen, Y. and Yau, M. K. (2001). Spiral bands in a simulated hurricane. Part I: Vortex Rossby wave verification, *J. Atmos. Sci.*, 58(15), 2128–2145.

Chin, P. C. (1965). Survey of objective methods used in Hong Kong for predicting typhoon movement, presented at the Inter-Regional Seminar on Advanced Tropical Meteorology, Manila, Philippines, pp. 1–14.

Chin, P. C. (1976). A diagnostic approach in tropical cyclone movement forecasting by objective techniques. Lecture delivered at the Ninth Session of the Typhoon Committee, Manila. 23–29 November 1976. *Royal Observatory: Hong Kong*, 63. p. 22. http://www.hko.gov.hk/publica/pubreprint_search.shtml?db_type=reprint&html_name=pubreprint.htm&search_field=1&search_for=A+diagnostic+approach+in+tropical+cyclone+movement+forecasting+&action=Search (Accessed 13 November 2017).

Chow, K. C., Chan, K. L. and Lau, A. K. H. (2002). Generation of moving spiral bands in tropical cyclones, *J. Atmos. Sci.*, 59(20), 2930–2950.

Cooper, G. A. and Falvey, R. J. (2008). Annual Tropical Cyclone Report, JTWC, Guam, Mariana Islands, Annual report.

DeMaria, M. (1985). Tropical cyclone motion in a nondivergent barotropic model, *Monthly Weather Review*, 113, 1199–1210.

DeMaria, M. (1996). The effect of vertical shear on tropical cyclone intensity change, *J. Atmos. Sci.*, 53(14), 2076–2088.

Demaria, M., Aberson, S. D., Ooyama, K. V. et al. (1992). A nested spectral model for hurricane track forecasting, *Monthly Weather Review*, 120(8), 1628–1643.

DeMaria, M., Knaff, J. A. and Connell, B. H. (2001). A tropical cyclone genesis parameter for the tropical Atlantic, *Weather and Forecasting*, 16(2), 219–233.

Diercks, J. W. and Anthes, R. A. (1976). Diagnostic studies of spiral rainbands in a nonlinear hurricane model, *J. Atmos. Sci.*, 33, 959–975.

Dillon, C. P. and Andrews, M. J. (1997). Annual Tropical Cyclone Report, JTWC, Guam, Mariana Islands, Annual report.

Dong, K. and Neumann, C. J. (1983). On the relative motion of binary tropical cyclones, *Monthly Weather Review*, 111(5), 945–953.

Dong, K. and Neumann, C. J. (1986). The relationship between tropical cyclone motion and environmental geostrophic flows, *Monthly Weather Review*, 114(1), 115–122.

Dvorak, V. F. (1984). Tropical cyclone intensity analysis using satellite data, NOAA, NESDIS 11.

Elsberry, R. L. (1995). Recent advancements in dynamical tropical cyclone track predictions, *Meteorol. Atmos. Phys.*, 56(1), 81–99.

Elsner, J. B. and Kara, A. B. (1999). *Hurricanes of the North Atlantic: Climate and Society.* Oxford University Press.

Emanuel, K. A. (1997). Some aspects of hurricane inner-core dynamics and energetics, *J. Atmos. Sci.*, 54(8), 1014–1026.

Emanuel, K. A. (1986). An air-sea interaction theory for tropical cyclones. Part 1: Steady-state maintenance, *J. Atmos. Sci.*, 43(6), 585–604.

Englebretson, R. E. (1992). The Joint Typhoon Warning Center (JTWC92) model, SAIC, Monterey, CA, AD-A258646.

Errico, R. and Baumhefner, D. (1987). Predictability experiments using a high-resolution limited-area model, *Monthly Weather Review*, 115(2), 488–504.

Feng, B. and Liu, J. N. K. (2004). Semo-Mamo, A 3-phase module to compare tropical cyclone satellite images using a modified Hausdorff distance, presented at the Machine Learning and Cybernetics, 2004, Shanghai, China, 6, 3808–3813.

Ferreira, R. N. and Schubert, W. H. (1997). Barotropic aspects of ITCZ breakdown, *J. Atmos. Sci.*, 54(2), 261–285.

Fett, R. W. (1968). Typhoon formation within the zone of the intertropical convergence, *Monthly Weather Review*, 96(2), 106–117.

Fett, R. W. and Brand, S. (1975). Tropical cyclone movement forecasts based on observations from satellites, *J. Appl. Meteorol.*, 14(4), 452–465.

Fiorino, M. and Elsberry, R. L. (1989). Some aspects of vortex structure related to tropical cyclone motion, *J. Atmos. Sci.*, 46, 975–990.

Fiorino, M., Harrison Jr, E. J. and Marks, D. G. (1982). A comparison of the performance of two operational dynamic tropical cyclone models, *Monthly Weather Review*, 110(7), 651–656.

Fulton, S. R. (2001). An adaptive multigrid barotropic tropical cyclone track model, *Monthly Weather Review*, 129(1), 138–151.

Gates, W. L., Pocinki, L. S. and Jenkins, C. F. (1995). Results of numerical forecasting with the barotropic and thermotropic atmospheric models, Airforce Cambridge Research Center, Bedford MA, Geophysical Research Paper No. AD0101943.

George, J. E. and Gray, W. M. (1976). Tropical cyclone motion and surrounding parameter relationships, *J. Appl. Meteorol.*, 15(12), 1252–1264.

Goerss, J. S. (2009). Impact of satellite observations on the tropical cyclone track forecasts of the navy operational global atmospheric prediction system, *Monthly Weather Review*, 137(1), 41–50.

Gray, W. M. (1968). Global view of the origin of tropical disturbances and storms, *Monthly Weather Review*, 96(10), 669–700.

Gray, W. M. (1979). Hurricanes: Their formation, structure and likely role in the tropical circulation, in Meteorology Over Tropical Oceans, *Roy. Meteor. Soc.*, pp. 155–218.

Gray, W. M. (1984). Atlantic seasonal hurricane frequency: Part I: El Niño and 30 mb quasi-biennial oscillation influences, *Monthly Weather Review*, 112, 1649–1668.

Gray, W. M. (1998). The formation of tropical cyclones, *Meteorl. Atmos. Phys.*, 67(1–4), 37–69.

Guinn, T. A. and Schubert, W. H. (1993). Hurricane spiral bands, *J. Atmos. Sci.*, 50(20), 3380–3403.

Haltiner, G. J. and Williams, R. T. (1980). *Numerical Prediction and Dynamic Meteorology.* Wiley, New York, 1980.

Hamuro, M., Kawata, Y., Matsuda, S. T. et al. (1969). Precipitation bands of Typhoon Vera in 1959 (Part I), *J. Meteor. Soc. Japan*, 47, 298–308.

Haque, S. M. A. (1952). The initiation of cyclonic circulation in a vertically unstable stagnant air mass, *Q. J. Roy. Meteor. Soc.*, 78(337), 394–406.

Harr, P. A. and Elsberry, R. L. (1991). Tropical cyclone track characteristics as a function of large-scale circulation anomalies, *Monthly Weather Review*, 119(6), 1448–1468.

Heta, Y. (1990). An analysis of tropical wind fields in relation to typhoon formation over the western Pacific, *J. Meteor. Soc. Japan*, 68, 65–76.

Hodur, R. M. (1982). Description and evaluation of NORAPS – The Navy Operational Regional Atmospheric Prediction System, *Monthly Weather Review*, 110, 1591–1602.

Holland, G. J. (1983). Tropical cyclone motion: Environmental interaction plus a beta effect, *J. Atmos. Sci.*, 40(2), 328–342.

Hope, J. R. and Neumann, C. J. (1970). An operational technique for relating the movement of existing tropical cyclones to past tracks, *Monthly Weather Review*, 98(12), 925–933.

Horsfall, F., DeMaria, M. and Gross, J. M. (1997). Optimal use of large-scale boundary and initial fields for limited-area hurricane forecast models, presented at the 22nd Conf. on Hurricanes and Tropical Meteorology, Fort Collins, CO, pp. 571–572.

Hovermale, J. B. and Livezey, R. E. (1977). Three-year performance characteristics of the NMC hurricane model, presented at the 11th Tech. Conf. on Hurricanes and Tropical Meteorology, Miami, FL, pp. 122–125.

Hubert, W. E. (1957). Hurricane trajectory forecasts from a non-divergent, non-geostrophic, barotropic model, *Monthly Weather Review*, 85(3), 83–87.

Isaksen, L. and Stoffelen, A. (2000). ERS scatterometer wind data impact on ECMWF's tropical cyclone forecasts, *IEEE T Geosci. Remote*, 38(4), 1885–1892.

Jarrell, J. D. and Somervell, W. L. (1970). A Computer Technique for Using Typhoon Analogs as a Forecast Aid, Navy Weather Research Facility, Norfolk, VA, Tech. Rep. No. A068800.

Jeffries, R. A. and Miller, R. J. (1993). Tropical Cyclone Forecasters Reference Guide 3. Tropical Cyclone Formation, NRL, Monterey, CA, 1993.

Jeffries, R. A., Sampson, C. R., Carr III, L. E. et al. (1993). Tropical Cyclone Forecasters Reference Guide. 5. Numerical Track Forecast Guidance, NRL, Monterey, CA, AD-A277 318.

Jones, S. C. (1995). The evolution of vortices in vertical shear. I: Initially barotropic vortices, *Q. J. Roy. Meteor. Soc.*, 121(524), 821–851.

Kanamitsu, M. (1989). Description of the NMC global data assimilation and forecast system, *Weather Forecast*, 4, 335–342.

Kasahara, A. (1961). A numerical experiment on the development of a tropical cyclone, *J. Meteor.*, 18(3), 259–282.

Kasiri, M. B., Aleboyeh, H. and Aleboyeh, A. (2008). Modeling and optimization of hetero-geneous photo-fenton process with response surface methodology and artificial neural networks, *Environ. Sci. Technol.*, 42(21), 7970–7975.

Keenan, T. D. (1982). A diagnostic study of tropical cyclone forecasting in Australia, *Aust. Met. Mag.*, 30, 69–80.

Kokhanovsky, A. A. and von Hoyningen-Huene, W. (2004). Optical properties of a hurricane, *Atmos. Res.*, 69(3), 165–183.

Kovordányi, R. and Roy, C. (2009). Cyclone track forecasting based on satellite images using artificial neural networks, *ISPRS J. Photogramm.*, 64(6), 513–521.

Kumar, S. and Prasad, K. (1973). An objective method for the tropical storm movement in Indian Seas, *Indian J. Meteor. Hydrol. and Geophys.*, 24, 31–34.

Kuo, H. L. (1965). On formation and intensification of tropical cyclones through latent heat release by cumulus convection, *J. Atmos. Sci.*, 22(1), 40–63.

Kurihara, Y. (1976). On the development of spiral bands in a tropical cyclone, *J. Atmos. Sci.*, 33, 940–958.

Kurihara, Y., Bender, M. A., Tuleya, R. E., and Ross, R. J. (1995). Improvements in the GFDL hurricane prediction system, *Monthly Weather Review*, 123(9), 2791–2801.

Kurihara, Y., Tuleya, R. E., and Bender, M. A. (1998). The GFDL hurricane prediction system and its performance in the 1995 hurricane season, *Monthly Weather Review*, 126(5), 1306–1322.

Lajoie, F. A. (1976). On the direction of movement of tropical cyclones, *Aust. Meteorol. Mag.*, 24(3).

Lajoie, F. A. and Nicholls, N. (1974). A relationship between the direction and movement of tropical cyclones and the structure of their cloud systems, Bureau of Meteorology, Melbourne, Australia, Technical Tech. Rep. No. 11.

Langland, R. H., Velden, C., Pauley, P. M. et al. (2009). Impact of satellite-derived rapid-scan wind observations on numerical model forecasts of Hurricane Katrina, *Monthly Weather Review*, 137(5), 1615–1622.

Lee, R. S. T. and Liu, J. N. K. (2000). Tropical cyclone identification and tracking system using integrated neural oscillatory elastic graph matching and hybrid RBF network track mining techniques, *IEEE T Neur. Network*, 11(3), 680–689.

Leslie, L. M. and Holland, G. J. (1995). On the bogussing of tropical cyclones in numerical models: A comparison of vortex profiles, *Meteorol. Atmos. Phys.*, 56(1), 101–110.

Leslie, L. M., LeMarshall, J. F., Morison, R. P. et al. (1998). Improved hurricane track forecasting from the continuous assimilation of high quality satellite wind data, *Monthly Weather Review*, 126(5), 1248–1258.

Li, X. and Wang, B. (1994). Barotropic dynamics of the beta gyres and beta drift, *J. Atmos. Sci.*, 51(5).

Lilly, D. K. (1960). On the theory of disturbances in a conditionally unstable atmosphere, *Monthly Weather Review*, 88(1), 1–17.

Ma, L. M. and Tan, Z. M. (2009). Improving the behavior of the cumulus parameterization for tropical cyclone prediction: Convection trigger, *Atmos. Res.*, 92(2), 190–211.

Marks, D. G. (1992). The Beta and Advection Model for Hurricane Track Forecasting, Washington, DC, Technical NWS NMC 70.

Mathur, M. B. (1988). Development of the NMC's high resolution hurricane model, presented at the 17th Conference on Hurricanes and Tropical Meteorology, Miami, FL, pp. 60–63.

Mathur, M. B. (1991). The National Meteorological Center's quasi-Lagrangian model for hurricane prediction, *Monthly Weather Review*, 119(6), 1419–1447.

Matsumoto, C. R. (1984). A Statistical Method for One-to Three-Day Tropical Cyclone Track Prediction., Air Force Inst. of Tech., Wright-Patterson Air Force Base, OH.

Mautner, D. A. and Guard, C. P. (1992). 1992 Annual Tropical Cyclone Report, JTWC, Guam, Mariana Islands, Annual report.

May, P. T. and Holland, G. J. (1999). The role of potential vorticity generation in tropical cyclone rainbands, *J. Atmos. Sci.*, 56(9), 1224–1228.

McBride, J. L. and Keenan, T. D. (1982). Climatology of tropical cyclone genesis in the Australian region, *J. Climatol.*, 2(1), 13–33.

Miller, B. I. and Chase, P. P. (1966). Prediction of hurricane motion by statistical methods, *Monthly Weather Review*, 94(6), 399–406.

Miller, B. I. and Moore, P. L. (1960). A comparison of hurricane steering levels, *Bull. Amer. Meteor. Soc.*, 41, 59–63.

Mnih, V. and Hinton, G. (2010). Learning to detect roads in high-resolution aerial images, *Computer Vision–ECCV 2010*, 210–223.

Molinari, J., Vollaro, D. and Corbosiero, K. L. (2004). Tropical cyclone formation in a sheared environment: A case study, *J. Atmos. Sci.*, 61(21), 2493–2509.

Møller, M. F. (1993). A scaled conjugate gradient algorithm for fast supervised learning, *Neural Networks*, 6(4), 525–533.

Montgomery, M. T. and Farrell, B. F. (1993). Tropical cyclone formation, *J. Atmos. Sci.*, 50(2), 285–310.

Montgomery, M. T. and Kallenbach, R. J. (1997). A theory for vortex rossby-waves and its application to spiral bands and intensity changes in hurricanes, *Q. J. Roy. Meteor. Soc.*, 123(538), 435–465.

Namias, J. (1955). Secular fluctuations in vulnerability to tropical cyclones in and off New England, *Monthly Weather Review*, 83(8), 155–162.

Neumann, C. J. (1972). An alternative to the HURRAN (hurricane analog) tropical cyclone forecast system, NHC, Miami, FL, Technical NWS SR-62.

Neumann, C. J. (1979). A guide to Atlantic and eastern Pacific models for the prediction of tropical cyclone motion, NHC, Miami, FL, NWS No. 932.

Neumann, C. J. (1992). The joint typhoon warning center (JTWC92) model, SAIC, Monterey, CA, SAIC Contract Rep. N 00014-90-C-6042 (Part 2).

Neumann, C. J. (1988). The National Hurricane Center NHC83 Model, NHC, Coral Gables, FL, Tech. Rep. No. NWS NHC 41.

Neumann, C. J. and Lawrence, M. B. (1975). An operational experiment in the statistical-dynamical prediction of tropical cyclone motion, *Monthly Weather Review*, 103, 665.

Neumann, C. J. and McAdie, C. (1991). A revised national hurricane center NHC83 model (NHC90), NHC, Coral Gables, FL, Tech. Rep. No. NWS NHC 44.

Nicholls, N. (1985). Predictability of interannual variations of Australian seasonal tropical cyclone activity, *Monthly Weather Review*, 113, 1144–1149.

Ooyama, K. V. (1982). Conceptual evaluation of the theory and modeling of tropical cyclone, *J Meteorol. Soc. Jpn.*, 60(1), 369–379.

O'Reilly, R. C. (1998). Six principles for biologically based computational models of cortical cognition, *Trends Cogn. Sci.*, 2(11), 455–462.

O'Reilly, R. C. and Munakata, Y. (2000). *Computational Explorations in Cognitive Neuroscience: Understanding the Mind by Simulating the Brain.* MIT Press.

Palmen, E. (1948). On the formation and structure of tropical hurricanes, *Geophysica*, 3, 26–38.

Pedersen, T. S. (1991). A comparison of the free ride and CISK assumptions, *J. Atmos. Sci.*, 48(16), 1813–1821.

Pedersen, T. S. and Rasmussen, E. (1985). On the cut-off problem in linear CISK-models, *Tellus A*, 37(5), 394–402.

Perrone, T. J. and Lowe, P. R. (1986). A statistically derived prediction procedure for tropical storm formation, *Monthly Weather Review*, 114(1), 165–177.

Pfeffer, R. L. (1958). Concerning the mechanics of hurricanes, *J. Atmos. Sci.*, 15, 113–120.

Pickle, J. (1991). Forecasting Short-Term Movement and Intensification of Tropical Cyclones Using Pattern-Recognition Techniques, Phillips Lab Hanscom AFB, MA, Environmental Research Paper No. AD-A256705.

Powell, M. D. (1990a). Boundary layer structure and dynamics in outer hurricane rainbands. Part I: Mesoscale rainfall and kinematic structure, *Monthly Weather Review*, 118, 891–917.

Powell, M. D. (1990b). Boundary layer structure and dynamics in outer hurricane rainbands. Part II: Downdraft modification and mixed layer recovery, *Monthly Weather Review*, 118, 918–938.

Prasad, K. and Rama Rao, Y. V. (2003). Cyclone track prediction by a quasi-Lagrangian limited area model, *Meteorol. Atmos. Phys.*, 83(3), 173–185.

Radford, A. M. (1994). Forecasting the movement of tropical cyclones at the Met. Office, *Meteorol. Appl.*, 1(4), 355–363.

Renard, R. J., Colgan, S. G., Daley, M. J. et al. (1973). Forecasting the motion of North Atlantic tropical cyclones by the objective MOHATT scheme, *Monthly Weather Review*, 101(3), 206–214.

Riehl, H. and Malkus, J. (1961). Some aspects of Hurricane Daisy, 1958, *Tellus*, 13(2), 181–213.

Riehl, H. and Shafer, R. J. (1944). The recurvature of tropical storms, *J. Atmos. Sci.*, 1, 42–54.

Rossby, C. G. (1948). On displacements and intensity changes of atmospheric vortices, *J. Mar. Res.*, 7(175), 71.

Rossby, C. G. (1949). On a mechanism for the release of potential energy in the atmosphere, *J. Atmos. Sci.*, 6, 164–180.

Rotunno, R. and Emanuel, K. A. (1987). An air–sea interaction theory for tropical cyclones. Part II: Evolutionary study using a nonhydrostatic axisymmetric numerical model, *J. Atmos. Sci.*, 44(3), 542–561.

Roy, C. and Kovordányi, R. (2012). Tropical cyclone track forecasting techniques – A review, *Atmospheric Research*, 104–105, 40–69.

Sampson, C. R., Jeffries, R. A. and Neumann, C. J. (1995). Tropical Cyclone Forecasters Reference Guide 4. Tropical Cyclone Motion, NRL, Monterey, CA, AD-A265 216.

Sanders, F. and Burpee, R. W. (1968). Experiments in barotropic hurricane track forecasting, *J. Appl. Meteor.*, 7, 313–323.

Sela, J. G. (1980). Spectral modeling at the National Meteorological Center, *Monthly Weather Review*, 108(9), 1279–1292.

Shapiro, L. J. (1982). Hurricane climatic fluctuations. Part II: Relation to large-scale circulation, *Monthly Weather Review*, 110(8), 1014–1023.

Shapiro, L. J. and Willoughby, H. E. (1982). The response of balanced hurricanes to local sources of heat and momentum, *J. Atmos. Sci.*, 39(2), 378–394.

Simpson, J., Ritchie, E., Holland, G. J. et al. (1997). Mesoscale interactions in tropical cyclone genesis, *Mon. Wea. Rev.*, 125(10), 2643–2661.

Smith, R. B. (1993). A hurricane beta-drift law, *J. Atmos. Sci.*, 50, 3213–3220.

Smith, R. K. (1980). Tropical cyclone eye dynamics, *J. Atmos. Sci.*, 37, 1227–1232.

Syōno, S. (1953). On the formation of tropical cyclones, *Tellus*, 5(2), 179–195.

Tomassini, M., LeMeur, D. and Saunders, R. W. (1998). Near-surface satellite wind observations of hurricanes and their impact on ECMWF model analyses and forecasts, *Monthly Weather Review*, 126(5), 1274–1286.

Tse, S. Y. W. (1970). Typhoon movement, *in* The Regional Training Seminar, Singapore, pp. 158–171.

Vanderman, L. W. (1962). An improved NWP model for forecasting the paths of tropical cyclones, *Monthly Weather Review*, 90(1), 19–22.

Veigas, K. W., Miller, R. G. and Howe, G. M. (1959). Probabilistic prediction of hurricane movements by synoptic climatology, Travelers Weather Research Center, Inc., Hartford, CT, *Occasional Papers in Meteorology*, no. 2.

Velden, C. S., Olander, T. L. and Wanzong, S. (1998). The impact of multispectral GOES-8 wind information on Atlantic tropical cyclone track forecasts in 1995. Part I: Dataset methodology, description, and case analysis, *Monthly Weather Review*, 126(5), 1202–1218.

Vigh, J. L. (2010). Formation of the hurricane eye. PhD dissertation, Colorado State University, Fort Collins, CO (Available online at http://hdl.handle.net/10217/39054).

Vigh, J., Fulton, S. R., DeMaria, M. et al. (2003). Evaluation of a multigrid barotropic tropical cyclone track model, *Monthly Weather Review*, 131(8), 1629–1636.

Villmann, T., Merényi, E. and Hammer, B. (2003). Neural maps in remote sensing image analysis, *Neural Networks*, 16(3), 389–403.

Wang, D., Liang, X., Duan, Y. et al. (2006). Impact of four-dimensional variational data assimilation of atmospheric motion vectors on tropical cyclone track forecasts, *Weather Forecast*, 21(4), 663–669.

Wang, Y. (1998). On the bogusing of tropical cyclones in numerical models: The influence of vertical structure, *Meteorol. Atmos. Phys.*, 65(3), 153–170.

Wang, Y. (2009). How do outer spiral rainbands affect tropical cyclone structure and intensity?, *J. Atmos. Sci.*, 66(5), 1250–1273.

Wang, Y. (2002). Vortex Rossby waves in a numerically simulated tropical cyclone. Part II: The role in tropical cyclone structure and intensity changes, *J. Atmos. Sci.*, 59(7), 1239–1262.

Wang, Y. and Wu, C. C. (2004). Current understanding of tropical cyclone structure and intensity changes – A review, *Meteorol. Atmos. Phys.*, 87(4), 257–278.

Weir, R. C. (1982). Predicting the Acceleration of Northward-Moving Tropical Cyclones Using Upper-Tropospheric Winds, JTWC, San Francisco, CA, Technical ADA127446.

Willoughby, H. E. (1978). A possible mechanism for the formation of hurricane rainbands, *J. Atmos. Sci.*, 35, 838–848.

Willoughby, H. E., Clos, J. A. and Shoreibah, M. G. (1982). Concentric eye walls, secondary wind maxima, and the evolution of the hurricane vortex, *J. Atmos. Sci.*, 39(2), 395–411.

Willoughby, H. E., Marks Jr, F. D. and Feinberg, R. J. (1984). Stationary and moving convective bands in hurricanes, *J. Atmos. Sci.*, 41(22), 3189–3211.

Xiao, Q., Zou, X., Pondeca, M. et al. (2002). Impact of GMS-5 and GOES-9 satellite-derived winds on the prediction of a NORPEX extratropical cyclone, *Monthly Weather Review*, 130(3), 507–528.

Xu, Y. and Neumann, C. J. (1985). A statistical model for the prediction of Western North Pacific tropical cyclone motion (WPCLPR). NHC.

Yanai, M. (1964). Formation of tropical cyclones, *Rev. Geophys.*, 2(2), 367–414.

Yang, Y. and Wang, J. (2005). An integrated decision method for prediction of tropical cyclone movement by using genetic algorithm, *Science in China Series D: Earth Sciences*, 48(3), 429–440.

Yu, H. and Kwon, H. J. (2005). Effect of TC-trough interaction on the intensity change of two typhoons, *Weather Forecast*, 20(2), 199–211.

Zehr, R. M. (1992). Tropical cyclogenesis in the western North Pacific., Technical NESDIS 61.

2 Tropical Cyclones
Effect on the Ionosphere

Natalia Perevalova

CONTENTS

2.1   INTRODUCTION

The ionosphere is a part of the Earth's upper atmosphere. It is a plasma layer that is formed due to solar radiation and is controlled by the geomagnetic field. The ionosphere processes are important components of space weather. Without taking ionospheric phenomena into account, it is impossible to robustly predict climate changes on Earth. As an ionized medium, the ionosphere has an impact on the functioning of many present-day systems of communication, navigation, power grids and space vehicles. Today, active space exploration requires accurate information on regular variations in ionospheric parameters and forecast of ionospheric plasma disturbances.

The dominant factors determining the ionospheric plasma behaviour are changes in solar radiation (solar activity) and variations in the Earth's magnetic field (geomagnetic activity). However, studies over the past 20 years have convincingly proved a correlation between the behaviour of ionospheric parameters and meteorological processes in the atmospheric lower layers (stratosphere, troposphere). Detailed reviews of meteorological effects in the ionosphere are presented in Hocke and Shlegel (1996), Kazimirovsky et al. (2003) and Laštovička (2006). The propagation of various-scale internal atmospheric waves (IAWs) (Gossard and Hooke, 1975; Hocke and Shlegel, 1996; Kazimirovsky et al., 2003; Laštovička, 2006) has been established to be one of the basic mechanisms to provide coupling between the ionosphere and the lower atmosphere. These IAWs are: planetary (periods: a few days), tidal (periods: a few hours), internal gravity (periods: 10–180 minutes), internal acoustic (periods: 1–10 minutes) and infrasonic (periods: several seconds through several minutes). The lower ionosphere is shown to be more sensitive to meteorological effects than other ionospheric regions. Here, meteorological processes can lead to essential (by a factor of two) changes in electron density that often exceeds its diurnal variations caused by solar radiation. Experimental studies have reliably established a connection between large-scale (several hours through 30 days) disturbances in the lower ionosphere and the stratosphere/troposphere characteristics (temperature, pressure, horizontal and vertical wind). This means that plasma behaviour in the lower ionosphere is, to a great extent, controlled by meteorological processes (Taubenheim, 1983; Kazimirovsky et al., 2003). Large-scale variations of the upper ionosphere parameters are most likely caused by solar and geomagnetic activities. At the same time, traveling ionospheric disturbances (TIDs) of medium and small scales that are permanently recorded in the upper ionosphere are the manifestation of acoustic and gravity waves in the neutral atmosphere. Here and below, the term 'neutral' refers to uncharged atoms and molecules (O_2, O, N_2, N, CO_2, H_2, H, etc.) which the Earth's atmosphere comprises. Sources of these waves may also be meteorological phenomena: atmospheric fronts, cyclones, hurricanes, thunderstorms, tornadoes, and so on.

In this chapter, we present results obtained from studying the ionospheric responses to such powerful meteorological disturbances as tropical cyclones (TCs). In Section 2.2, we describe the basic notions used, possible mechanisms for the TC effect on the ionosphere, as well as the methods to detect these effects in the ionosphere. Section 2.3 presents the TC-induced phenomena in the lower ionosphere. In Section 2.4, we discuss the TC effects in the upper ionosphere. Section 2.5 contains final notes and conclusions.

2.2 BASIC NOTIONS AND POSSIBLE MECHANISMS FOR TC EFFECT ON THE IONOSPHERE

2.2.1 IONOSPHERE CHARACTERISTICS

The near-Earth space is a complex system (Figure 2.1a) whose basic elements are: atmosphere (Earth's gas shell), ionosphere (ionized gas layer in the upper atmosphere) and magnetosphere (the area where the Earth's magnetic field dominates).

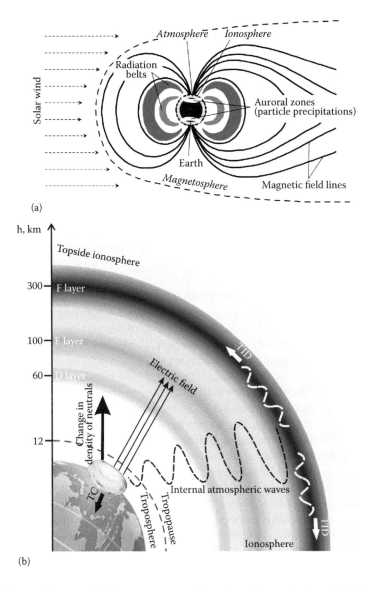

(a)

(b)

FIGURE 2.1   The near-Earth space structure (a) and possible mechanisms for TC effect on the ionosphere (b).

Each of these regions has its own structure and features, and interaction between these different regions, as well as with solar radiation and the interplanetary magnetic field, determines environmental conditions in the near-Earth space.

The ionosphere spans altitudes from 50–1000 km above the Earth's surface. Free electrons and ions are generated at these heights due to atmospheric gas ionization, mainly by solar ultraviolet and X-ray radiations. The electron density (Ne) is one of the main characteristics of the ionosphere. The electron height distribution is heterogeneous. The solar ionizing fluxes with different wavelengths are absorbed at different heights in the atmosphere. This leads to stratification of the ionosphere. It is possible to identify three basic layers (regions) (Figure 2.1b), referred to as the D, E and F2 layers. The D layer has an ionization maximum height, $h_{mD} \sim 60-70$ km, and maximum electron density being $Ne_{mD} \sim 10-10^3$ sm^{-3}; for the E layer these parameters are $h_{mE} \sim 100-120$ km, $Ne_{mE} \sim 10^3-10^5$ sm^{-3}; and for the F2 layer, $h_{mF2} \sim 250-300$ km, $Ne_{mF2} \sim 5 \cdot 10^5$ sm^{-3}. F2 is the most powerful ionospheric layer; it is the ionization main maximum. Above the main maximum, Ne gradually diminishes. The region above 600 km is termed the topside ionosphere.

The 'total electron content' (TEC) parameter is also used to describe the ionospheric plasma. TEC is the total number of electrons along a path between two points. It is calculated as integral of Ne along a ray. A special measurement unit, 1 TECU $= 10^{16}$ m^{-2}, was introduced for TEC. Also important in the ionosphere are the temperature of ions (Ti) and of electrons (Te) that characterize the energy of these particles.

Charged particles (electrons and ions) determine the ionosphere's unique properties. The ionized layer plays an important role in the Earth's global electrical circuit: some currents flowing from the magnetosphere are closed in the ionosphere. The ionospheric plasma is affected and controlled by the geomagnetic field. At the same time, the concentration of neutral atmospheric particles significantly exceeds the plasma density even in the F2 region. Therefore, the neutral atmosphere has a considerable effect on the ionosphere's behaviour.

When detecting TC ionospheric effects, one should take into account the background behaviour of ionization, as well as the disturbances caused by other sources. Global distribution of ionization, diurnal and seasonal variations in the behaviour of ionospheric parameters at various heights has been extensively studied. One of the large-scale ionospheric structures affecting the possibility to detect TC responses in the ionosphere is the equatorial anomaly (EA) of ionization. EA forms during local afternoons on both sides of the magnetic equator. Figure 2.2 shows the 28 August 2005 TEC distribution map for 20:00 UT (14:00 local time [LT] at 90°W). The map clearly shows two regions of higher TEC values corresponding to the northern and southern EA crests. Under quiet geomagnetic conditions, the ionization northern crest is at 8–10°N. During magnetic storms, the crest displaces northward and may reach 20–22°N (Astafyeva et al., 2007; Perevalova et al., 2008). The EA crest slope represents the region of the TEC gradient's greatest values. Here, the probability of ionospheric plasma irregularities being generated is high, and it is almost impossible to detect the TC-caused disturbances against these irregularities. As an example, Figure 2.3 gives the TEC variations dI(t) obtained at the KYW1 GPS station (see Section 2.2 for details) on 23–31 August 2005 and filtered to extract variations with periods 02–20 min. On 23–31 August, strong magnetic storms were registered, at the same time as TC Katrina was occurring in the Atlantic Ocean. TEC responses to these sources are discussed in Section 4.2; here we focus on the effects of EA. Grey colour in Figure 2.3 marks the EA activity time within 60–120°W. The TEC variation intensity bursts were recorded in individual days. Most bursts coincide with the EA activity periods. During the night the EA disappears, and the ionization spatial-temporal variations feature a weak variability thus facilitating detection

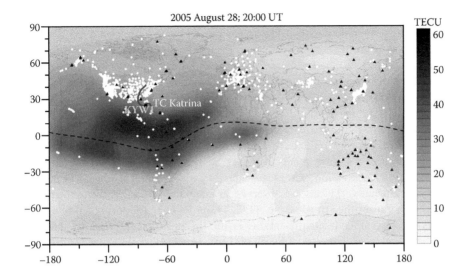

FIGURE 2.2   28 August 2005 TEC distribution map for 20:00 UT (14:00 LT at 90°W). Dots mark the positions of GPS stations. The KYW1 station is presented by the black dot. Triangles mark the positions of VS ionosondes. The thick grey line shows the TC Katrina trajectory. The black dotted line indicates the geomagnetic equator.

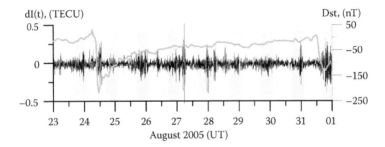

FIGURE 2.3   The 02–20 min TEC variations (dI(t)) obtained at the KYW1 station on 23–31 August 2005 (black line). Grey colour marks the equatorial anomaly activity time within 120–60°W. The thick grey line shows the Dst index behaviour.

of TC effects: the 27–29th August TEC bursts, outside the EA activity periods, are effects of TC Katrina (see Section 4.2.2). Bursts at nighttime on 24 and 31 August are caused by geomagnetic storms.

The important factors masking TC effects in the ionosphere are solar flares, geomagnetic storms and the solar terminator. These processes are the sources for various-scale disturbances of ionospheric plasma. A solar flare causes Ne to rapidly increase in the ionosphere over the entire lit Earth's surface simultaneously (Davies, 1990). The geomagnetic storm effect on the ionosphere is very complex. For our purposes, the storm-generated ionospheric wave disturbances are important. After the sudden commencement of a magnetic storm, 40–60 min period large-scale waves appear in the ionosphere almost synchronously in the Northern and Southern Hemispheres around the geomagnetic poles. The waves propagate equatorward to 35–30°. This propagation is accompanied by an increase in the 1–10 min oscillation intensity (Perevalova et al., 2008). In Figure 2.3, one can see the TEC response to two powerful magnetic storms, on 24 August 2005 (Dst index dropped to –216 nT) and on 31 August 2005 (Dst = –131 nT), respectively. The Dst and Kp indices characterize the geomagnetic activity level (World Data Centre for Geomagnetism, n.d.). The solar terminator (ST) separates the day and the night sides of the Earth. It moves east to west according to the Earth's rotation. The morning ST corresponds to sunrise, the evening ST corresponds to sunset. ST passage causes a radical ionosphere rearrangement: ionization increases at sunrise and decreases at sunset. These characteristic diurnal variations are observed in many ionospheric measurements. The terminator causes the strongest Ne, TEC, Te and Ti changes during a day, and one of its most significant manifestations is the disappearance of EA northern and southern crests after the evening ST passage. In addition, the solar terminator is a source of wave disturbances in the ionospheric plasma. These disturbances have 15–30 min periods and 100–300 km wavelengths. Depending on the season and latitude, the terminator waves can be observed 2–3 h prior to or 1–3 h after ST arrival (Afraimovich et al., 2009).

Considering the strong variability of ionospheric parameters, detecting ionospheric responses to TCs requires a careful analysis of the geophysical situation. Detecting and identifying responses is possible under quiet heliogeophysical conditions during the local nighttime, when the ionization background spatial-temporal variations feature a weak variability.

2.2.2   RESEARCH TECHNIQUES TO STUDY TROPICAL CYCLONE EFFECTS IN THE IONOSPHERE

Various techniques are applied to study TC effects in the ionosphere. Direct rocket and satellite measurements provide immediate recording of the ionospheric plasma parameters (Ne, Ti, Te).

Until the mid-1990s, vertical (VS) and oblique (OS) soundings with ionosondes were the most prevalent techniques to diagnose the disturbances originating in the ionosphere from various effects (including TC). Today, more than 100 VS ionosondes are in operation (Figure 2.2). A VS ionosonde records a height-frequency matrix for the amplitudes of the sounding radio signal reflected from the ionosphere (ionogram). These ionograms allow the electron density height profile to be recovered, and the maximum height and the electron density in the ionization maximum to be computed. The peak frequency of the radio wave that is reflected from the upper ionosphere at vertical incidence is known as critical frequency, fo. It is the measure of maximal electron density Ne_{max} in the layer. The highest ionospheric layer, F2, has the greatest critical frequency foF2 that varies within 5–12 MHz. The radio waves with frequencies greater than foF2 propagate through the ionosphere without reflecting. OS is a modification of the VS technique, in which the ionosonde transmitters and receivers are spaced apart. The sounding signal propagates along the path between transmitter and receiver reflecting from the ionosphere. OS at very low and low frequencies (VLF/LF) (15–50 kHz) enables investigating the lower ionosphere (D region). To detect travelling ionospheric disturbances, Doppler frequency shift (DFS) measurements of radio signal at VS or OS are often used. One can also estimate foF2 from transionospheric sounding data, by using signals of low-orbit (low-orbit tomography) or high-orbit (high-orbit tomography) navigation satellites.

Transionospheric sounding is based on recording high-frequency radio signals that have passed from a space emitter to a ground-based receiver. This method enables TEC to be estimated: when propagating in the ionosphere, radio signals experience a delay whose value is proportional to TEC. Both natural space radio sources and satellite-based transmitters are used as emitters. When recording the signals of low-orbit satellites, one measures either the wave polarization plane rotation angle (due to the wave's Faraday rotation); the Doppler frequency shift; or the phase variations of the received signal. The operating frequency range is 100–400 MHz. Using global navigation satellite systems (GNSS) signals takes a special place in transionospheric sounding. It is difficult to represent modern life without GNSS such as American GPS system, Russian GLONASS system, European Galileo system and Chinese Beidou system. GNSS has two basic operating frequencies: f1~1600 MHz and f2~1200 MHz. The measured GNSS navigation parameters are the signal propagation time (code measurements) and the carrier phase (phase measurements). TEC can be computed from both code and phase navigation measurements. In the practice of ionospheric studies, the phase measurements at two operating frequencies are most often used, as they have low noise (Hofmann-Wellenhof et al., 2008). The data temporal resolution is within 1–30 s. Special multipurpose (for geodesy, positioning, meteorology, space weather, geophysical applications) networks (such as the International GNSS service network, GEONET network in Japan, CORS network in the United States, EUREF network in Europe, ARGN, SPRGN networks in Australia and many others) of ground-based dual-frequency GNSS receivers present us opportunity for sounding the ionosphere (Figure 2.2). Now, there are more than 5000 such continuously operating GNSS stations in the world, which provide a high spatial-temporal data resolution and detection of rather weak TEC disturbances. Below, we use "GPS" instead of "GNSS" because GPS receivers dominated until 2010 and the GNSS results presented here were obtained from GPS data.

One of the effective ground-based instruments to diagnose TIDs is the Super Dual Auroral Radar Network (SuperDARN) of 35 radars that point into the polar regions of the Earth (http://superdarn.jhuapl.edu). These radars operate in the short-wave range (8–20 MHz) (Chisham et al., 2007). The coverage sector of any individual radar is ~52°. TID effects in the data from SuperDARN radars are in quasiperiodic temporal variations of the sounding signal parameters. More than 10 countries actively participate in the SuperDARN project. There are 23 radars in the Northern Hemisphere and 12 in the Southern one.

2.2.3 TROPICAL CYCLONES

A TC represents an atmospheric disturbance like a vortex, typically 300–1000 km across, with the air low pressure in the centre and with a storm wind (see also Chapter 1 by Roy, this volume). The ground pressure in the TC centre is 950–960 hPa on average dropping below 900 hPa in certain cases. A TC may extend up to 10–12 km high. TCs differ from extratropical cyclones in size (TCs are smaller), in more extreme pressure gradients, and in higher wind speeds.

There is a small zone ('eye') free from clouds and with scant winds in the TC centre. The zone is 5–50 km across. The wind reaches its peak values in the narrow band around the 'eye': this defines the cyclone wind speed. In the Northern Hemisphere, the wind in a TC is counterclockwise, while in the southern one it is clockwise. Air flows into a cyclone due to a pressure gradient at the base of the TC. At the 'eye' boundary, an approximate equilibrium is established between the pressure gradient force and the opposing centrifugal and Coriolis forces. With no possibility to move farther in to the TC centre, the air is displaced upward, transporting warmth and moisture which leads to the formation of powerful cumulonimbus clouds, and precipitation of shower rains with thunderstorms.

A TC undergoes several stages in its evolution (Pielke and Pielke, 1997; Pokrovskaya and Sharkov, 2006; Roy, this volume), starting as a tropical depression (wind speed in the cyclone being V~15–18 m/s), and developing into a tropical storm (V~18–23 m/s), a strong tropical storm (V~23–33 m/s), hurricane (V>33 m/s) and eventually an extratropical disturbance (the area of low ground pressure at mid latitudes, V<15 m/s). In South East Asia, tropical cyclones at hurricane stage are termed 'typhoons'. The TC average duration is about 9 days, with the maximal being up to 4 weeks.

The TC formation regions are between 5 and 20° of latitude north and south of the equator. Within ±5°, TCs are recorded very rarely because of the small torque caused by the Coriolis force. TCs form only over the overheated marine surface. Over the land, a TC collapses rapidly due to the increased inflow of the dry continental air into the cyclone low layers. On average, 80–120 TCs form on the globe annually, of which about 30 occur in South Eastern Asia, and about 20 in the Southern Hemisphere. TCs most often originate in summer and in autumn, when the ocean surface is particularly warmed, while they are observed extremely rarely in winter and in spring.

Upon forming, a TC moves first east to west (toward common air transport in the tropical region) at 10–20 km/h (3–6 m/s), deviating toward higher latitudes. A TC that reached the tropics (20–30°) rounds the subtropical atmospheric anticyclone from the west and, passing to the mid latitudes, it reverses its direction of travel from westerly to easterly and deviates toward higher latitudes. Thus, the TC characteristic trajectory is like a parabola with its vertex lying to the west. When leaving the tropics, the TC characteristics and traveling speed approach the features of extratropical cyclones.

2.2.4 Possible Mechanisms for TC Effects on the Ionosphere

TCs, like powerful tropospheric vortices, refer to the most destructive atmospheric phenomena. The TC mean kinetic energy (10^{16}–10^{18} J) (Chernogor, 2006; Maclay et al., 2008) is comparable to the energy of major (8–9 magnitude) earthquakes. There are some possible channels of a TC effect on the Earth's ionosphere: generation of acoustic and gravity waves, disturbance of the quasi-stationary electric field, generation of electromagnetic radiation and change in the density of neutral particles (atoms and molecules) in the atmosphere.

2.2.4.1 TC-Caused Generation of Acoustic and Gravity Waves

Experimental work (Pfister et al., 1993; Wang et al., 2006) has repeatedly demonstrated that TCs can be the source of acoustic-gravity waves (AGWs, periods 1–10 min) and internal gravity waves (IGWs, periods 10–180 min) in the troposphere and the stratosphere (Pfister et al., 1993; Wang et al., 2006), the properties of which have been studied through both theoretical computations and numerical simulation (Pogoreltsev, 1996; Kuester et al., 2001; Kim et al., 2005; Chernogor, 2006; Kunitsyn et al., 2007). According to estimates, the pressure variations in TC-caused IGWs start to exceed the atmospheric noise level (0.5–0.7 Pa) when the wind speed in the cyclone becomes more than 35–40 m/s, that is, a TC is at the hurricane stage (Chernogor, 2006). Theoretical computations have shown that AGWs with periods of several minutes are to be observed near the local source in the atmosphere (Chernogor, 2006; Kunitsyn et al., 2007). At large horizontal distances from the local source, IGWs with periods from tens of minutes to 2–3 h will be recorded, caused by rapid AGW attenuation due to the processes of viscosity and of thermal conductivity in the atmosphere. Height distribution of temperature and of background wind significantly impacts the upward propagation of the AGW/IGW. Vertical gradients of temperature and of wind speed can lead to the wave reflection and refraction. The issues of AGW/IGW vertical propagation have been actively discussed in the literature (see, for example, Vadas and Nicolls, 2012; Gavrilov and Kshevetskii, 2013; Godin, 2015), but they are still far from their solution.

A distinguishing feature of acoustic and gravity waves is the amplitude increase with height due to the atmosphere density depletion (see Figure 2.1b). Under favourable conditions, such waves can reach ionospheric heights, and can lead to noticeable ionospheric plasma disturbances. On the one hand, the wave propagation modulates the parameters of the neutral atmosphere and the ionosphere. In this case, AGWs/IGWs most often manifest themselves as traveling ionospheric disturbances (Gossard and Hooke, 1975; Hocke and Shlegel, 1996; Kazimirovsky et al., 2003; Laštovička, 2006). On the other hand, the wave energy dissipation caused by various factors – molecular viscosity, thermal conductivity, turbulence, ionic deceleration, saturation and instability – leads to warming of the environment. This additional warming causes a whole chain of processes, including change in density and conductivity of ionospheric plasma, geomagnetic field variations, formation of irregularities in the topside ionosphere, and affect not only the ionosphere but also the Earth's magnetosphere.

2.2.4.2 Formation of the Additional Electric Field within TC Areas

Electric fields originating over the TC regions may lead to ionospheric plasma irregularity generation. One mechanism for the quasi-stationary electric field disturbance in TC assumes the emergence of an external current in the lower atmosphere (Sorokin et al., 2005). The production and ionization of aerosols and water drops over the sea surface is accelerated by the presence of strong winds in the TC. Intensive vertical convection in the cyclone transporting the charged particles leads to the formation of an external current. This causes the modification of the currents in the atmosphere-ionosphere electric circuit as well as the perturbation of the electric field in the ionosphere. The horizontal dimension of the external current generation region on the Earth's surface is hundreds of kilometres. Theoretical estimates have shown that the electric field disturbances in the ionosphere may reach 10–20 mV/m (Sorokin et al., 2005). The increase in the atmospheric current in a TC may also cause geomagnetic field variations.

Another mechanism for electric field disturbance is associated with the generation of the additional local electric field formed by a ground ionized layer (Pulinets et al., 2000). As noted above, a TC leads to an increase in ionization: many atmospheric particles get an electric charge. Due to various mobility of different-sign ions affected by the natural Earth's electric field, E_0, a ground layer with a local (additional) electric field E_1 forms near the Earth's surface. The E_1 field, depending on the direction, either strengthens or weakens the basic E_0 field both near the ground and in the ionosphere. E_0 modifications generate irregularities in the ionospheric plasma. Notable changes in the electric field at ionospheric heights are shown to be observed only from large-scale ground sources having an extent of more than 100 km and electric field strength values of 500–1000 V/m.

The changes in the electron density and the formation of the ionospheric plasma irregularities may be a consequence of the additional electric field formation in the ionosphere. Besides, such an electric field, weakening insignificantly, may penetrate the magnetosphere and affect the motion of the charged particles in the Earth's radiation belts (radiation belts are the part of the magnetosphere containing high energy electrons and protons, see Figure 2.1a). Under certain conditions, this will facilitate precipitation of the charged particles from a radiation belt into the upper atmosphere (Pulinets et al., 2000; Sorokin et al., 2005; Chernogor, 2006).

2.2.4.3 Generation of Electromagnetic Radiation

TCs are accompanied by intensive storm activity. During the TC existence (5–6 days), the number of lightning discharges can reach 1000–10,000 (Chernogor, 2006). Thunderstorms are sources for acoustic and electromagnetic radiation in a wide frequency range. Reaching the ionosphere, acoustic and electromagnetic waves affect significantly its parameters. The ultra-low frequency (ULF) and VLF electromagnetic waves along the magnetic field lines may penetrate into the magnetosphere thus causing the geomagnetic field pulsations and particle precipitation. Many studies (Kelly et al., 1985; Mikhailova et al., 2002; Hoffman and Alexander, 2010) address investigating various manifestations of thunderstorms. However, the thunderstorm effects in the ionosphere are a special major subject that goes beyond the scope of this chapter.

2.2.4.4 Change in Density of Neutral Atmospheric Particles

The TC's effect on the ionospheric plasma is also possible through a change in the density of neutral particles (atoms and molecules) in the atmosphere (Liu et al., 2006; Vanina-Dart et al., 2007, 2008). Because of TC effects, turbulence intensifies in the lower atmosphere, and an increase in the turbulent diffusion coefficient D_t occurs, as well as a decrease in the molecular diffusion coefficient D_m. Such changes lead to an increase in the turbopause altitude (the turbopause represents the troposphere boundary where D_t and D_m values are equal). Model estimates have shown that the turbopause rise causes an increase in the density of neutral particles at 100–150 km (Liu et al., 2006). The latter leads to an increase in the electron recombination rate. This, therefore, results in the electron density decrease in the ionosphere.

2.3 TROPICAL CYCLONE EFFECTS IN THE LOWER IONOSPHERE

Due to a small electron density, the lower ionosphere (region D) is most complicated to study. Therefore, there are few data on TC effects in the lower ionosphere. Ne measurements in this region are most often taken by rockets and satellites. Rocket sounding at the Thumba observatory (India) recorded an Ne decrease in the ionosphere's region D at about 1000 km from the 8502-02A tropical cyclone centre on 29 May 1985 (Vanina-Dart et al., 2008). The 8502-02A TC originated in the Arabian Sea, west of the Indian coast on 27 May 1985. On 29 May, the TC merged into a strong storm: the cyclone centre pressure dropped to 990 mbar, the wind speed was 26 m/s. Rocket measurements of the electron density profile were taken on 29 May at 11:55 UT. The Ne greatest decrease (by a factor of 3–4) was observed at 71 ± 3 km. Upon analyzing the databank of the Ne rocket measurements at the Thumba observatory, and from satellite monitoring tropical cyclogenesis in the Indian and Pacific Oceans in 1985 and 1988, Vanina-Dart et al. (2007) established that the electron density at 65–80 km on the TC days is, on average, lower than on the TC-free days. Figure 2.4 is developed on the basis of data presented in Vanina-Dart (2007). The Ne greatest decrease was observed close to 70 km. Over 80 km, TC effect is weak. Ionospheric responses to TC were recorded at up to 8000 km from the cyclone trajectory. Ne decrease in region D may be caused by the neutral particle content (ozone in particular) redistribution, under the TC effect: Ne decreases as ozone increases (Vanina-Dart et al., 2007, 2008).

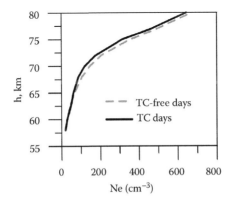

FIGURE 2.4 The electron density (Ne) averaged profiles at Thumba (India) during the days with TCs and during TC-free days.

The method of VLF/LF radio sounding is also used for studying disturbances in the ionospheric D region. Rozhnoi et al. (2014) analysed in detail the VLF/LF signal sensitivity to eight TCs in the Pacific Ocean over 2010–2013: Dianmu (August 2010), Malou (September 2010), Damrey (July 2012), Haikui (August 2012), Kai-tak (August 2012), Bolaven (August 2012), Soulik (July 2013) and Utor (August 2013). The author used the data from three VLF/LF receiving stations in the Russian Far East. For six TCs, negative nighttime anomalies were recorded in the VLF/LF signal amplitude that were, most probably, caused by TC activity. The anomalies were observed for 1–2 days, when the TCs moved under the radio paths. Spectral analysis revealed the fluctuations with 7–16-minute and 15–55-minute periods that correspond to the internal gravity wave periods.

2.4 TROPICAL CYCLONE EFFECTS IN THE UPPER IONOSPHERE

Recording TC effects in the upper and topside ionosphere can be performed through various techniques. One can identify three types of effect caused by a TC influence in the upper ionosphere: Ne variation, TID formation, and electric field disturbances.

2.4.1 ELECTRON DENSITY VARIATIONS IN THE UPPER IONOSPHERE

Electron density variations in the upper ionosphere are recorded by variations of either critical frequency, foF2 or TEC. According to the vertical sounding data (Bauer, 1958; Shen, 1982; Liu et al., 2006; Rice et al., 2012) and the tomographic sounding data (Vanina-Dart, 2011), a possible response of the upper ionosphere to TC is that the foF2 change may reach 10–50%. In some cases, foF2 was observed to increase as the TC approached (Bauer, 1958; Vanina-Dart et al., 2011; Rice et al., 2012), whereas in others it decreased (Bauer, 1958; Vanina-Dart et al., 2011). Such variances may be explained by different positions of measuring stations relative to the TC trajectory (Bauer, 1958). In Shen (1982), foF2 is shown to decrease directly over the hurricane centre, but it increases in the adjacent regions. A strong vertical convection in the troposphere, caused by a hurricane, may lead to a change in the local circulating system in the stratosphere and the mesosphere. This, in the end, will lead to the changes in the electron density and in foF2.

An interesting effect was recorded in the foF2 and TEC behaviour after TC landfall (Liu et al., 2006; Mao et al., 2010; Yu et al., 2010). As shown in Liu et al. (2006) and Yu et al. (2010), before its landfall, TC can lead (in some cases) to an foF2 short-term increase. The foF2 began to decrease one day after hurricane landfall. The minimal foF2 values are observed about two days later. The foF2 deviation from the median can reach 6–20% (Liu et al., 2006). At the same time, in the surroundings of ionosondes, the ground pressure and the height of 500 mbar surface rise. The magnitude and duration of the foF2 deviations correlate with the TC intensity. A similar effect was observed in the GPS-TEC variations over TC Matsa in July/August 2005 (Mao et al., 2010). One day before this TC landfall, a TEC increase by 5 TECU relative to monthly average values was recorded. After Matsa crossed the coastline, the TEC started to decrease. A day later, TEC reached its minimal value, 1 TECU less than the monthly average level. Guha et al. (2016) observed anomalous

GPS-TEC decreases (1.5–3.8 TECU) from the monthly mean value on the landfall days for TCs Mahasen (May 2013), Hudhud (October 2014) and Vongfong (October 2014). The foF2 and TEC decrease effect might be associated with the turbopause height increase (Section 2.4.4) (Liu et al., 2006). Because the turbulence intensification is more effective over the land than over the ocean, the foF2 and TEC decrease is recorded after the TC landfall.

2.4.2 Traveling Ionospheric Disturbances in the Upper Ionosphere

Most TIDs in the upper atmosphere are considered to be the manifestations of AGWs/IGWs. Detecting TIDs associated with the TC effect on the upper ionosphere (F-region) becomes complicated due to a strong variability of ionospheric parameters and a small amplitude of the ionospheric response to TC. Early studies (1975–2005) focused on recording individual TIDs. With short observational series this did not provide any complete confidence that the detected disturbances were caused by the TC effect. However, a TC acts during a comparatively long time (for several days). If it generates AGW/IGW during that time, this will lead to an increase in the intensity of the entire spectrum of variation in the ionospheric parameters, or in its individual bands. Therefore, when detecting TC effects in the upper ionosphere, special attention is now paid to search for an increase in the intensity of variation within the different ranges of periods.

2.4.2.1 Tropical Cyclones in the Pacific Ocean

The ionosphere F-region wave TIDs caused by tropical cyclones in the northwestern part of the Pacific Ocean have been repeatedly observed through various techniques, as will be outlined below.

According to the Doppler frequency shift data, the F-region response represented wave TIDs with periods of 1.4–9.7 minutes (Okuzawa et al., 1986), of 13–14 minutes (Huang et al., 1985), and of 20–90 minutes (Bertin et al., 1975; Hung and Kuo, 1978; Xiao, S-G. et al., 2006, 2012; Xiao, Z. et al., 2007). The response in the ionosphere is not always observed. In Huang et al. (1985), 12 typhoons over 1982–1983 were studied; the F-region response was detected only for two events (the 22–30 July 1982 typhoon Andy, and the 23–25 July 1983 typhoon Wayne). Xiao et al. (2007) recorded TIDs in the F-region during 22 of 32 typhoons that occurred between 1987 and 1992. TIDs are shown to be always observed at landfall. An increase in the TID period and amplitude over time is noted. TID transition into an intensive F-spread was repeatedly observed that testifies to the presence of small-scale irregularities in the ionosphere. Trajectory computations performed by Bertin et al. (1975) showed that the TID generation regions, most likely, were in the troposphere in the cyclone formation regions. Horizontal distances of probable sources from the TID recording locations were 1000–2000 km. All the recorded TIDs propagated in the direction opposite to the neutral wind direction in the ionosphere's F-region. These results confirm theoretical estimates (Hocke and Shlegel, 1996; Sorokin et al., 2005), according to which the neutral wind in the ionosphere is a filter for AGWs/IGWs.

The TEC TIDs excited by typhoons Dujuan (September 2015) (Kong et al., 2017), Meranti (September 2016) (Chou et al., 2017), Rammasun (July 2014) and Matmo

(July 2014) (Song et al., 2017) were detected using ground-based GNSS networks in Taiwan and China. According to these authors, the TID characteristics agree to the AGW theory: the estimated TID horizontal phase velocity was 240 m/s (Kong et al., 2017), 106–220 m/s (Chou et al., 2017), and 118–132 m/s (Song et al., 2017). Yang et al. (2016) recorded scintillations of GPS signals over the area of 13–23°N and 110–120°E during passage of typhoon Tembin (August 2012) from the ground-based GPS stations and from GPS receivers on board COSMIC satellites (GPS radio-occultation data).

The variations of ionospheric parameters during TCs Damrey (September 2005), Saola (September 2005), Longwang (September/October 2005), Kirogi (October 2005), Tembin (November 2005) and Bolaven (November 2005), all of which occurred in the northwest Pacific Ocean, were studied from the data of VS ionosondes and GPS-stations located in the East Siberian and Far Eastern regions of Russia, in China, and in Australia (Perevalova and Polekh, 2009; Perevalova and Ishin, 2011; Polyakova and Perevalova, 2013) (Figure 2.5). The period from 17 September 2005 to 30 November 2005 featured sufficiently quiet geomagnetic

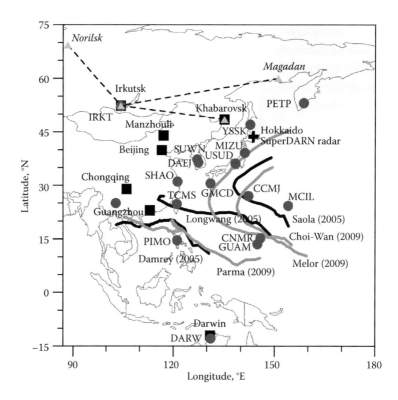

FIGURE 2.5 TC effect recording instruments in the northwestern Pacific: VS ionosondes (squares), OS ionosondes (triangles), GPS stations (dots), the Hokkaido SuperDARN radar (cross). The dashed lines show the OS paths. Grey colour indicates the SuperDARN radar coverage sector. The solid lines mark the trajectories of TCs Damrey, Saola, Longwang (in September 2005) and of TCs Choi-Wan, Melor, Parma (in September 2009).

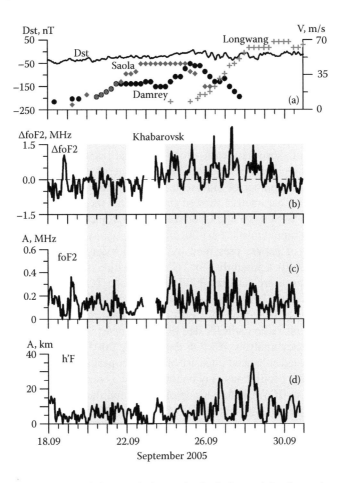

FIGURE 2.6 Variations of the Dst index and of wind speed in the cyclones Damrey (dots), Saola (rhombs), Longwang (crosses) (a). Ionospheric parameter disturbances over Khabarovsk: foF2 deviations from the daily median (b), intensity of 4 h oscillations in variations of foF2 (c), and h'F (d). Grey colour marks the periods of the increase in the TEC disturbance intensity at the GPS stations in the region.

conditions, and an absence of significant magnetic storms (Figure 2.6a shows the Dst variations in September 2005). The ionospheric disturbances most probably connected with the cyclonic activity were recorded at almost all the monitoring stations, the few exceptions being due to remoteness of the station from the TC trajectory, lack of sufficient observation data, or highly noisy data. Figure 2.6b–d presents examples of ionospheric parameter variations received with the Khabarovsk VS ionosonde in September 2005: the foF2 deviations from the daily median, the intensity of 4-h oscillations in the foF2 variations and of F-region minimal height h'F. According to the ionosonde data, the ionospheric disturbances exhibited themselves in the foF2 (Figure 2.6b) and h'F significant deviations from the median values, as well as in the 1.5–6 h foF2 and h'F variation strength increases (Figure 2.6c–d). No significant deviations were recorded in the maximum ionization height (h_{mF2}) behaviour. At the same time, an increase in the intensity of 02–20-, 20–60- and 60–90-min TEC oscillations was observed at the GPS stations in the region (Polyakova and Perevalova, 2013), as shown in Figure 2.6c–d. The highest amplitude of the TEC variation was observed when the wind speed in the TC was maximal. The TEC disturbances were more intense when several cyclones acted simultaneously.

Oinats et al. (2012) analysed the Hokkaido SuperDARN oblique backscatter radar data obtained in September 2008 and 2009. In presenting their results, they discuss the association of those data with strong TCs that occurred in the north-western Pacific during the observation period. Despite generally quiet geomagnetic conditions, a significant number of multiscale TID signatures were revealed. The estimates for the TID parameters showed that some of the recorded TIDs may have been caused by the Sinlaku and Choi-Wan strong tropical cyclones, both of which happened at that time (Oinats et al., 2012).

In the series of papers presented by Chernigovskaya et al. (2010, 2014) and Beletsky et al. (2010), a possibility for TC manifestation in the ionosphere and atmosphere, at a distance greater than 3000 km from the TC trajectories, was studied. Chernigovskaya et al. (2010, 2014) used the data of the MOF for ionosonde oblique-sounding signals (see Section 2.2) on the Magadan-Irkutsk, Khabarovsk-Irkutsk and Norilsk-Irkutsk radio paths located in the northeastern region of Russia (Figure 2.5). The frequency analysis of MOF data for the 2005–2010 spring and autumn periods has revealed enhanced intensity of the short-period (tens of minutes – hours) oscillations. These oscillations may be interpreted as manifestations of large-scale TIDs generated by IGWs with periods of 1–5 h and the TCs in the northwestern Pacific are considered as potential sources of these. Figure 2.7 shows the example of MOF spectra variations on the Khabarovsk-Irkutsk path in September 2009. Kp index variations characterizing geomagnetic conditions are given at the bottom. Vertical arrows mark the instants of local meteorological fronts' transiting in the region of the path's midpoint. Horizontal arrows show the TC Choi Wan, Parma and Melor action periods. A significant increase in the 1–5 h wave disturbances was recorded on the oblique Magadan-Irkutsk, Norilsk-Irkutsk and Khabarovsk-Irkutsk paths in the 2005–2010 autumn periods. These considerable increases do not appear to be associated with manifestations of helio-geomagnetic activity, and do not always coincide in time with the passage of local meteorological fronts; but they were observed during TC activities in the northwestern Pacific and might be associated with TC effect on the atmosphere. The TID intensity decreased as the midpoints of the oblique radio paths moved away from potential sources. The TID propagation velocity was estimated as 90–170 m/s. TIDs were also observed during the spring equinox under quiet helio-geomagnetic conditions, and in the absence of active TCs in the northwest Pacific, but the spring TID intensities were much lower than those of the autumn ones.

Thus, there is evidence (Perevalova and Polekh, 2009; Beletsky et al., 2010; Chernigovskaya et al., 2010, 2014; Perevalova and Ishin, 2011) of possible association between TIDs, observed at large distances from TC trajectories, and the activity of those cyclones. However, the possibility to record TC ionospheric effects at large distances requires, in our opinion, further experimental validation. One should

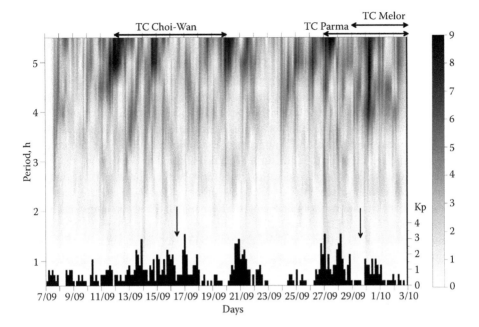

FIGURE 2.7 Maximum observed frequencies (MOF) current spectra at the Khabarovsk-Irkutsk path in September 2009. Below: The Kp index behaviour. Horizontal arrows indicate the TC effective periods. Vertical arrows mark the instants of the meteorological fronts passing in the region of the path's midpoint.

consider that, at large distances from the cyclone trajectory, the observed effects can be caused not by a TC effect, but by local sources of disturbance (for example, diurnal reconfiguration in the atmosphere, weather fronts, thunderstorms, particle precipitation, earthquakes, etc.).

2.4.2.2 Tropical Cyclones in the Atlantic Ocean

GPS stations are also actively used when studying ionospheric responses to TCs in the Atlantic Ocean. Ionospheric effects of TCs Katrina, Rita and Wilma (Perevalova and Ishin, 2011; Polyakova and Perevalova, 2011; Bondur and Pulinets, 2012; Zakharov and Kunitsyn, 2012) have been studied in greatest detail.

TC Katrina (23–31 August 2005) was one of the most destructive natural catastrophes in the history of the United States. Detecting ionospheric responses to TC Katrina was complicated by noise in the data caused by two geomagnetic storms (Figure 2.3): 24–26 August 2005 (Dst ~ –216 nT, Kp = 9) and 31 August–05 September 2005 (Dst ~ –131 nT, Kp = 7). However, on 27–29 August, when Katrina was in the hurricane phase (the wind speed in the cyclone varied 55–77 m/s, the pressure dropped to 902 mbar), geomagnetic conditions appeared quiet enough to enable us to detect Katrina-caused ionospheric disturbances. TCs Rita (18–26 September 2005) and Wilma (15–25 October 2005) were both active in the absence of geomagnetic storms (Dst ~ –50 nT, Kp ≤ 4). TC Rita was in the hurricane phase on 20–24 September (the wind speed reached 78 m/s, the pressure being 895 mbar); TC Wilma was in the hurricane stage on 19–26 October. On 19 October, the pressure at the Wilma centre reached an extremely low value of 882 mbar. The wind speed V was ≈ 78 m/s.

The electron density increase in the F-layer maximum over the centre of TC Katrina was recorded on 28 September 2005, based on the analysis of height profiles of the electron density (Ne) obtained from the GPS-sounding (Bondur and Pulinets, 2012). The detected Ne growth, according to Bondur and Pulinets (2012), may be caused by the electric field created because of the convection increase in the hurricane.

To compare the behaviour of ionospheric and meteorological parameters, Polyakova and Perevalova (2011) suggested a technique for building and comparatively analysing maps of TEC disturbance intensity and of meteorological parameters (temperature, pressure, speed of zonal and meridional winds). To build the spatial distribution of the TEC disturbance intensity, 'receiver-satellite' ray trajectories at the ionization maximum height are mapped. The position of each point is designated by a circle whose radius is proportional to the filtered TEC absolute value recorded at the given ray at the given instant. Meteoparameter maps are built from the NCEP/NCAR reanalysis archives (http://www.esrl.noaa.gov/psd). Using this technique allowed us to detect the TEC disturbances caused by TC Katrina, Rita and Wilma, and to trace their dynamics. Figure 2.8 exhibits the spatial-temporal distributions of the TEC variations for the periods 02–20 min and 20–60 min. This is compared with the maps of ground pressure and of the meridional wind speeds on the TC highest activity days. Thick black lines show the TC trajectories.

For all three TCs, a significant increase in intensity of TEC variations for both period ranges was registered at the 'receiver-satellite' rays passing near the cyclone trajectory. This increase was observed under quiet geophysical conditions, in local night and evening hours during the cyclones' maximal activity. We may note that no TEC disturbances were recorded directly over the TC 'eye'. As the cyclone weakened and travelled away from the station, the recorded TEC disturbance intensity decreased. The TEC disturbance behaviour testifies that an ionospheric plasma irregularity region forms at the ionospheric heights over the cyclone trajectory. It occurs when the cyclone reaches the hurricane phase and follows the cyclone. TEC disturbance variations were expressed more strongly in the long-period (20–60 min) than in the short-period (02–20 min) disturbances. Long-period TEC disturbances covered a large territory. The horizontal extent of the large-scale TEC disturbances was about 2000 km.

At the individual 'receiver-satellite' rays, due to favourable observation conditions, the TEC variations caused by the TC Katrina were sometimes higher than that generated by EA or magnetic storms (Figure 2.3 and Figure 2.8). The large TEC variation amplitude was obtained for the satellites that are low above the horizon (at low elevation angles) and when the TC was within immediate proximity of the GPS-station. As was shown in Polyakova and Perevalova (2013), recording TEC variations at low elevation angles, with the aid of special antennas and two-frequency phase measurements, is an effective instrument for detecting weak ionospheric disturbances generated by TCs, as well as those caused by distant earthquakes.

The ionosphere response to TCs Rita and Wilma was significantly weaker than that to TC Katrina (Polyakova and Perevalova, 2011; Perevalova et al., 2011). To reveal the cause of this, we analysed the height profiles of the pressure, of temperature, and

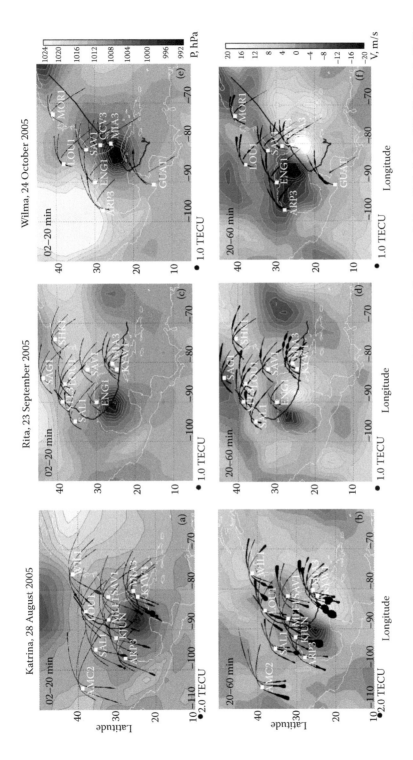

FIGURE 2.8 02–20 min and 20–60 min TEC intensity variation spatial distribution (black lines of variable width) during TCs Katrina (a–b), Rita (c–d), Wilma (e–f). Top: Filling shows the ground pressure P distribution. Bottom: Filling presents meridional wind speed V distribution. The positive meridional wind is northward. The squares mark the GPS station locations. Left bottom: The circle determining the TEC disturbance scale is shown.

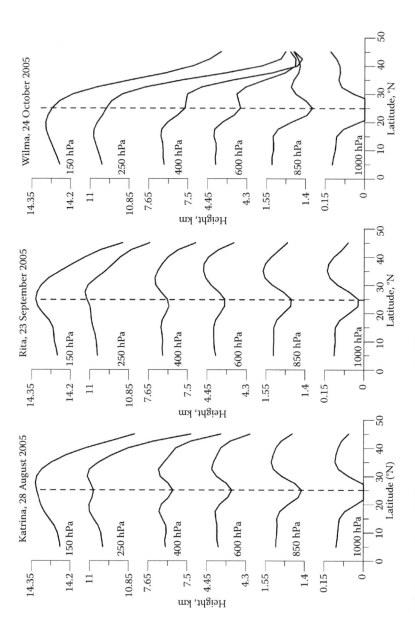

FIGURE 2.9   TC Katrina, Rita and Wilma height extents from the NCEP/NCAR reanalysis data: The height-latitude profiles for geopotential height of six levels of pressure (1000, 850, 600, 400, 250, 150 hPa) at the longitude that passed through the TC centre. Vertical lines mark the positions of the TC centre.

of wind speeds over the three TCs. As an example, Figure 2.9 shows the height-latitude profiles for geopotential height of six levels of pressure at the longitude that passed through the TC centre during high-activity days of each of the three cyclones. The analysis showed that Katrina was the highest cyclone: it was traced up to 250 hPa (~11 km), whereas TC Rita and Wilma were traced only up to 400 hPa (~8 km). Due to this, TC Katrina could have a stronger effect on the ionosphere.

To improve the reliability of detection and identification of the ionospheric responses to TCs, Zakharov and Kunitsyn (2012) and Zakharov and Budnikov (2012) used cluster analysis to process ground-based GPS network data. As well as TCs Katrina, Rita and Wilma, the authors also investigated three other large hurricanes that impacted the U.S. Atlantic coast: Ivan (2–24 September 2004), Dean (13–23 August 2007) and Ike (1–14 September 2008). For all six TCs, they detected wave structures having speeds of 300–1200 m/s. These values correspond to the IAW velocity. The spectrum analysis for these structures showed that a sharp decrease in the wave spectrum roll-off factor was observed during an abrupt change in TC intensity. This testifies to an increase in environmental turbulence associated with the TC wave's effect on the ionosphere. Also, an association between the wave structures and orographic objects (islands, mountains) along the TC trajectory was revealed. The wave generation in this case is probably related to the emergence of vortex formations in the region of the TC flow around an obstacle. At the same time, there was no statistically significant correlation between the wave structure emergence and the TC centre ('eye') (Zakharov and Kunitsyn, 2012).

The ionospheric disturbances caused by TC Odette in December 2003 were studied at the Arecibo Observatory in Puerto Rico by Bishop et al. (2006), using the incoherent scatter radar (ISR), the ionosonde located at Ramey and the GPS radio-occultation data. They reported that the ISR observed anomalously high velocities of the plasma drift in the F-region, during the period when Odette was at less than 600 km from the observatory. The velocities exceeded those observed in ISR under similar geomagnetic conditions by a factor of almost two. At that time, drift velocity wave variations with a 90-min average period were observed at 367 km. The ionosonde recorded intensive mid-latitude F-spread during the night after the TC's nearest passage. The GPS radio-occultation measurements recorded an AGW activity increase in the stratosphere and scintillations in the F-region.

2.4.3 ELECTRIC FIELD DISTURBANCES IN THE TOPSIDE IONOSPHERE

The 'Kosmos-1809' (1986–1993) and 'Interkosmos-24' (1989–1995) satellites have been used to measure the quasi-stationary electric field and ULF-VLF electromagnetic oscillations in the topside ionosphere (600–100 km) (Isaev et al., 2002; Mikhailova et al., 2002; Isaev et al., 2010).

Abnormally high ULF-VLF intensity values of the electric field variations were recorded by 'Interkosmos-24' in September 1990 around the Pacific Ocean (Mikhailova et al., 2002). These values were observed at the topside ionosphere heights (1500–2500 km) in the afternoon, during the TCs Lowel, Marie, Ed, Gene, Polo, Dot, Norbert, Flo, Odile, and Rachel C within 120–280°E and 7°S–15°N.

From the 'Kosmos-1809' measurements, the effect of an increase in the quasi-stationary electric field over the emergence and evolution regions of 10 TCs was noted (Isaev et al., 2002, 2012). The data that were obtained demonstrated that TCs are accompanied by localized disturbances of the quasi-stationary electric field in the topside ionosphere, at different stages of their evolution (from depression to hurricane). The detected disturbances reached ~20 mV/m at 950 km high; this value agrees with the results of theoretical estimates (Sorokin et al., 2005). Dynamic processes in a TC cause a substantial plasma density increase in the night ionosphere, stretching across a broad (greater than 1000 km) region that extends from the geomagnetic equator toward the TC. Under daytime conditions, TCs may be a cause of an increase in the plasma equatorial maxima in the localized regions west of TC trajectories.

2.5 CONCLUSION

Studying TC ionospheric effects corroborate the idea that the Earth's atmosphere represents an integrated dynamic system (Hocke and Shlegel, 1996; Kazimirovsky et al., 2003; Laštovička, 2006). The destructive tropospheric vortex effect is not only confined to the ground layers, but also noticeably changes the ionospheric plasma behaviour. In most cases, TC effects in the ionosphere are weaker than the disturbances caused by geomagnetic storms and major solar flares, but they are magnitudinally comparable with the disturbances from minor solar flares, solar terminator, equatorial anomaly of ionization and earthquakes. The last circumstance makes it necessary to carefully analyse geophysical conditions when detecting the ionosphere responses to TCs.

Experimental data generally confirm the results of theoretical and model computations. TCs are the sources of acoustic and gravity waves, and of the electric fields affecting the ionospheric plasma. In the lower ionosphere over the TC area, a decrease in the electron density (most strongly expressed near ~70 km) is recorded. In the upper ionosphere, TCs can cause either an increase or a decrease in the electron density, as well as electric field variations. Over the cyclone trajectory in the upper ionosphere, an ionospheric plasma irregularity region emerges. This region forms when the cyclone reaches the hurricane phase, has the horizontal extent of about 2000 km, and follows the cyclone. Recorded irregularities in the form of travelling ionospheric disturbances have periods from 10 minutes to several hours. The issue of the TC's effective radius on the ionosphere requires a special study. The problem is that at large distances from the TC trajectory, any disturbances that are observed could be caused by many various local sources.

Using dense networks of GNSS (GPS/GLONASS/Galileo/Beidou) receivers has opened new possibilities in studying ionospheric responses to TCs, and in determining their characteristics. The GNSS provides continuous global monitoring of the ionospheric plasma with high spatial and temporal resolutions. Using the techniques of spectral, correlative and cluster analyses may provide separation of disturbances from various sources, and identification of ionospheric responses to TCs. With other techniques of ionospheric measurements, and with meteorological data, these methods are leading to progress in studying the effects of powerful tropospheric vortices on the Earth's upper atmosphere.

ACKNOWLEDGEMENTS

The author is grateful to the Scripps Orbit and Permanent Array Center (SOPAC) for the data from the GPS receiver global network; to the National Centers for Environmental Prediction (NCEP) and National Center for Atmospheric Research (NCAR) for the meteorological data from the NCEP-NCAR Reanalysis Archives. NCEP reanalysis data is provided by the NOAA/OAR/ESRL PSD, Boulder, Colorado, from their website at http://www.esrl.noaa.gov/psd. The author thanks A.S. Polyakova and A.V. Oinats for most useful discussions and assistance in preparing the manuscript. Also, the author appreciates Yu. M. Kaplunenko's assistance in the English text editing.

REFERENCES

Afraimovich, E.L., Edemskiy, I.K., Voeykov, S.V., Yasukevich, Yu.V. and Zhivetiev, I.V. (2009). Spatial-temporal structure of the wave packets generated by the solar terminator. *Adv. Space Res.*, 44 (7), 824–835.

Astafyeva, E.I., Afraimovich, E.L. and Kosogorov, E.A. (2007). Dynamics of total electron content distribution during strong geomagnetic storms. *Adv. Space Res.*, doi:10.1016/j.asr.2007.03.0062007.

Bauer, S.J. (1958). An apparent ionospheric response to the passage of hurricanes. *J. Gephys. Res.*, 63, 265–269.

Beletsky, A.B., Mikhalev, A.V., Chernigovskaya, M.A., Sharkov, E.A. and Pokrovskaya, I.V. (2010). Study of possibility of manifestation of tropical cyclone activity in the atmospheric airglow. *Earth Research from Space*, 4, 41–49 (in Russian).

Bertin, F., Testud, J. and Kersley, L. (1975). Medium scale gravity waves in the ionospheric F-region and their possible origin in weather disturbances. *Planet. Space Sci.*, 23, 493–507.

Bishop, R.L., Aponte, N., Earle, G.D., Sulzer, M., Larsen, M.F. and Peng, G.S. (2006). Arecibo observations of ionospheric perturbations associated with the passage of tropical storm Odette. *J. Geophys. Res.*, 111, A11320, doi:10.1029/2006JA011668.

Bondur, V.G. and Pulinets, S.A. (2012). Effect of mesoscale atmospheric vortex processes on the upper atmosphere and ionosphere of the Earth. *Izvestiya, Atmospheric and Oceanic Physics*, 48 (9), 871–878.

Chernigovskaya, M.A., Kurkin, V.I., Oinats, A.V., Poddelsky, I.N. (2014). Ionosphere effects of tropical cyclones over the Asian region of Russia according to oblique radio-sounding data. *Proc. of SPIE*, 9292, 92925E-1.

Chernigovskaya, M.A., Kurkin, V.I., Orlov, I.I., Sharkov, E.A. and Pokrovskaya, I.V. (2010). Study of coupling of the ionospheric parameter short period variations in the northeastern region of Russia with manifestation of tropical cyclones. *Earth Research from Space*, 5, 32–41 (in Russian).

Chernogor, L.F. (2006). The tropical cyclone as an element of the Earth-Atmosphere-Ionosphere-Magnetosphere system. *Journal of Space Science and Technology*, 12 (2/3), 16–36 (in Russian).

Chisham, G., Lester, M., Milan, S.E. et al. (2007). A decade of the Super Dual Auroral Radar Network (SuperDARN): Scientific achievements, new techniques and future directions. *Surveys in Geophysics*, 28 (1), 33–109.

Chou, M. Y., Lin, C.C.H., Yue, J., Tsai, H.F., Sun, Y.Y., Liu, J.Y., and Chen, C.H. (2017). Concentric traveling ionosphere disturbances triggered by Super Typhoon Meranti (2016), *Geophys. Res. Lett.*, 44, 1219–1226.

Davies, K. (1990). *Ionospheric Radio*. London: Peter Peregrinus.

Gavrilov, N.M., Kshevetskii, S.P. (2013). Numerical modeling of propagation of breaking non-linear acoustic-gravity waves from the lower to the upper atmosphere. *Adv. Space Res.*, 51, 1168–1174.

Godin, O.A. (2015). Wentzel-Kramers-Brillouin approximation for atmospheric waves. *J. Fluid Mech.*, 777, 260–290.

Gossard, E.E. and Hooke, W.H. (1975). *Waves in the Atmosphere.* Elsevier: New York.

Guha, A., Paul, B., Chakraborty, M. and De, B.K. (2016). Tropical cyclone effects on the equatorial ionosphere: First result from the Indian sector. *J. Geophys. Res. Space Physics*, 121, 5764–5777.

Hocke, K. and Shlegel, K. (1996). A review of atmospheric gravity waves and traveling iono-spheric disturbances: 1982–1995. *Ann. Geophys.*, 14, 917–940.

Hoffmann, L. and Alexander, M.J. (2010). Occurrence frequency of convective gravity waves during the North American thunderstorm season. *J. Geophys. Res. Atmospheres*, 115 (20), D20111.

Hofmann-Wellenhof, B., Lichtenegger, H. and Wasle, E. (2008). *GNSS – Global Navigation Satellite Systems. GPS, GLONASS, Galileo, and More.* Springer: Wien, New York.

Huang, Y.-N., Cheng, K. and Chen, S.-W. (1985). On the detection of acoustic gravity waves generated by typhoon by use of real time HF Doppler frequency shift sounding system. *Radio Sci.*, 20, 897–906.

Hung, R.J. and Kuo, J.P. (1978). Ionospheric observation of gravity waves associated with Hurricane Eloise. *J. Geophys.*, 45, 67–80.

Isaev, N.V., Kostin, V.M., Belyaev, G.G., Ovcharenko, O.Ya., Trushkina, E.P. (2010). Disturbances of the topside ionosphere caused by typhoons *Geomagnetism and Aeronomy*, 50 (2), 243–255.

Isaev, N.V., Sorokin, V.M., Chmyrev, V.M., Serebryakova, O.N. and Yashchenko, A.K. (2002). Disturbance of the electric field in the ionosphere by sea storms and typhoons. *Cosmic Research*, 40 (6), 547–553.

Kazimirovsky, E., Herraiz, M. and De La Morena, B.A. (2003). Effects of the ionosphere due to phenomena occurring below it. *Surv. Geophys.*, 24, 139–184.

Kelly, V.S., Siefring, C.L., Pfaff, R.F. et al. (1985). Electrical measurements in the atmosphere and the ionosphere over an active thunderstorm. 1. Campaign overview and initial iono-spheric results. *J. Geophys. Res.*, 90 (A10), 9815–9824.

Kim, S.-Y., Chun, H.-Y. and Baik, J.-J. (2005). A numerical study of gravity waves induced by convection associated with Typhoon Rusa. *Geophys. Res. Lett.*, 32, L24816, doi:10.1029/2005GL024662.

Kong, J., Yao, Y., Xu, Y., Kuo, C., Zhang, L., Liu, L., Zhai, C. (2017). A clear link connecting the troposphere and ionosphere: Ionospheric reponses to the 2015 Typhoon Dujuan. *J. Geod.* doi:10.1007/s00190-017-1011-4.

Kuester, M.A., Alexander, M.J. and Ray, E.A. (2001). A Model Study of Gravity Waves over Hurricane Humberto. *J. Atmosphere Sci.*, 65 (10), 3231–3246.

Kunitsyn, V.E., Suraev, S.N. and Akhmedov, R.R. (2007). Modeling of atmospheric prop-agation of acoustic gravity waves generated by different surface sources. *Moscow University Physics Bulletin*, 62 (2), 122–125.

Laštovička, J. (2006). Forcing of the ionosphere by waves from below. *J. Atmos. Sol. Terr. Phys.*, 68, 479–497.

Liu, Y.-M., Wang, J-S. and Suo, Y-C. (2006). Effects of typhoon on the ionosphere. *Adv. Geosci.*, 29, 351–360.

Maclay, K.S., DeMaria, M. and Haar, T.H.V. (2008). Tropical cyclone inner-core kinetic energy evolution. *Monthly Weather Rev.*, 136, 4882–4898.

Mao, T., Wang, J-S., Yang, G-L., Yu, T., Ping, J-S. and Suo, Y-C. (2010). Effects of typhoon Matsa on ionospheric TEC. *Chinese Sci. Bull*, 55 (8), 712–717.

Mikhailova, G.A., Mikhailov, Yu.M. and Kapustina, O.V. (2002). Variations of ULF-VLF electric fields in the external ionosphere over powerful typhoons in Pacific Ocean. *Adv. Space Res*, 30 (11), 2613–2618.

Oinats, A.V., Kurkin, V.I., Nishitani, N. and Chernigovskaya, M.A. (2012). Meteorological effects in the ionosphere by SuperDARN Hokkaido HF radar. *Actual Problems in Remote Sensing of the Earth from Space*, 9 (4), 113–120 (in Russian).

Okuzawa, T., Shibata, T., Ichinose, T., Takagi, K., Nagasawa, C., Nagano, I., Mambo, M., Tsutsui, M. and Ogawa, T. (1986). Short-period disturbances in the ionosphere observed at the time of typhoons in September 1982 by a network of HF doppler receivers. *J.-Geomag. Geoelectr.*, 38, 239–266.

Perevalova, N.P., Afraimovich, E.L., Voeykov, S.V. and Zhivetiev, I.V. (2008). Parameters of large scale TEC disturbances during strong magnetic storm on October 29, 2003. *J. Geophys. Res.*, 113, A00A13, doi:10.1029/2008JA013137.

Perevalova, N.P. and Ishin, A.B. (2011). Effects of tropical cyclones in the ionosphere from data of sounding by GPS signals. *Izvestiya, Atmospheric and Oceanic Physics*, 47 (9), 1070–1081.

Perevalova, N.P. and Polekh, N.M. (2009). An investigation of the upper atmosphere response to cyclones using ionosonde data in Eastern Siberia and the Far East. *Proceedings of SPIE 2009*, 7296, 72960J1-72960J11.

Perevalova, N.P., Polyakova, A.S., Ishin, A.B. and Voeykov, S.V. (2011). Comparative analysis of ionospheric and meteorological parameter variations over tropical cyclones Rita (18–26.09.2005) and Wilma (15–25.10.2005). *Actual Problems in Remote Sensing of the Earth from Space*, 8 (1), 303–312 (in Russian).

Pfister, L., Chan, K.R., Bui, T.P., Bowen, S., Legg, M., Gary, B., Kelly, K., Proffitt, M. and Starr, W. (1993). Gravity waves generated by a tropical cyclone during the STEP tropical field program: A case study. *J. Geophys. Res.*, 98 (D5), 8611–8638.

Pielke Jr., R.A. and Pielke Sr., R.A. (1997). *Hurricanes. Their Nature and Impacts on Society.* John Wiley & Sons, Chichester, New York.

Pogoreltsev, A.I. (1996). Production of electromagnetic field disturbances due to the interaction between acoustic gravity waves and the ionospheric plasma. *Journal of Atmospheric and Terrestrial Physics*, 58 (10), 1125–1141.

Pokrovskaya, I.V. and Sharkov, E.A. (2006). *Tropical Cyclones and Tropical Disturbances of the World Ocean, Chronology and Evolution.* Version 3.1 (1983–2005). Poligraph Services, Moscow.

Polyakova, A.S. and Perevalova, N.P. (2011). Investigation into impact of tropical cyclones on the ionosphere using GPS sounding and NCEP/NCAR Reanalysis data. *Advances in Space Research*, 48, 1196–1210.

Polyakova, A.S. and Perevalova, N.P. (2013). Comparative analysis of TEC disturbances over tropical cyclone zones in the North-West Pacific Ocean. *Advances in Space Research*, 52, 1416–1426.

Pulinets, S.A., Boyarchuk, K.A., Khegai, V.V., Kim, V.P. and Lomonosov, A.M. (2000). Quasielectrostatic model of atmosphere–thermosphere–ionosphere coupling. *Adv. Space Res.*, 26 (8), 1209–1218.

Rice, D.D., Sojka, J.J., Eccles, J.V. and Schunk, R.W. (2012). Typhoon Melor and ionospheric weather in the Asian sector: A case study. *Radio Science*, 47, RS0L05, doi:10.1029/2011RS004917.

Rozhnoi, A., Solovieva, M., Levin, B., Hayakawa, M., and Fedun, V. (2014). Meteorological effects in the lower ionosphere as based on VLF/LF signal observations. *Nat. Hazards Earth Syst. Sci.*, 14, 2671–2679.

Shen, C-S. (1982). The correlation between the typhoon and the foF2 of ionosphere. *Chinese Journal of Space Science,* 2 (4), 335–340 (in Chinese).

Song, Q., Ding, F., Zhang, X., Mao, T. (2017). GPS detection of the ionospheric disturbances over China due to impacts of Typhoons Rammasum and Matmo. *J. Geophys. Res. Space Physics*, 122 (1), 1055–1063.

Sorokin, V.M., Isaev, N.V., Yaschenko, A.K., Chmyrev, V.M. and Hayakawa, M. (2005). Strong DC electric field formation in the low latitude ionosphere over typhoons. *Atmos. Solar. Terr. Phys.*, 67, 1269–1279.

Taubenheim, J. (1983). Meteorological control of the D-region. *Space Sci. Rev.*, 34, 397–411.

Vadas, S.L., Nicolls, M.J. (2012). The phases and amplitudes of gravity waves propagating and dissipating in the thermosphere: Theory. *J. Geophys. Res: Space Phys.*, 117 (5), A05322.

Vanina-Dart, L.V., Pokrovskaya, I.V. and Sharkov, E.A. (2008). Response of the lower equatorial ionosphere to strong tropospheric disturbances. *Geomagnetism and Aeronomy*, 48 (2), 245–250.

Vanina-Dart, L.V., Romanov, A.A. and Sharkov, E.A. (2011). Influence of a tropical cyclone on the upper ionosphere according to tomography sounding data over Sakhalin Island in November 2007. *Geomagnetism and Aeronomy*, 51 (6), 774–782.

Vanina-Dart, L.V., Sharkov, E.A. and Pokrovskaya, I.V. (2007). The solar activity influence on the equatorial lower ionosphere reply during the active phase of tropical cyclones. *Earth Research from Space*, 6, 3–10 (in Russian).

Wang, L., Alexander, M.J., Bui, T.P. and Mahoney, M.J. (2006). Small-scale gravity waves in ER-2 MMS/MTP wind and temperature measurements during CRYSTAL-FACE. *Atmos. Chem. Phys.*, 6, 1091–1104.

World Data Centre for Geomagnetism, n.d. World Data Centre for Geomagnetism, Kyoto: Home Page http://swdcwww.kugi.kyoto-u.ac.jp (Accessed 27 April 2017).

Xiao, S-G., Hao, Y-Q., Zhang, D-H. and Xiao, Z. (2006). A case study on the detailed process of the ionospheric responses to the typhoon. *Chin. J. Geophys*, 49 (3), 546–551.

Xiao, S-G., Shi, J-K., Zhang, D-H., Hao, Y-Q. and Huang, W-Q. (2012). Observational study of daytime ionospheric irregularities associated with typhoon. *Sci China Tech. Sci.*, 55 (5), 1302–1304.

Xiao, Z., Xiao, S., Hao, Y. and Zhang, D. (2007). Morphological features of ionospheric response to typhoon. *J. Geophys. Res.*, 112, A04304, doi:10.1029/2006JA011671.

Yang, Z., and Liu, Z. (2016). Observational study of ionospheric irregularities and GPS scintillations associated with the 2012 tropical cyclone Tembin passing Hong Kong. *J. Geophys. Res. Space Physics*, 121, 4705–4717.

Yu, T., Wang, Y., Mao, T., Wang, J., Wang, S., Shuai, F., Su, W., and Li, J. (2010). A case study of the variation of ionospheric parameter during typhoons at Xiamen. *Acta Meteorologica Sinica*, 4, 569–576 (in Chinese).

Zakharov, V.I. and Budnikov, P.A. (2012). The application of cluster analysis to the processing of GPS_interferometry data. *Moscow University Physics Bulletin*, 67 (1), 25–32.

Zakharov, V.I. and Kunitsyn, V.E. (2012). Regional features of atmospheric manifestations of tropical cyclones according to ground-based GPS network data. *Geomagnetism and Aeronomy*, 52 (4), 533–545.

3 Typhoon and Sea Surface Cooling

Xidong Wang

3.1 INTRODUCTION

The typhoon (in this chapter, tropical storms, tropical cyclones and hurricanes are collectively called typhoons; the term tropical cyclone [TC] is more generally used elsewhere in this volume) is one of the most destructive natural disasters for humanity, as well as severely impacts the natural ecosystems (Pielke et al., 2008; Altman et al., 2013). A global average of 86 storms acquires tropical storm strength or greater (greater than 17 ms^{-1} sustained winds); also, an average of 46.9 category 1 or stronger typhoons (greater than 33 ms^{-1} sustained winds) per year occurred between 1980 and 2010. Using EM-DAT data during 1980–2000, researchers with the United Nations Development Programme identified that four developed nations and 29 developing nations were significantly exposed to tropical cyclones (Shultz et al., 2005). During approximately the last two centuries, typhoons have been responsible for the deaths of about 1.9 million people worldwide (Adler, 2005). It is estimated that 10,000 people per year perish due to typhoons (Adler, 2005), while estimates of the current damages from typhoons for the entire world are US$13 billion per year (Mendelsohn et al., 2012).

A typhoon is developed and maintained by drawing energy from the underlying warm ocean surface (see also Roy and Kovordanyi, Chapter 1 this volume). It can only form over waters of 26°C or higher and its intensity is very sensitive to sea surface temperature (SST) (Tuleva and Kurihara, 1982; Emanuel, 1986). Treating a typhoon as a Carnot heat engine, Emanuel (1986) suggested that the maximum potential intensity (MPI) of a typhoon was primarily determined by the underlying SST. Typhoon-induced SST cooling ranging from 1 to 6°C (Liepper, 1967; Emanuel, 1986), or even as much as 9°C (Sakaida et al., 1989), has been known since the 1960s (Leipper, 1967).

More recently, satellite images have shown that SST dropped more than 9 and 11°C, respectively, in responses to the passage of typhoons Kai-Tak (Lin et al., 2003) and Ling-Ling (Shang et al., 2001). Previous studies also suggested that ocean conditions ahead of typhoon passage can significantly impact the SST cooling induced by the typhoon itself, which may in turn affect the typhoon intensity (Shay et al., 2000; Goni and Trinanes, 2003; Emanuel et al., 2004; Lin et al., 2005; Scharroo et al., 2005; Oey et al., 2007; Wu et al., 2007; Shang et al., 2008; Wang et al., 2011; Balaguru et al., 2012).

During its passage across the Gulf of Mexico in September 2004, Typhoon Ivan hovered over cyclonic eddies in the ocean, and the elevated level of the upwelling-induced cooling exceeded 5°C (Walker et al., 2005; Halliwell et al., 2008). This was a clear example of negative feedback, in which upper-ocean cooling affects the ocean mixed layer and heat fluxes that feed into the typhoon, and eventually reduces the typhoon's intensity (Jacob and Shay, 2003; Shay and Uhlhorn, 2008). In contrast, the ocean mixed layer is not significantly cooled by typhoon-induced mixing in a region of warm eddies. During the passage of typhoon Opal across the Gulf of Mexico in 1995, the ocean mixed layer was found to have cooled less than 1°C in a warm core eddy detached from the Gulf of Mexico Loop Current based on buoy measurements and satellite SST images (Shay et al., 2000). Additionally, SST cooling may be greatly suppressed due to the presence of a thicker warm layer when a typhoon encounters warm ocean features (Wu et al., 2008).

Besides the self-feedback effect, the sea surface cold wakes can exist in the ocean for days to weeks and have a continuing effect upon the atmosphere and ocean, and can even contribute to climate variability and change. Firstly, these cold wakes have a large effect on the seasonal cycle of SST. Typhoons occurring in the first half of the season disrupt the seasonal warming trend, which is not resumed until 30 days after the typhoon's passage. Conversely, typhoon occurrences in the second half of the season bring about a mean temperature drop of approximately 0.5°C, from which the ocean does not recover due to the seasonal cooling cycle (Dare and McBride, 2011). Secondly, the persistence of SST anomalies for 1–2 months after the typhoon's passage could potentially impact upon larger-scale atmospheric circulations (Hart et al., 2007). Thirdly, recovery of the cold wake to pretyphoon conditions requires a net increase in the vertically integrated enthalpy of the sea water layer affected by the typhoon. In statistical equilibrium, this must be balanced by a lateral heat flux out of the region of typhoon activity, which could be a feasible way to drive poleward heat transport (Emanuel, 2001). This can reflect the effect of SST cooling on the global oceans and the climate system.

In this chapter, the main aim is to document the universal features of the SST cooling and to offer a systematic analysis of the effect of upper ocean structure on sea surface cooling regionally and globally. Some cases are discussed to illustrate or emphasize points made. The rest of this chapter is organized as follows. In Section 3.2, the statistics of typhoon-induced SST cooling are presented and the effect of ocean subsurface structure on the cooling also is introduced. Section 3.3 introduces the feedback effect of typhoon self-induced cooling on typhoon development and intensification. Section 3.4 gives a brief summary.

3.2 TYPHOON-INDUCED SEA SURFACE COOLING (SST) COOLING

3.2.1 STATISTICAL FEATURES OF TYPHOON-INDUCED SST COOLING

In this section we review the general characteristics of typhoon-induced SST cooling in the regional and global ocean scale. Chiang et al. (2011) showed a statistical characteristic of maximum cooling during 1958–2008 in northern South China Sea (Figure 3.1). The maximum cooling of 31 typhoons ranges from 1–8°C. The average maximum cooling is 4.31 ± 2.03°C. The maximum cooling represents a negative correlation with the typhoon moving speed (Figure 3.1).

Using global SST and typhoon track data during 1981–2008, Dare and McBride (2011) studied the response of SST to a typhoon. They found that the occurrence time of maximum SST cooling ranged from day –1 to day +7 relative to typhoon passage. The most universal appearance of the maximum SST cooling was 1 day after typhoon passage. Corresponding to the day of maximum cooling, a global composite average response was an SST reduction of 0.9°C. The global average maximum cooling for a weak typhoon was approximately –0.7°C while that for the intense typhoon (sustained wind speed exceeding 44 ms^{-1}) was –1.4°C. In addition, the influence of typhoon translation speed on sea surface cooling was also investigated. The discovery was that the slowly moving typhoon (less than 8 km h^{-1}) had an average maximum cooling of –1.1°C, compared with –0.75°C for the rapidly moving typhoon.

Vincent et al. (2012) used a global ocean general circulation model with 1/2° resolution to investigate global sea surface cooling induced by a typhoon during 1998–2007. The model successfully reproduces the observed spatial distribution of typhoon-induced cooling (Figure 3.2). The average cooling within typhoon-active regions is about 1°C in most ocean basins, with maximum amplitude of about 2°C in the northwest Pacific region where most of the strongest typhoons usually occur. The modelling average cooling is overestimated by almost 1°C in the northeast Pacific basin, which may be attributed to a shallower than observed thermocline in this region. The observational cooling magnitude monotonically increases with wind intensity up to a Category 2 typhoon on the Saffir–Simpson scale, and then saturated from Category 2 to 5.

3.2.2 EFFECT OF OCEAN STATE ON TYPHOON-INDUCED SST COOLING

With advances in satellite remote sensing, the oceanic meso-scale process can be effectively detected by satellite altimetry. Recent observational findings have shown that a great number of typhoons intensified rapidly when they passed through a positive sea surface height anomaly (SSHA) region, indicating a warmer and deeper mixed layer (Shay et al., 2000; Lin et al., 2005; Scarroo et al., 2005). Generally speaking, this rapid intensification is attributed to the typhoon self-induced cooling, suppressed by the warmer and thicker mixed layer. Based on 13 years of satellite altimetry, *in situ* and climatological temperature/salinity data, the best typhoon track data of the U.S. Joint Typhoon Warning Center (JTWC), together with an ocean mixed layer model, 30 western North Pacific Category 5 typhoons occurring during the typhoon season from 1993 to 2005 were systematically examined by Lin et al. (2008).

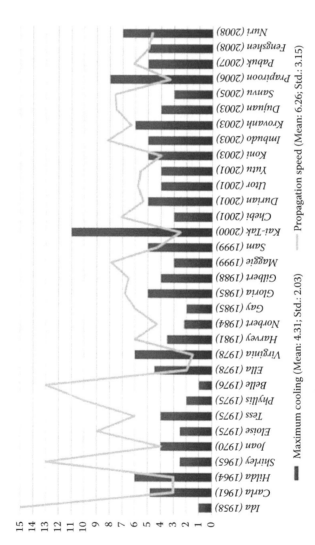

FIGURE 3.1   Maximum cooling (8°C) and translation speed (ms⁻¹) of typhoons. The 1958–1988 periods in various regions are from Bender et al. (1993). The 1998–2008 periods are for the northern SCS: maximum cooling from TMI SST images and translation speed from JTWC best-track data. (Adapted from Chiang, T. L., Wu, C. R. and Oey, L.Y. (2011). Typhoon Kai-Tak: An ocean's perfect storm. *J. Phys.Oceanogr.*, 41, 221–233.)

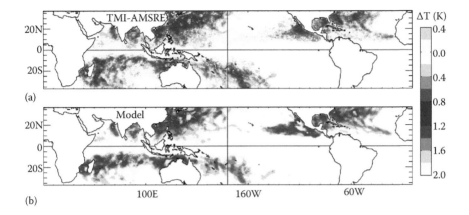

FIGURE 3.2 Spatial maps of average cold wake amplitude over 1998–2007 for (a) TMI-AMSRE observations and (b) model experiment. These maps are produced as follows: the maximum amplitude of cold wake during a typhoon passage over each grid point within 200 km of each typhoon position are averaged for all typhoon passing over the same grid point within the same typhoon season; these seasonal maps are then averaged over the 1998–2007 period. (Adapted from Vincent, E. M., Lengaigne, M., Vialard, J., Madec, G., Jourdain, N. C., and Masson, S. (2012). Assessing the oceanic control on the amplitude of sea surface cooling induced by tropical cyclones. *J. Geophys. Res.*, 117, C05023, doi:10.1029/2011JC007705.)

The comparison of Figures 3.3a,b shows that under the same translation speed and wind forcing, the typhoon-induced SST cooling is evidently larger under the climatological condition (Figure 3.3a) than under the condition with the positive-SSHA features (Figure 3.3b).

As seen in Figure 3.3b, given translation speeds of 3, 4 and 5 ms^{-1}, typhoon-induced SST cooling never falls into the shaded region throughout the intensification period from category 1 to 5. Even with the intense 65 ms^{-1} wind, the typhoon-induced SST cooling is still only around 1.2°C when traveling at 5 ms^{-1} (Figure 3.3b). The results indicate that the presence of the positive-SSHA features can effectively limit the typhoon self-induced cooling in the western North Pacific South Eddy Zone (127–170°E, 21–26°N, SEZ). In the gyre central region (121–170°E, 10–21°N, GCR), more pronounced cooling is detected when the typhoon is over the negative-SSHA feature (Figure 3.4b) than under the climatological condition (Figure 3.4a). Note that the typhoon-induced SST cooling falls to the shaded region only when the typhoon translation speed is 1–2 ms^{-1} under climatology (Figure 3.4a). Though cooling is more pronounced in the negative-SSHA features, these negative-SSHA features are nonetheless found to show a less shoaling situation from an existing thick, warm layer in the gyre central region. Even after becoming shallow from the negative-SSHA features, the warm layer thickness is relatively large. Figure 3.4b shows that if a typhoon is fast-moving (i.e., the translation speed is 6–10 ms^{-1}), the self-induced cooling will not exceed the threshold limit since the faster the typhoon travels, the induced self-induced cooling is less pronounced. In the gyre, the background climatological warm layer is already very deep, with a depth of 26°C isotherm (D26) extending to around 100 m. Over the positive-SSHA features in the gyre region, the

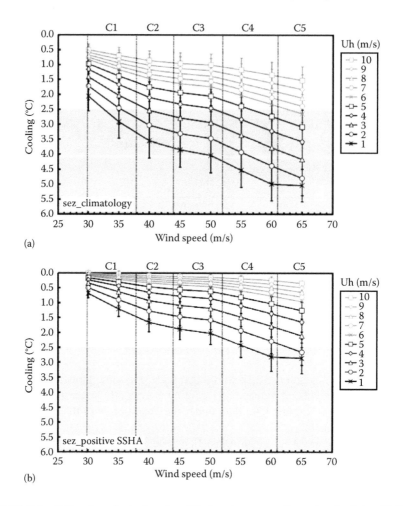

FIGURE 3.3 Estimation of a typhoon's self-induced ocean cooling using the Price–Weller–Pinkel ocean mixed layer model for the SEZ cases. The typhoon-induced cooling is estimated progressively with increase in wind forcing (every 5 ms^{-1} interval) from category 1 to 5 (denoted as C1-C5) at various translation speeds (1, 2, 3,…10 ms^{-1}). Region of self-induced cooling exceeding the 2.5°C threshold is shaded in grey, indicating unsuitable condition for intensification. (a) Estimation based on the climatological profiles along the typhoon tracks. (b) Estimation based on the in situ profiles inside the positive-SSHA features. (Adapted from Lin, I. I., Wu, C. C., Pun, I. F., and Ko, D. S. (2008). Upper-ocean thermal structure and the western North Pacific category 5 typhoons. Part I: Ocean features and the category 5 typhoons' intensification, *Mon. Weather Rev.*, 136, 3288–3306, doi:10.1175/2008MWR2277.1.)

warm layer can extend even deeper, corresponding to D26 of an average of about 130 m. Under this situation, the corresponding typhoon-induced SST cooling is very limited. Only when the translation speeds are between 1 and 2 ms^{-1} would the typhoon-induced SST cooling exceed the threshold value (Figure 3.4c).

Using a global ocean model, Vincent et al. (2012) investigated the oceanic influence on typhoon-induced cooling. The ocean model realistically sampled the ocean

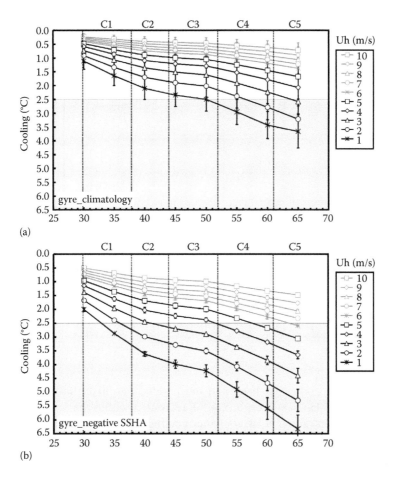

FIGURE 3.4 As in Figure 3.3, but for the self-induced ocean cooling estimation for the gyre cases. (a) Estimation based on the climatological profiles along the typhoon tracks. (b) Estimation based on the in situ profiles inside the negative SSHA features. *(Continued)*

response to more than 3000 typhoons over the last 30 years. This provided a comprehensive and global quantification of the sensitivity of TC-induced cooling amplitude to pretyphoon ocean state. Their study derived a physically grounded ocean parameter, the cooling inhibition (CI) index. CI can be used to describe the oceanic inhibition to a wind-induced cooling because it integrated two relevant parameters for the cooling amplitude: the initial mixed layer depth, and the strength of the thermal stratification below it. The results showed that using the CI improved the statistical hindcasts of the cold wake amplitude by 40%. This also indicated that the upper ocean thermal structure significantly impacted the amplitude and pattern of sea surface cooling.

In regions of high fresh water input, where significant stratification due to salinity sets within a deep isothermal layer, a so-called barrier layer between

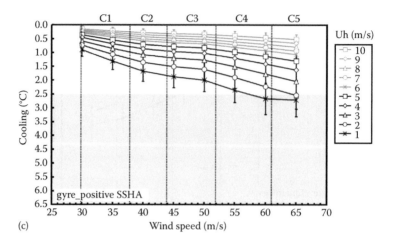

FIGURE 3.4 (CONTINUED) As in Figure 3.3, but for the self-induced ocean cooling estimation for the gyre cases. (c) Estimation based on the in situ profiles inside the positive-SSHA features. (Adapted from Lin, I. I., Wu, C. C., Pun, I. F., and Ko, D. S. (2008). Upper-ocean thermal structure and the western North Pacific category 5 typhoons. Part I: Ocean features and the category 5 typhoons' intensification, *Mon. Weather Rev.*, 136, 3288–3306, doi:10.1175/2008MWR2277.1.)

the base of the isothermal layer and the base of the mixed layer can appear. When there is a barrier layer, mixing is restricted within the shallower mixed layer, and the entrainment of cool thermocline water into the mixed layer is reduced. Several studies have suggested an active role of the barrier layer in typhoon-induced SST cooling (Ffield, 2007; Sengupta et al., 2008; Wang et al., 2011; Balaguru et al., 2012). Both the array for real-time geostrophic ocean-ography (Argo) measurements, and satellite SST data revealed that there was less-than-usual surface cooling during the passage of typhoons Kaemi in July 2006 and Cimaron in October 2006, both in the tropical western North Pacific (Wang et al., 2011). It was found that SST cooling induced by a typhoon can be greatly suppressed by a barrier layer with a thickness of 5–15 m. Such a barrier layer could reduce the entrainment cooling by 0.4–0.8°Cd^{-1} during a typhoon's passage, according to a diagnostic mixed layer model. The pre-existing barrier layer led to a reduction in typhoon-induced surface cooling, and favoured typhoon development. The average SST cooling under barrier layer condition is 0.4–0.8°C less than that under the no-barrier-layer condition. To address whether the effect of barrier layer on typhoon-induced SST cooling holds true in general, Balaguru et al. (2012) analyzed a decade of TC tracks from 1998 to 2007 in the major tropical barrier layer regions, which included a total of 587 typhoons in the northwestern and southwestern tropical Pacific, northwestern tropical Atlantic, and northern tropical Indian Ocean basins. They concluded that, due to the barrier layer effect, the mean SST cooling induced by typhoons is reduced by 36% (Figure 3.5).

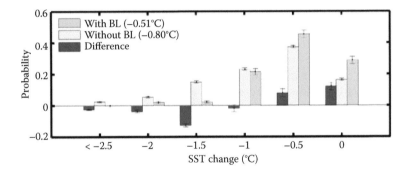

FIGURE 3.5 The comparison between typhoon-induced SST cooling with barrier layer and without barrier layer condition. (Adapted from Balaguru, K., Chang, P., Saravanan, R., Leung, L. R., Xu, Z., Li, M. K., and Hsieh, J. S. (2012). Ocean barrier layers' effect on tropical cyclone intensification. *PNAS*, doi:10.1073/pnas.1201364109.)

3.3 FEEDBACKS OF SST COOLING TO TYPHOON INTENSIFICATION

The typhoon-induced SST cooling plays a critical negative feedback role in the intensification of a typhoon (Emanuel, 1999; Schade and Emanuel, 1999; Bender and Ginis, 2000; Cione and Uhlhorn, 2003; Emanuel et al., 2004; Lin et al., 2005; Zhu and Zhang, 2006; Wu et al., 2007). Typhoon intensity is most sensitive to the SST cooling under the typhoon eye. Typhoon-induced cooling can limit the evaporation, therefore resulting in the negative feedback of air–sea. Recent observational findings may provide more highlights on this feedback effect.

SST variability between the typhoon inner-core environment and the ambient ocean environment ahead of the typhoon was investigated using airborne expendable bathythermograph (AXBT) observations and buoy-derived archived SST data during 1975–2002 (Cione and Uhlhorn, 2003). Results showed that relatively modest changes in typhoon inner-core SST can dramatically alter air–sea fluxes within the high-wind inner-core typhoon environment. Initial estimates showed that SST changes on the order of 1°C led to surface enthalpy flux changes of 40% or more. The magnitude of SST change (ambient minus inner core) was statistically linked to subsequent changes in typhoon intensity for the 23 typhoons. A relationship between reduced inner-core SST cooling and typhoon intensification is constructed. Under certain circumstances, the variability associated with inner-core SST change appeared to be an important factor directly linked to the intensity change process.

The influence of oceanic changes on typhoon activity was investigated using observational estimates of SST, air–sea fluxes and ocean subsurface thermal structure during 1998–2007 (Lloyd and Vecchi, 2011). SST conditions were examined before, during and after the passage of typhoons, through Lagrangian composites along typhoon tracks across all ocean basins, with particular focus on the North Atlantic. A non-monotonic SST intensity relationship was discovered. The SST-intensity response for intensifying and decaying typhoons indicated that decaying typhoons have larger SST cooling. The larger SST cooling was associated with more markedly non-monotonic events, while smaller SST cooling was linked to mono-tonic ones. Given that ocean feedback is important for typhoon intensity, the conclusion drawn is that the strongest typhoons would most frequently develop in areas where ocean cooling is inhibited by large-scale oceanic conditions. The results also implied that large-scale oceanic conditions were a control on typhoon intensity, since they controlled oceanic sensitivity to atmospheric forcing.

3.4   SUMMARY

In this review, we have focused on the typhoon-induced SST cooling, the effect of subsurface thermohaline structure on the cooling, and its negative feedback effect on typhoon development and intensification. After a typhoon has passed through, cold wakes are left behind, and are typically more pronounced to the right of the typhoon tracks. The magnitude of the SST cooling depends on several factors such as wind intensity, translation speed and oceanic subsurface structure. The pre-existing oceanic meso-scale anomaly and salinity barrier layer can significantly modulate amplitude of the SST cooling, and in turn can adjust track and intensification of the typhoon. When the typhoon-induced SST cooling exists in the eyewall region, the surface fluxes of latent and sensitive heat to the typhoon are reduced, limiting the intensification of the typhoon. Besides the self-feedback effect, the sea surface cold wakes can exist in the ocean for days to weeks after the typhoon has passed, and can generate a continual effect upon the atmosphere and ocean, even climate variability and change.

In summary, the interaction between a typhoon and self-induced cooling involves multiscale processes and a variety of factors. Integrating these influences can be difficult in practice, and under current observational systems it remains unclear what the ultimate limits of typhoon forecasting are. Thus, developing an optimal observational network of multisensors for monitoring the spatial evolution of the upper ocean is important for typhoon prediction, and also to improve knowledge of upper oceanic environments during a typhoon's passage.

ACKNOWLEDGEMENTS

This study is co-sponsored by grants from the National Natural Science Foundation of China (41776004) and the Fundamental Research Funds for the Central Universities (2016B12514).

REFERENCES

Adler, R. F. (2005). Estimating the Benefit of TRMM Tropical Cyclone Data in Saving Lives. *American Meteorological Society, 15th Conference on Applied Climatology*, Savannah, GA.

Altman, J., Doležal J., Černý T., and Song, J. S. (2013). Forest response to increasing typhoon activity on the Korean peninsula: Evidence from oak tree-rings. *Global Change Biology*, 19, 498–504.

Balaguru, K., Chang, P., Saravanan, R., Leung, L. R., Xu, Z., Li, M. K., and Hsieh, J. S. (2012). Ocean barrier layers' effect on tropical cyclone intensification. *PNAS*, doi:10.1073 /pnas.1201364109.

Bender, M. A. and Ginis, I. (2000). Real-case simulations of hurricane–ocean interaction using a high resolution coupled model: Effects on hurricane intensity. *Mon. Wea. Rev.*, 128, 917–946.

Bender, M. A., Ginis, I., and Kurihara, Y. (1993). Numerical simulations of tropical cyclone–ocean interaction with a high-resolution coupled model. *J. Geophys. Res.*, 98 (D12), 23 245–23263.

Chiang, T. L., Wu, C. R., and Oey, L. Y. (2011). Typhoon Kai-Tak: An ocean's perfect storm. *J. Phys.Oceanogr.*, 41, 221–233.

Cione, J. J. and Uhlhorn, E. W. (2003). Sea surface temperature variability in hurricanes: Implications with respect to intensity change. *Mon. Wea. Rev.*, 131, 1783–1795.

Dare, R. A. and McBride, J. L. (2011). Sea surface temperature response to tropical cyclones. *Mon. Weather Rev.*, 139, 3798–3808. doi:10.1175/MWR-D-10-05019.1.

Emanuel, K. A. (1986). An air–sea interaction theory for tropical cyclones. Part 1: Steady-state maintenance. *J. Atmos. Sci.*, 43, 585–604.

Emanuel, K. A. (1999). Thermodynamic control of hurricane intensity. *Nature*, 401, 665–669.

Emanuel, K. A. (2001). The contribution of tropical cyclones to the oceans' meridional heat transport. *J. Geophys. Res.*, 106, 14771–14781.

Emanuel, K. A., DesAutels C., Holloway, C., and Korty, R. (2004). Environmental control of tropical cyclone intensity. *J. Atmos. Sci.*, 61, 843–858.

Ffield, A. (2007). Amazon and Orinoco River plumes and NBC Rings: Bystanders or participants in hurricane events? *J. Clim.*, 20, 316–333.

Goni, G. J. and Trinanes, J. A. (2003). Ocean thermal structure monitoring could aid in the intensity forecast of tropical cyclones. *Eos, Trans. Amer. Geophys. Union*, 84, 573–580.

Halliwell, Jr. G. R., Shay, L. K., Jacob, S. D., Smedstad, O. M., and Uhlhorn, E. W. (2008). Improving ocean model initialization for coupled tropical cyclone forecast models using GODAE nowcasts. *Mon. Wea. Rev.*, 136, 2576–2591.

Hart, R. E., Maue, R. N., and Watson, M. C. (2007). Estimating local memory of tropical cyclones through MPI anomaly evolution. *Mon. Wea. Rev.*, 135, 3990–4005.

Jacob, S. D. and Shay, L. K. (2003). The role of oceanic mesoscale features on the tropical cyclone–induced mixed layer response: A case study. *J. Phys. Oceanogr.*, 33, 649–676.

Jansen, M. F., Ferrari, R., and Mooring, T. A. (2010) Seasonal versus permanent thermocline warming by tropical cyclones. *Geophys. Res. Lett.*, 37, L03602, doi:10.1029 /2009GL041808.

Leipper, D. (1967). Observed ocean conditions and Hurricane Hilda, 1964. *J. Atmos. Sci.*, 24, 182–196.

Lin, I., Liu, W. T., Wu, C., Wong, G. T. F., Hu, C., Chen, Z., Liang, W., Yang, Y., and Liu, K. (2003). New evidence for enhanced ocean primary production triggered by tropical cyclone. *Geophys. Res. Lett.*, 30, 1781, doi:10.1029/2003GL017141.

Lin, I. I., Wu, C. C., Emanuel, K. A., Lee, I. H., Wu, C. R., and Pun, I. F. (2005). The interaction of super typhoon Maemi (2003) with a warm ocean eddy. *Mon. Wea. Rev.*, 133, 2635–2649.

Lin, I. I., Wu, C. C., Pun, I. F., and Ko, D. S. (2008). Upper-ocean thermal structure and the western North Pacific category 5 typhoons. Part I: Ocean features and the category 5 typhoons' intensification, *Mon. Weather Rev.*, 136, 3288–3306, doi:10.1175/2008 MWR2277.1.

Lloyd, I. D. and Vecchi, G. A. (2011). Observational evidence of oceanic controls on hurricane intensity. *J. Clim.*, 24, 1138–1153, doi:10.1175/2010JCLI3763.1.

Mendelsohn, R., Emanuel, K., Shun, C., and Bakkensen, L. (2012). The impact of climate change on global tropical cyclone damage. *Nature Climate Change*, 2, 205–209.

Oey, L. Y., Ezer, T., Wang, D. P., Yin, X. Q., and Fan, S. J. (2007). Hurricane-induced motions and interaction with ocean currents. *Cont. Shelf Res.*, 27, 1249–1263.

Pielke, R. A., Gratz, J., Landsea, C. W., Collins, D., Saunders, M. A., and Musulin, R. (2008). Normalized hurricane damage in the United States: 1900–2005. *Natural Hazards Review*, 9, 29–42.

Price, J. F. (1981). Upper ocean response to a hurricane. *J. Phys. Oceanogr.*, 11, 153–175.

Sakaida, F., Kawamura, H., and Toba, Y. (1998). Sea surface cooling caused by typhoons in the Tohuku area in August 1989. *J. Geophys. Res.*, 103(C1), 1053–1065.

Schade, L. R., and Emanuel, K. A. (1999) The ocean's effect on the intensity of tropical cyclones: Results from a simple coupled atmosphere–ocean model. *J. Atmos. Sci.*, 56, 642–651.

Scharroo, R., Smith, W. H. F., and Lillibridge, J. L. (2005). Satellite altimetry and the intensification of hurricane Katrina. *EOS, Trans. Amer. Geophys. Union*, 86, 366–367.

Sengupta, D., Goddalehundi, B. R., and Anitha, D. S. (2008). Cyclone-induced mixing does not cool SST in the post-monsoon north Bay of Bengal. *Atmos Sci Letts.*, 9, 1–6.

Shang, S.-L., Li, L., Sun, F., Wu, J., Hu, C., Chen, D., Ning, X., Qiu, Y., Zhang, C., and Shang, S. (2008). Changes of temperature and bio-optical properties in the South China Sea in response to typhoon Lingling, 2001. *Geophys. Res. Lett.*, 35, L10602. doi:10.1029/2008GL033502.

Shay, L. K., Goni, G. J., and Black, P. G. (2000). Effects of a warm oceanic feature on Hurricane Opal. *Mon. Wea. Rev.*, 128, 1366–1383.

Shay, L. K. and Uhlhorn, E. W. (2008). Loop current response to Hurricanes Isidore and Lili. *Mon. Wea. Rev.*, 136, 3248–3274.

Shultz, J. M., Russell, J., and Espinel, Z. (2005). Epidemiology of tropical cyclones: The dynamics of disaster, disease, and development. *Epidemiologic Reviews*, 27, 21–35, PMID 15958424, doi:10.1093/epirev/mxi011.

Sriver, R. L. and Huber, M. (2007). Observational evidence for an ocean heat pump induced by tropical cyclones. *Nature*, 447, 577–580, doi:10.1038/nature05785.

Tuleya, R. E. and Kurihara, Y. (1982). A note on the sea surface temperature sensitivity of a numerical model of tropical storm genesis. *Mon. Wea. Rev.*, 110, 2063–2068.

Vincent, E. M., Lengaigne, M., Madec, G., Vialard, J., Samson, G., Jourdain, N. C., Menkes, C. E., and Jullien, S. (2012). Processes setting the characteristics of sea surface cooling induced by tropical cyclones, *J. Geophys. Res.*, 117, C02020, doi:10.1029/2011JC007396.

Vincent, E. M., Lengaigne, M., Vialard, J., Madec, G., Jourdain, N. C., and Masson, S. (2012). Assessing the oceanic control on the amplitude of sea surface cooling induced by tropical cyclones. *J. Geophys. Res.*, 117, C05023, doi:10.1029/2011JC007705.

Walker, N., Leben, R. R., and Balasubramanian, S. (2005). Hurricane forced upwelling and chlorophyll a enhancement within cold core cyclones in the Gulf of Mexico. *Geophys. Res. Letters*, 32, L18610, doi: 10.1029/2005GL023716.

Wang X. D., Han, G. J., Qi, Y. Q., and Li, W. (2011). Impact of barrier layer on typhoon-induced sea surface cooling. *Dyn Atmos Oceans*, 52, 367–385.

Wu, C. C., Lee, C. Y., and Lin, I. I. (2007). The effect of the ocean eddy on tropical cyclone intensity. *J. Atmos. Sci.*, 64, 3562–3578.

Wu, C.-R., Chang, Y. L., Oey, L.-Y., Chang, J., Chang, C. W.-J., and Hsin, Y.-C. (2008). Air-sea interaction between tropical cyclone Nari and Kuroshio. *Geophys. Res. Lett.*, 35, L12605, doi:10.1029/2008GL033942.

Zhu, T. and Zhang, D. L. (2006). The impact of the storm-induced SST cooling on hurricane intensity. *Advances in Atmospheric Sciences*, 23, 1, 14–22.

4 Indian Cyclones and Earthquakes

Their Impact on Structures

Parthasarathi Mukhopadhyay
and Sekhar Chandra Dutta

CONTENTS

4.1   INTRODUCTION

The incredible advancement of science and technology, and their multitude efforts to make this planet a safer place, have not been able to thwart the devastating effects of natural disasters. Succumbing to the vagaries of nature has become a regular global concern. Though the number of natural disasters is 56.17% of the total number of disasters reported globally in the last 15 years, they account for, respectively, 92.63% of all the people reported killed, and 99.96% of all the people affected, by disasters in the same period (Kynman et al., 2007; Zetter, 2012). Approximately 97.82% of the amount of disaster-estimated damage since the turn of the century is also due to natural disasters, including avalanches, landslides, droughts, food insecurity, earthquakes, tsunamis, extreme temperatures, floods, forest and scrub fires, insect infestations, volcanic eruptions and windstorms (Figure 4.1). Of these, earthquakes, tsunamis and windstorms together are responsible for 67.36% of the global damage, whereas the damage caused by technological disasters is markedly less than its number.

The further disturbing fact is that almost one in seven of the people worldwide who were killed, and one in four persons affected by disasters between 1992 and 2011, are from the Indian subcontinent (Table 4.1). India is one of the worst-affected countries where disaster-driven poverty, deprivation and death are common features of life. In fact, within these 20 years, disasters reportedly killed 0.12 million people and affected another 978.98 million in India alone (Table 4.1). Real-life experiences tell us that these figures are likely to greatly underreport the actual numbers. The Orissa Super Cyclone of 1999, the Bhuj Earthquake of 2001, tsunamis triggered by the Indian Ocean Earthquake in 2004, the Kashmir Earthquake of 2005 and the Very Severe Cyclonic Storm Sidr of 2007 all claimed heavy tolls of life and properties. Not surprisingly, much current research effort is devoted to minimizing the effect of these devastating forces of nature.

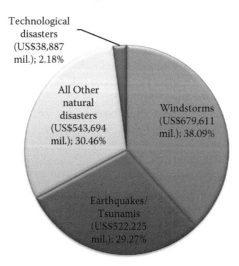

FIGURE 4.1   Damage caused by different types of disasters during 1997–2011 in millions of U.S. dollars (Kynman et al., 2007; Zetter, 2012).

TABLE 4.1

**Total Number of Persons Reported Killed and Affected
by Disasters in the Indian Subcontinent (1992–2011) (Zetter, 2012)**

Country/Territory	Persons Reported Killed	Persons Reported Affected
Afghanistan	14,775	13,080,997
Bangladesh	22,244	136,341,975
Bhutan	263	86,628
India	120,406	978,978,557
Maldives	143	28,963
Nepal	6,756	3,974,048
Pakistan	89,408	73,028,173
Sri Lanka	36,576	11,917,110
Subcontinent	290,571	1,217,436,451
World	2,121,051	5,048,729,028

4.2 REGIONAL SETTING

Different contexts have given rise to different interpretations of the term 'Indian subcontinent'. Considering the contemporary political-economic trends and keeping in mind the sociocultural backgrounds of the region, the geographic extent constituting the South Asian Association of Regional Cooperation (SAARC) is considered the Indian subcontinent. Thus, it may be defined as the peninsula, islands and inland areas constituting Afghanistan, Bangladesh, Bhutan, India, Maldives, Nepal, Pakistan and Sri Lanka, bounded by the Himalayas to the north, the Bay of Bengal to the east, the Arabian Sea to the west and the Indian Ocean to the south (Figure 4.2).

Geologically, the subcontinent rests substantially on the Indian tectonic plate, which is characterized by high seismicity and deformation. The vulnerability of the area to seismic tremors has been demonstrated many times throughout history. In addition, an extensive portion of the more than 5000 km-long coast of peninsular India is less than 20 m above mean sea level (Figure 4.2). This minimal topography is more pronounced in the deltaic regions of the rivers flowing into the Bay of Bengal and the Arabian Sea. Considering their area, the Ganges-Brahmaputra Delta (consisting of portions of India and Bangladesh, flowing into the Bay of Bengal), the Mahanadi River Delta (flowing into the Bay of Bengal in India) and the Indus River Delta (flowing into the Arabian Sea in Pakistan) are three major deltaic regions, the first of these being the largest delta of the world. Calamities, such as monsoon floods and tropical cyclones, are perennial features of these deltaic and coastal regions, and are discussed further in other chapters in this volume.

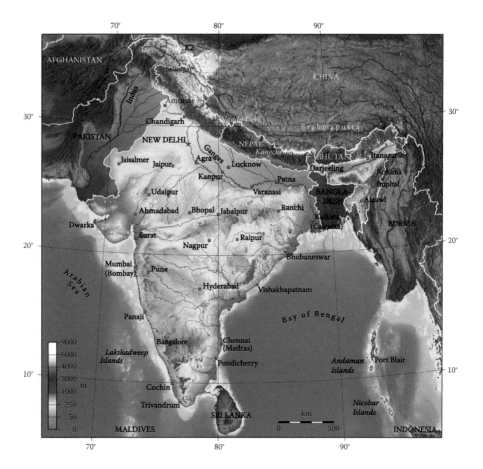

FIGURE 4.2   Regional setting of the Indian subcontinent (Wikipedia, 2012).

4.3   GENERAL PROFILE OF THE INDIAN BUILDING STRUCTURES

The Census of India 2011 (Government of India, 2011) reported surveying 330.84 million houses, out of which 304.88 million were verified to be in occupation of people. Whereas 53.2% of these census houses were in 'good' physical condition, 41.5% were said to be 'liveable', and the condition of 5.3% were reported to be 'dilapidated'. The conditions of houses described as 'good', 'liveable' and 'dilapidated' are based on the perception of the respondents of the census questionnaire and, therefore, are qualitative in nature and to an extent subjective. The physical condition of a house not only determines the level of comfort enjoyed by its occupants, but a 'good' physical condition also largely assures relative safety against natural disasters. Knowledge of strength properties of the predominant building materials used in construction can further qualify the subjective assessment. Distribution of predominant building materials of the Indian census houses in roofs, walls and floors has been provided in Table 4.2. Indian building structures built out of these materials can be broadly classified into three categories:

- *Engineered buildings*: These are constructed with 'engineered' materials whose methods of construction and strength properties are specified in the relevant standards and codes. These include materials like concrete, burnt brick, machine made tiles, and so on.

TABLE 4.2

Distribution of Indian Census Houses by Predominant Building Materials, as in 2011 (Government of India, 2011)

Parameter	Number of Houses (in millions)	Percentage
Total Number of Houses	304.88	100.0
Roof		
Grass/thatch/bamboo/wood/ mud, etc.	46.99	15.4
Plastic/polythene	2.07	0.7
Handmade tiles	40.28	13.2
Machine made tiles	26.43	8.7
Burnt brick	20.25	6.6
Stone/slate	26.98	8.8
G.I./metal/asbestos sheets	50.34	16.5
Concrete	90.24	29.6
Other materials	1.30	0.4
Wall		
Grass/thatch/bamboo, etc.	28.94	9.5
Plastic/polythene	1.10	0.4
Mud/unburnt brick	66.45	21.8
Wood	2.78	0.9
Stone not packed with mortar	10.44	3.4
Stone packed with mortar	33.04	10.8
G.I./metal/asbestos sheets	2.33	0.80
Burnt brick	146.55	48.1
Concrete	10.98	3.6
Other materials	2.26	0.7
Floor		
Mud	138.69	45.5
Wood/bamboo	2.58	0.8
Burnt brick	7.86	2.6
Stone	23.98	7.9
Cement	98.06	32.2
Mosaic/floor tiles	32.27	10.6
Other materials	1.46	0.5

- *Non-engineered buildings*: These are made with 'non-engineered' materials whose methods of construction and strength properties are not specified in the standards/codes. These refer to materials like grass, thatch, bamboo, mud, handmade tiles, unburnt brick, stone masonry without mortar, and so on.
- *Semi-engineered buildings*: These are constructed either by combination of an 'engineered' material with a 'non-engineered' one, or by using an 'engineered' material in dimension, quantity and method, not in conformity to relevant standards/codes. A roof consisting of galvanized iron (G.I.) sheets of known thickness, fixed on bamboo roof structures, may be considered as a typical example of a 'semi-engineered' roofing element, constructed in combination of respectively 'engineered' and 'non-engineered' building materials. A half-brick thick unreinforced external load bearing wall of burnt bricks in mud-lime mortar is another typical example of a 'semi-engineered' walling element, where an 'engineered' material is used in a manner not specified in the codes.

Based on the above understanding, it may be concluded from Table 4.2 that the number of 'engineered' buildings is low in India because unfortunately there are not many Indian buildings whose roofs, walls and floors are all built as per the provisions of the building codes. This is because, in most cases, artisans are involved in building construction instead of engineers and architects. Such semi- and non-engineered building structures are less capable to resist the thrusts of both seismic as well as wind loads, and are prone to get devastated even during moderate earthquakes or tropical depressions.

4.4 MECHANISM OF THE NATURAL DISASTERS

The growth of agro-based economies on the Indian peninsula, due to two tropical monsoons per year, and the simultaneous devastation of life and property due to typically between three and five tropical cyclones during May to November, are respectively the most welcome and notorious weather phenomena of the subcontinent. In addition to these, India is visited by low to moderate magnitude earthquakes at regular intervals. Further, the subcontinent gets devastated by the tremors of one or two major earthquakes in a decade. It is, therefore, considered worthwhile to examine the mechanism of these natural disasters.

4.4.1 TROPICAL CYCLOGENESIS IN THE NORTH INDIAN OCEAN BASIN

Variation in surface area of the continents and oceans and in incoming solar radiation produce consequent variations in density of the atmosphere. This leads toward formation of high and low atmospheric pressure belts. This relative pressure difference generates wind movement. Tropical cyclogenesis or formation of tropical cyclones in the North Indian Ocean Basin refers to one such atmospheric closed circulation, which is formed when an organized system of anti-clockwise revolving wind develops over it. Despite limitation of knowledge regarding tropical cyclogenesis, the most accepted six preconditions considered to be suitable for it are as follows (Dutta and Mukhopadhyay, 2012).

- Genesis of tropical cyclones is found to take place at a minimum distance of 500 km from the equator. The Trade Winds, deflected by the Coriolis force, need at least this distance to gain momentum to be converted into a cyclone.
- Presence of high humidity in the lower to middle levels of the troposphere over the Basin helps in creating disturbances within it, thereby fulfilling one of the conditions for formation of tropical cyclones. This is because a dehydrated atmosphere is unfavourable for uninterrupted formation of cyclones.
- It is reported that low wind shear over the Basin is conducive in cyclone formation as high wind shear tends to interrupt formation of thunderstorms, which are important to the core of a cyclone.
- Feeble disturbance near the Basin surface, with sizeable spin and low-level inflow, helps in cyclogenesis by pulling clouds inward, rotating the clouds with the aid of the Coriolis force.
- The atmosphere over the basin cools fast enough with height, releasing huge amounts of heat of condensation that acts as another source of heat for initiating tropical cyclones.
- Solar radiation, which is the source of all energy on earth, is highest during late summer in the North Indian Ocean basin. The tropical cyclone season in the basin, therefore, starts in May when heated ocean water fuels the heat flow cycle of the tropical cyclone.

The origins, behaviour and consequences of tropical cyclones are discussed in further detail in Chapters 1 and 2 of this volume; here, the focus is on the potential impact of these storms on built structures in India.

The wind speeds found in tropical cyclones vary considerably, as does the rate of progress of an individual storm across the country. These variables in turn will affect the amount and nature of the associated devastation. The Indian Wind Code (Bureau of Indian Standards, 1989) has divided the country into six wind zones, based on maximum expected basic wind speed, worked out for a 50-year return period. The values in ascending order are, respectively, 33 ms^{-1} (118.8 kmh^{-1}), 39 ms^{-1} (140.4 kmh^{-1}), 44 ms^{-1} (158.4 kmh^{-1}), 47 ms^{-1} (169.2 kmh^{-1}), 50 ms^{-1} (180 kmh^{-1}) and 55 ms^{-1} (198 kmh^{-1}). Table 4.3 enumerates cyclones making landfall on the east and west coasts of the Indian peninsula since 1970, whose recorded sustained peak wind speed was above 150 kmhr^{-1}.

4.4.2 Subduction of the Indo-Australian Plate under the Eurasian Plate

The Indian subcontinent lies at the northwestern end of the Indo-Australian tectonic plate, which comprises the continental crusts of the Indian Plate and the Australian Plate, and the oceanic crust of the Indian Ocean. Before fusing with the Australian Plate, the Indian Plate was part of the supercontinent of Gondwanaland. Around 140 million years ago, the supercontinent disintegrated and one of the resulting pieces, the Indian

TABLE 4.3

Cyclonic Storms Making Landfall in the Indian Peninsula with Recorded Sustained Peak Wind Speed above 150 km/h since 1970 (Indian Meteorological Department, n.d.)

Name (IMD Designation)	3-min Sustained Peak Wind Speed	Origin	Landfall	Duration
		East Coast (Bay of Bengal)		
1970 Bhola Cyclone	51.4 m/s (185 km/h)	Central Bay of Bengal	Coast of Bangladesh (erstwhile East Pakistan)	Nov 03–13, 1970
1977 Andhra Pradesh Cyclone	53.6 m/s (193 km/h)	Bay of Bengal	Chirala, Andhra Pradesh (A.P.), India	Nov 14–20, 1977
1989 Kavali Cyclone (later development of Typhoon Gay)	66.7 m/s (240 km/h)	Gulf of Thailand; later emerged into Bay of Bengal after crossing Thailand	Kavali, A.P. India	Nov 01–09, 1989
1990 Andhra Pradesh Cyclone (BOB 01)	65.3 m/s (235 km/h)	Bay of Bengal	40 km southwest of Machilipatnam, A.P., India	May 04–09, 1990
1991 Bangladesh Cyclone (BOB 01)	65.3 m/s (235 km/h)	Bay of Bengal	Chittagong coast of Bangladesh across Sandweep Island	Apr 24–30, 1991
Very Severe Cyclonic Storm (VSCS) BOB 03	45.8 m/s (165 km/h)	South China Sea; later emerged over Bay of Bengal after crossing Malay Peninsula	30 km north of Karaikal, Puducherry, India	Dec 01–04, 1993
VSCS BOB 01	59.7 m/s (215 km/h)	Southeastern Bay of Bengal	Teknaf, Cox's Bazar, Bangladesh	Apr 02–May 02, 1994
May 1997 Bangladesh Cyclone (VSCS BOB 01)	45.8 m/s (165 km/h)	A near equatorial trough in the Indian Ocean	Chittagong, Bangladesh	May 14–20, 1997
1999 Orissa Cyclone (Super Cyclonic Storm BOB 06)	72.2 m/s (260 km/h)	South China Sea; later emerged over Andaman Sea after crossing Malay Peninsula	Near Paradip, Orissa, India	Oct 25–Nov 03, 1999

(Continued)

TABLE 4.3 (CONTINUED)

Cyclonic Storms Making Landfall in the Indian Peninsula with Recorded Sustained Peak Wind Speed above 150 km/h since 1970 (Indian Meteorological Department, n.d.)

Name (IMD Designation)	3-min Sustained Peak Wind Speed	Origin	Landfall	Duration
2000 Sri Lanka Cyclone (VSCS BOB 06)	45.8 m/s (165 km/h)	Central Bay of Bengal	Near Trincomalee, Sri Lanka	Dec 23–28, 2000
VSCS Sidr	59.7 m/s (215 km/h)	Bay of Bengal	Bangladesh coast around 100 km east of the Sagar Islands, West Bengal, India	Nov 11–16, 2007
VSCS Phailin	59.7 m/s (215 km/h)	Gulf of Thailand; later emerged into Andaman Sea after passing over Malay Peninsula	Near Gopalpur, Orissa, India	Oct 8–14, 2013
VSCS Hudud	51.4 m/s (185 km/h)	Andaman Sea	Near Visakhapatnam, A.P., India	Oct 7–14, 2014
West Coast (Arabian Sea)				
1975 Saurashtra Cyclone	44.4–50 m/s (160–180 km/h)	Arabian Sea	15 km northwest of Porbandar, Gujarat, India	Oct 19–24, 1975
1998 Gujarat Cyclone (VSCS ARB 02)	45.8 m/s (165 km/h)	Arabian Sea	Near Porbandar, Gujarat, India	Jun 05–09, 1998
1999 Pakistan Cyclone (VSCS ARB 01)	54.2 m/s (195 km/h)	Arabian Sea	Karachi, Pakistan	May 16–22, 1999
2001 India Cyclone (VSCS ARB 01)	59.7 m/s (215 km/h)	East of Somalia, Arabian Sea	Saurashtra, Gujarat, India	May 21–28, 2001

Note: Since 1988, IMD adopted 3-min sustained peak wind speed and began rating cyclones with speed above 118 km/h as 'very severe cyclonic storms'. In 1998, IMD introduced the category 'super cyclonic storms' for those with speeds above 222 km/h. The practice of naming North Indian cyclonic storms only started in 2004.

Plate, started moving northeast, reaching speeds of up to 18–20 cm yr^{-1} during the late Cretaceous period (Kumar et al., 2007) before slowing to about 5 cm yr^{-1} when it collided with the massive Eurasian Plate about 50 million years ago.

Because of this, the thickness of the Indian Plate is half that of other components of the former Gondwanaland (Kumar et al., 2007). In the process, the Indo-Australian Plate and the Eurasian Plate have formed a convergent or compressional boundary and the former is undergoing subduction below the latter (Figure 4.3). This colossal collision is responsible for formation of the Himalayas and the Tibetan Plateau. The inter-tectonic zone between these two major tectonic plates has become a subduction zone and has remained seismically very active. Table 4.4 lists significant Indian earthquakes of the last century, the majority of which occurred in the Western and Eastern Himalayas and at their foothills. In fact, prior to 2001, it was generally conceived by the Indian earthquake researchers that earthquakes in India are restricted to the inter-tectonic zones. Accordingly, Zone I in the Indian seismic zone map (Bureau of Indian Standards, 1986) was designated as free from earthquake. But a series of intra-tectonic earthquakes culminating in the catastrophic 2001 Bhuj Earthquake led to a major change of the Indian seismic zone map. It was upgraded to show four instead of previous five seismic zones, in the process abolishing the entire Zone I (Bureau of Indian Standards, 1986, 2002).

The reason behind these intra-tectonic earthquakes may be explained by the suggestion that the Indo-Australian Plate is in the process of splitting into two separate Indian and Australian sub-plates, primarily due to differential movement of the eastern and western parts of the Indo-Australian Plate (Delescluse et al., 2012). Such propositions are further confirmed by the occurrence of two major intra-tectonic Indian Ocean earthquakes in and around Sumatra, Indonesia and the Nicobar Islands, India in 2004 (M_w 9.3) and 2012 (M_w 8.5) respectively, and many frequent medium and minor intra-tectonic tremors occurring in the area since then and still continuing.

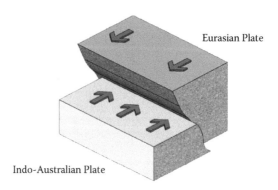

Eurasian Plate

Indo-Australian Plate

FIGURE 4.3   Relative inter-tectonic motion at the subduction zone between Indo-Australian Plate and Eurasian Plate.

TABLE 4.4

Significant Indian Earthquakes of the Last 100 Years (Indian Meteorological Department, n.d.)

Date	Location	Epicenter		Magnitude (M_w)
		Latitude	Longitude	
Apr 4, 1905	Kangra, Himachal Pradesh	32.3°N	76.3°E	8.0
Jul 8, 1918	Srimangal, Assam	24.5°N	91.0°E	7.6
Jul 2, 1930	Dhubri, Assam	25.8°N	90.2°E	7.1
Jan 15, 1934	Bihar-Nepal border	26.6°N	86.8°E	8.3
Jun 26, 1941	Andaman Islands	12.4°N	92.5°E	8.1
Oct 23, 1943	Assam	26.8°N	94.0°E	7.2
Aug 15, 1950	India (Arunachal Pradesh)—China border	28.5°N	96.7°E	8.5
Jul 21, 1956	Anjar, Gujarat	23.3°N	70.0°E	7.0
Dec 10, 1967	Koyna, Maharashtra	17.37°N	73.5°E	6.5
Jan 19, 1975	Kinnaur, Himachal Pradesh	32.38°N	78.49°E	6.2
Aug 6, 1988	Manipur—Myanmar border	25.13°N	95.15°E	6.6
Aug 21, 1988	India (Bihar)—Nepal border	26.72°N	86.63°E	6.4
Oct 20, 1991	Uttarkashi, Uttarakhand	30.75°N	78.86°E	6.6
Sep 30, 1993	Killari, Latur-Osmanabad, Maharashtra	18.07°N	76.62°E	6.3
Mar 29, 1999	Chamoli, Uttarakhand	30.41°N	79.42°E	6.8
Jan 26, 2001	Bhuj, Gujarat	23.40°N	70.28°E	7.7
Sep 14, 2002	Diglipur, North Andaman Islands	13.01°N	93.15°E	6.5
Dec 26, 2004	Indian Ocean at India (Nicobar Islands)—Indonesia (west coast of Sumatra) border	3.34°N	96.13°E	9.3
Oct 8, 2005	North Kashmir (Pakistan occupied)	34.49°N	73.15°E	7.6
Oct 28, 2008	Pakistan	30.7°N	67.4°E	6.3
Aug 10, 2009	Andaman Islands	14.1°N	92.9°E	7.7
Jan 18, 2011	Southwestern Pakistan	28.9°N	64.0°E	7.4
Sep 18, 2011	India (Sikkim)—Nepal border	27.8°N	88.1°E	6.9
Apr 11, 2012	Indian Ocean at India (Nicobar Islands)—Indonesia (west coast of Sumatra) border	2.3°N	93.0°E	8.5
Sep 24, 2013	Pakistan	27.0°E	65.7°E	7.4
Mar 21, 2014	Nicobar Islands region	7.5°N	94.2°E	6.3

4.5 A FEW MAJOR NATURAL DISASTERS SINCE 1965

In the following sections, brief accounts of five major tropical cyclones and five major earthquakes of the last half-century that devastated the subcontinent are given.

4.5.1 MAJOR TROPICAL CYCLONES OF THE NORTH INDIAN OCEAN

Table 4.3 illustrates the fact that wind depressions generated in the Arabian Sea are usually less destructive than their Bay of Bengal counterparts. Accordingly, the five examples presented below each had their landfall at different places along the eastern coast of the Indian peninsula. The sustained wind speeds of the first two cyclones are within the uppermost range (180–198 kmh^{-1}) that the code (Bureau of Indian Standards, 1989) expects India to experience, and that of the last three are even beyond that.

4.5.1.1 1970 Bhola Cyclone (185 kmh^{-1})

This very severe cyclonic storm made landfall along the coasts of Bangladesh (still at that time known as East Pakistan, and part of the federation that included Pakistan to the west of the subcontinent) on 12 November, 1970 and, at its peak, reached wind speeds of up to 185 kmh^{-1}. This was much less than that of the 1991 Bangladesh Cyclone (235 kmh^{-1}), but the 1970 cyclone, also known as the 'Bhola Cyclone', was deadlier and is considered one of the deadliest tropical cyclones that has ever rocked Bangladesh and West Bengal in India. A population of 4.7 million was affected, and 300,000–500,000 people were killed; 280,000 cattle and 500,000 poultry were lost; 400,000 houses and 3500 schools were damaged; and 99,000 fishing boats, used in inland waters and marine waters, were destroyed (Frank and Husain, 1971). The storm wiped out numerous villages in Patuakhali, Bakerganj, Noakhali and other northern offshore islands of the erstwhile East Pakistan. However, the 1970 Bhola Cyclone was significant for the scientific community, as this was the first time that there was an opportunity to document a cyclone-induced storm surge in East Pakistan (Eskander and Barbier, 2014).

4.5.1.2 1977 Andhra Pradesh Cyclone (193 kmh^{-1})

This was a Super Cyclonic Storm, also known as the 1977 Andhra Pradesh cyclone, that originated in the Bay of Bengal and made landfall on 19 November, 1977. It hit the coasts of Andhra Pradesh, Southern India, with wind speed above 193 kmh^{-1}, killing 10,000 people and rendering another 3.4 million homeless (Indian Meteorological Department, n.d.). The worst affected areas were in the Krishna River delta region, where a two-storey high storm surge came over the island of Diviseema. Fields of paddy and cash crops, along with hundreds of villages, were submerged by the ensuing floods. Landslides disrupted major transportation, and in particular, the railway lines of the Waltair-Kirandal route were destroyed.

4.5.1.3 1991 Bangladesh Cyclone (235 kmh^{-1})

This Super Cyclonic Storm, originating in the Bay of Bengal, is also known as the 1991 Bangladesh Cyclone or the Tropical Cyclone Marian. It struck the Chittagong District of Bangladesh on the night of 29 April, 1991, with a peak wind speed of 235 kmh^{-1}, and killed almost 138,000 people, leaving about 10 million people homeless (Bern et al., 1993). Despite a prior cyclone warning being issued, few people moved to cyclone shelters built after the Bhola Cyclone (see further discussion of the ineffectiveness of this warning, and the consequential high loss of life, in Ferrer-Julià et al., Chapter 5 this volume). The worst affected areas were the coastal lowlands from Chittagong to Cox's Bazaar, and the islands of Sandwip and Kutubdia.

4.5.1.4 1999 Orissa Cyclone (260 kmh^{-1})

A tropical depression initially formed in the South China Sea, crossed the Malay Peninsula, and gradually emerged over the Andaman Sea on 25 October, 1999.

This depression later gained momentum, developing into the strongest tropical cyclonic storm ever recorded in the North Indian Ocean, and struck the Indian state of Orissa with peak wind speeds of 260 kmh⁻¹. This deadly cyclone was responsible for killing 9885 people, destroying crops over 165,000 hectares of land, destroying approximately 275,000 homes and leaving 1.67 million people homeless (Indian Meteorological Department, n.d.). The non- and semi-engineered residential structures of coastal Orissa suffered maximum damage. Their failure modes ranged from the roof being blown off to complete building collapse. Inadequate design and poor detailing played a significant role in triggering and increasing damage, and failure of roof connections and gable end walls, and collapse of steel roof trusses were observed to be particularly important factors in the failure of the commercial and industrial structures (Simon et al., 2001).

4.5.1.5   2007 Cyclone Sidr (215 kmh⁻¹)

The Very Severe Cyclonic Storm Sidr of 15 November, 2007 struck the Bangladesh coast with a peak speed of 215 kmh⁻¹, killing 3295 people and affecting millions. Material damage was severe, with over 563,877 houses destroyed and a further 940,438 houses being partially damaged (Mukhopadhyay and Dutta, 2012). It was the deadliest cyclonic storm of the Bay of Bengal portion of the North Indian Ocean basin in the first decade of the new millennium. The non-engineered residences of Khulna and Barisal divisions of Bangladesh were particularly impacted. These primarily had large overhang pitched roofs with low acute roof angles. The dominant roof cladding material was G.I. sheets of non-uniform quality, while the major building skeleton including columns and beams were also found to be made of wood or bamboo. The walling consisted of either 125 mm thick second class brick wall, or various combinations of G.I. sheets, tin sheets, wooden planks, bamboo mesh, and so on. Damage to the roof structure was the most common damage feature and, in many instances, partially damaged or undamaged entire roof structures were observed lying on the ground in the aftermath of the storm (Mukhopadhyay and Dutta, 2008).

4.5.2   MAJOR INDIAN EARTHQUAKES

The case studies presented here are selected in a manner so that both inter-tectonic and intra-tectonic examples are represented. In fact, as researchers are less acquainted with the intra-tectonic earthquakes, three out of the five examples presented represent such instances. Of the two inter-tectonic earthquakes, one example is taken from each of the Eastern and the Western Himalayas.

4.5.2.1   1967 Koyna Earthquake (M_w 6.5)

This M_w 6.5 earthquake took place on 10 December, 1967, near the 103 m concrete gravity Koyna Dam, Maharashtra, Western India, and had an intensity of VIII (MM) at Koynanagar (Berg et al., 1969). The earthquake caused widespread damage, killing about 200 people and injuring more than 1500 people. Before construction of the Koyna Dam and its reservoir in 1962, this area was considered to be free from seismicity. Accordingly, the area was designated as Zone 0 in the then Indian seismic map (Indian Standards Institute, 1962), which had seven zones from Zones 0 – VI. Since the Koyna Dam Project became operational, the area became significantly seismic. This was reflected in the 1966 revision of the seismic code (Indian Standards Institute, 1967), where the area around Koyna and the then Bombay were designated as Zone I in the seven-scale seismic map. In fact, the epicentre, fore- and after-shocks were all located near or under the Koyna Dam reservoir (Gupta, 2002). This earthquake has since been identified by researchers as a typical example of a reservoir-induced earthquake, where earthquakes are triggered by construction of artificial reservoirs. The 1967 Koyna Earthquake prompted further revision of the Indian seismic map (Indian Standards Institute, 1971), where the number of seismic zones was reduced from seven to five, named as Zones I–V. A new Western Coast Zone III was introduced that included Bombay; and a small area around Koynanagar was upgraded to Zone IV.

4.5.2.2   1993 Killari Earthquake (M_w 6.3)

The districts of Latur and Osmanabad of Maharashtra, Central India, were severely shaken by an earthquake on the morning of 30 September, 1993. It killed about 10,000 people, and completely destroyed stone/mud structures in about 20 villages covering an area about 15 km wide, centred 5 km west of a village named Killari in Latur District (Jain et al., 1994). The maximum intensity level of shaking was about MM VII – IX. The core affected area did not have any modern buildings or major industries; otherwise, the death toll would likely have increased many times. The Killari Earthquake was an addition to the small but distinct list of intra-tectonic Indian earthquakes, at locations which were originally considered non-seismic. In fact, this list includes the 1819 Rann of Kutch Earthquake in Gujarat, Western India; the 1843 Bellari Earthquake in Karnataka, Southern India; the 1900 Coimbatore Earthquake in Tamil Nadu, Southern India; the 1956 Anjar Earthquake in Gujarat; the 1967 Koyna Earthquake in Maharashtra (introduced above); and the 1970 Bharuch Earthquake in Gujarat.

4.5.2.3   2001 Bhuj Earthquake (M_w 7.7)

The earthquake that impacted the Bhuj in Gujarat, Western India, on 26 January, 2001, was one of the country's most destructive earthquakes ever recorded. This M_w 7.7 earthquake caused extensive liquefaction and slope failures over extensive areas, without producing any primary surface fault rupture. The earthquake injured over 167,000 people and killed another 13,800; over 230,000 one- and two-storey masonry houses collapsed, and over 980,000 more were damaged (Jain et al., 2002). Further, several hundred reinforced concrete (RC) frame buildings collapsed, including some 11-storey structures. The tremor of this intra-tectonic earthquake was felt as far as Kolkata, about 1900 km east of the epicentre. The human impact of the Bhuj Earthquake is unparalleled because it was the first major earthquake to hit an urban area of India in more than 50 years. As mentioned above, its death toll led to the revision of the Indian seismic zone map, with the removal of the entire seismicity-free Zone I (Bureau of Indian Standards, 1986) and its replacement with the present map (Bureau of Indian Standards, 2002) showing only Zones II–V.

4.5.2.4   2005 Kashmir Earthquake (M_w 7.6)

A strong M_w 7.4 earthquake occurred on 8 October 2005 with the epicentre situated in the Muzaffarabad region of Pakistan Occupied Kashmir (POK), which is 170 km west-northwest of Srinagar, the capital of Kashmir in India. Though the intensity of this event was similar to that of the 2001 Bhuj Earthquake, its fundamental difference with the previous intra-tectonic quake lies in the fact that the Kashmir Earthquake occurred in a known subduction zone, with active thrust faults, in the area where the Eurasian and Indian tectonic plates are colliding. The quake caused widespread destruction in POK, Pakistan's North-West Frontier Province, and western and southern parts of the state of Jammu & Kashmir (J&K) in the Indian side of the Line of Control. The official death toll was 47,700 in Pakistan and 1329 in India. The Indian casualty list included 72 army personnel. The earthquake further affected the lives of about 4 million people and collapsed 37,607 masonry buildings (Langenbach, 2007). The buildings that suffered maximum damage were constructed using random rubble masonry and bricks laid in clay mud mortar, with corrugated G.I. sheet roofing. Use of weak mortar, the absence of bonding in stones across the stone walls, which caused nonalignment between the inner and outer walls, and separation of the walls at the corners all contributed to the total collapse of the buildings.

4.5.2.5 2011 Sikkim Earthquake (M_w 6.9)

In the evening of 18 September, 2011, Sikkim, a small Indian state at the foothills of the Eastern Himalayas, experienced an M_w 6.9 earthquake that had its origin near the Sikkim-Nepal border. This particular event, considered the largest mid-to-deep crustal earthquake of the region as recorded by the Himalayan Nepal Tibet Seismic experiment (Sheehan et al., 2002), was followed by a number of aftershocks, three of whose magnitudes were more than M_w 4.2 (Rajendran et al., 2011). The combined impact of the principal quake along with the aftershocks, a typical feature of this region, was disastrous. The catastrophic effect of the shock was most felt in North Sikkim, where many existing structures became debris. Sikkim had been previously impacted by a moderate M_w 5.7 earthquake in 2006. The main damage features of the masonry walls, due to the 2006 quake, were generation of shear cracks in infill walls especially from openings, separation of wall junctions, out-of-plane collapse and pounding of buildings (Mukhopadhyay et al., 2009). The 2011 Sikkim Earthquake surpassed all the damage features that were witnessed during the moderate quake of 2006, and in addition damaged many reinforced cement concrete (RCC) framed structures as well. Further, two notable trends were observed: firstly, the government buildings suffered more than their private counterparts; and secondly, damage was observed more in the newer structures than the older ones. These observations suggest a combination of the following reasons, namely, (1) low-grade quality of material, (2) mediocre workmanship, and (3) second-rate technical supervision (Dutta et al., 2015). However, it was notable that the majority of the non-engineered buildings, made with a combination of wood and bamboo, survived.

4.6 TYPICAL DAMAGE FEATURES

The brief case studies of the natural disasters presented in the previous section show a pattern in the damage features of the building structures affected. One can easily differentiate between the devastation of a fishers' settlement in the Orissa coast, with the damage to structures and toppling of trees in the city of Dhaka. Similarly, there exists a difference in the ferocity of the impact of an earthquake upsetting the hill towns of

Sikkim with that experienced on the plains of Bhuj. However, some common structural failure emerges as typical. These are discussed below, to provide an overall view of the impacts of such natural hazards on structures in the Indian subcontinent. Further, a few exceptional cases have also been cited to illustrate indigenous artisanal solutions.

4.6.1 FAILURE OF ROOF STRUCTURES

The most obvious pattern of building failure due to cyclonic windstorms is loss of claddings and damage to rafters of sloped roof structures (Figure 4.4). The causes are usually inadequate thickness of the cladding materials, inappropriate fastening devices and insufficient frequencies of fasteners in the known areas of greater wind suction (Mukhopadhyay and Dutta, 2012). The dominant roof cladding materials, for the non-engineered structures, are thatching, burnt clay tiles, and G.I. sheets of non-uniform quality. The cladding materials are generally arranged over wooden or bamboo rafters. In some public structures like bus stands, and so on, GI sheet clad steel framed structures are also used. The absence of proper connections between the rafters, and between the roof structures and the vertical load-bearing members, results in flying away of roof structures and sometimes collapse of the entire building (Figure 4.5).

4.6.2 FAILURE OF UNREINFORCED MASONRY JUNCTIONS

Generation of shear cracks (Figure 4.6), and out-of-plane failure of unreinforced masonry (URM) walls due to absence of proper connections at masonry junctions, seem to be two typical damage features due to earthquakes in the subcontinent. Figure 4.7 illustrates the second case, where the 500-mm thick north and south rubble walls of a Buddhist prayer hall had undergone out-of-plane collapse, as it stood across the north-south direction of the ground motion of the 2011 Sikkim Earthquake. However, the east and west walls remained almost intact. A proper connection between the two sets of walls would have extended the collapse time of the walls in the weaker direction, for as long as those in the stronger direction did not also fail. Among the different code-recommended methods for augmenting junction strength of load-bearing mutually orthogonal URM walls, introduction of toothed joints, or insertion of steel dowels at corners and T-junctions are two important ones (Bureau of Indian Standards, 2004; IAEE and NICEE, 2004). These may also be reinforced by encasing with polypropylene bands, leading to a quantifiable increase in the lateral strength of junctions, over that obtained by reinforcing through toothed junctions or by steel dowels (Dutta et al., 2013).

4.7 POSITIVE INDIGENOUS EXPERIENCES

Amidst these catastrophes, instances have been encountered where non-engineered or semi-engineered buildings survived the natural disasters. Such examples, though few in number, were true for both cyclones and earthquakes. Lessons learned from such survivals are discussed below as they may provide crucial input for better performance of such structures in the future.

FIGURE 4.4 Loss of roof claddings and damage of rafters due to the impact of 2007 Cyclone Sidr, Bagerhat District, Khulna Division, Bangladesh.

FIGURE 4.5   Flying away of roof structures and collapse of entire house due to 2007 Cyclone Sidr, Nalchithi municipality, Jhalakathi District, Barisal Division, Bangladesh.

FIGURE 4.6   Shear cracks generating from the door-window openings in infill walls during 2011 Sikkim Earthquake, Siliguri Special Correctional Home, Darjeeling District, West Bengal.

4.7.1   Survival during the Tropical Cyclones

After the 2007 Cyclone Sidr, a single-storied non-engineered residence was discovered at Ghola Para, on the Sagar Islands of India, that had suffered no damage. The building is near square in overall plan form, with a roof consisting of four slopes forming a pyramid at the centre. The roof tiles are well connected, with regularly spaced bamboo trusses; and the bamboo roof structure, in its turn, is also well-connected to the mud walls (Mukhopadhyay and Dutta, 2012). Figure 4.8 demonstrates an indigenous way followed in the Sagar Islands, where roof claddings are tied to the roof structure through the application of nylon fishing nets. The nylon nets did not allow the roof to blow away during Sidr. In the Khulna Division of Bangladesh, two non-engineered double-storied residential buildings were observed,

FIGURE 4.7 Out-of-plane collapse of masonry walls during 2011 Sikkim Earthquake in Mani Lhakhang Prayer Hall of stone masonry at Rongyek near Sikkim State Jail, East Sikkim.

one at Rayenda, Swarankhola (Mukhopadhyay and Dutta, 2012) and another at Molemberia (Figure 4.9), which survived Cyclone Sidr. The overall plan of these buildings is also square with four pitched roofs forming a pyramid at the centre. The buildings were found to have many bolts at junctions between different mutually perpendicular wooden structural members.

4.7.2 SURVIVAL DURING THE EARTHQUAKES

A significant observation during the 1993 Killari Earthquake was that some of the thatched houses, with wooden vertical posts and rafters connected with coir rope ties, and walls made of bamboo woven mats, suffered only minor cracks (Jain et al., 1994).

FIGURE 4.8 Indigenous way of tying down roof claddings with the application of fishing nets as practiced at Ghola Para, Sagar Islands, India.

FIGURE 4.9   An undamaged non-engineered building near Molemberia within Bagerhāt District of the Khulna Division of Bangladesh, used as a shelter during Cyclone Sidr.

Elsewhere, and more recently, during the 2005 Kashmir Earthquake, it was interesting to note that no collapse was reported for traditional timber-laced brick constructions, locally known as Dhajji-dewari and Taq systems of construction (Langenbach, 2007). The timber frames subdivide brick walls into many smaller panels, arresting formation of large cracks. They also introduce frictional dampening over many cycles, with slow rates of degradation. Further, during both 2006 and 2011 Sikkim earthquakes, it was noted that a few traditional residences, made of combinations of wood, bamboo and rattan with corrugated G.I. sheet roofing, did not suffer any damage (Figure 4.10). The natural characteristics of these traditional materials

FIGURE 4.10   An undamaged non-engineered building in the vicinity of the damaged Mani Lhakhang Prayer Hall, East Sikkim, observed during the 2011 Sikkim Earthquake.

of being light in weight and fibrous in morphology make them highly flexible. Buildings made with such materials become inherently resistant to the lateral shock waves of earthquake forces because of their large deformation accommodating capability.

4.8 CONCLUDING REMARKS

It is a matter of deep concern that the major casualties of natural disasters in the Indian subcontinent have particularly been people from the lowest socioeconomic strata of society, who can hardly manage to have an engineered shelter. In fact, an overwhelming percentage of roofing and walling materials of the different housing units of the subcontinent are non-engineered. The impact of tropical cyclones and earthquakes, in terms of casualty, injury and property loss, is tremendous. In this context, the major challenge to the subcontinent lies in improving the existing construction practices, without making any sea change in materials and procedure yet introducing crucial technical inputs, adequate for survival in the event of such natural disasters.

The length of the coasts of peninsular India is around 5423 km and that inclusive of the islands is around 7517 km (Kumar, V.S. et al., 2006). Devastation of the vernacular buildings in the entire coastal region is an annual feature, but the extensive damage caused to many parts of the eastern coast of the peninsula and the Andaman and Nicobar Islands, by the Indian Ocean Earthquake of December 2004, and its associated tsunami, was most unexpected. One of the many long-term follow-up actions that resulted from this incident was amendment of the Environment (Protection) Act, 1986, in 2011, which initiated a set of new Coastal Regulation Zone (CRZ) rules. A 'no-development zone' from the high-tide line was fixed at 100 m, and the floor space index (FSI) was increased to 2.5.* The CRZ was extended to 12 nautical miles (about 22 km) seaward. New industries, and expansion of old industries, are prohibited in the CRZs, with some exceptions, such as facilities for generating power by non-conventional energy sources, and so on. Discharge of untreated waste and effluents is also prohibited, as is dumping of city or town waste including construction debris, industrial solid waste, and so on. However, lack of experience of tsunamis is a major issue which calls for further research in this area. This should include investigation of factors that may play important roles in the performance of buildings within the CRZ, like orientation and shape of the buildings, location of doors or windows, orientation of the masonry at junctions, nature of bonding, and so on. Further, the allowable distance between buildings constructed in the CRZ may be another issue to be explored in a region-specific manner.

A paucity of funds is thwarting the task of upgrading the huge number of non- and semi-engineered buildings in the developing economies of the subcontinent.

---

* FSI or floor area ratio (FAR) is the quotient obtained by dividing the combined floor area of all the floors of a building by the area of the plot on which the building is built. Thus, FSI or FAR is a unitless quantity, being a ratio of two areas. The information "FSI of a plot of $\times$ m^2 area is 2.5" suggests that the combined area of all the floors of a building built on this plot can be maximum $2.5 \times$ m^2.

The situation warrants development of a tool for quick assessment and grading of damageability of existing buildings, in the eventuality of future cyclones and earthquakes, to enable prioritization of needs. A dearth of technical personnel, on the other hand, requires it to be a simple methodology, which may be implemented by persons passing high school grades with fundamental knowledge of mathematics. In this context, the Rapid Visual Screening (RVS) procedure, which was first proposed in 1988 by the Federal Emergency Management Agency (FEMA) in the United States, is worth mentioning. It consists of survey-based simple procedures to assess buildings that may be damaged in the event of earthquakes, and prioritize them with respect to their extent of vulnerability (FEMA, 2002).

The RVS of a building consists of a scoring system, based on certain visually identifiable quantitative and qualitative parameters, with an aim to generate a 'Structural Score' related to the probability of the building getting damaged. However, the FEMA-recommended RVS scheme does not include earthquake-affected non-engineered buildings in particular, and cyclone-affected buildings in general. Inspired by FEMA, two different RVS procedures for both cyclone- and earthquake-affected non- and semi-engineered structures in the Indian subcontinent have been developed. The proposed strategy for cyclone-affected buildings includes 8 building types and 11 score-modifying parameters (Mukhopadhyay and Dutta, 2012); and that for the earthquake-affected structures consists of 9 building types and 19 score-modifying parameters (Mukhopadhyay, 2014). These parameters, some quantitative and some qualitative in nature, are visually identifiable. Furthering such research work for improving performance of the non-engineered building structures is being felt as the need of the hour.

ACKNOWLEDGEMENTS

The authors acknowledge the authorities of Bengal Engineering and Science University, Shibpur and Indian Institute of Technology, Bhubaneswar, for sponsoring reconnaissance-based damage surveys undertaken, respectively, at Bangladesh and Sikkim. They sincerely thank Dr. Md. Zoynul Abedin, Professor, Department of Civil Engineering, Bangladesh University of Engineering and Technology for extending logistic support during survey of the impact of 2007 Cyclone Sidr. They also thank Mr. Sonam Daju Bhutia, Additional Chief Architect, Building and Housing Department, Government of Sikkim for providing support during damage survey after the 2011 Sikkim Earthquake.

REFERENCES

Berg, G.V., Das, Y.C., Gokhale, K.V.G.K. et al. (1969). The Koyna, India, Earthquakes. *Proceedings of the Fourth World Conference on Earthquake Engineering, Vol. III*, Santiago de Chile, Jan 13–18, 1969, University of Chile, Santiago, J2-44–J2-57.
Bern, C.J., Sniezek, G.M., Mathoor, M.S. et al. (1993). Risk factors for mortality in the Bangladesh cyclone of 1991. *Bulletin of the World Health Organization*, 71(1), 73–38.
Bureau of Indian Standards (1986). *Criteria for Earthquake Resistant Design of Structures (Fourth Revision) (Reaffirmed 1998), IS: 1893–1984*, Bureau of Indian Standards (BIS), New Delhi.

Bureau of Indian Standards (1989). *Code of Practice for Design Loads (Other Than Earthquake) for Buildings and Structures, Part 3 Wind Loads* (Second Revision) (Reaffirmed 1987), IS: 875 (Part 3) – 1987, BIS, New Delhi, 1989.

Bureau of Indian Standards (2002). *Criteria for Earthquake Resistant Design of Structures, Part 1, General Provisions and Buildings (Fifth Revision), IS 1893 (Part 1): 2002*, BIS, New Delhi.

Bureau of Indian Standards (2004). *Earthquake Resistant Design and Construction of Buildings – Code of Practice (Second Revision) (Incorporating Amendment Nos. 1 & 2) (Reaffirmed 1998), IS 4326: 1993*. BIS, New Delhi.

Delescluse, M., Chamot-Rooke, N., Cattin, R. et al. (2012). April 2012 intra-oceanic seismicity off Sumatra boosted by the Banda-Aceh megathrust. *Nature*, 490, 240–244.

Dutta, S.C. and Mukhopadhyay, P. (2012). Occurrence of tropical cyclones: The North Indian Ocean. In *Improving Earthquake and Cyclone Resistance of Structures: Guidelines for the Indian Subcontinent*, TERI, New Delhi, 161–173.

Dutta, S.C., Mukhopadhyay, P. and Goswami, K. (2013). Augmenting strength of collapsed unreinforced masonry junctions: The principal damage feature due to moderate Indian earthquakes. *Natural Hazards Review*, 14(4), 281–285.

Dutta, S.C., Mukhopadhyay, P.S., Saha, R. et. al. (2015). 2011 Sikkim Earthquake at Eastern Himalayas; Lessons learnt from performance of structures. *Soil Dynamics and Earthquake Engineering*, 75, 121–129.

Eskander, S.M.S.U. and Barbier, E.B. (2014). *Long-Run Impacts of the 1970–74 Series of Disasters in Bangladesh.* https://papers.ssrn.com/sol3/papers.cfm?abstract_id=2548744 (accessed Sep 2017).

Federal Emergency Management Agency (FEMA) (2002). *Rapid Visual Screening of Buildings for Potential Seismic Hazards: A Handbook*, 2nd ed. (CD-ROM), FEMA 154, Department of Homeland Security, Washington, DC.

Frank, N.L. and Husain, S.A. (1971). The deadliest tropical cyclone in history? *Bulletin of the American Meteorological Society*, 52(6), 438–444.

Government of India (2011). *Census of India, 2011.* New Delhi, Registrar General and Census Commissioner, India.

Gupta, H.G. (2002). A review of recent studies of triggered earthquakes by artificial water reservoirs with special emphasis on earthquakes in Koyna, India. *Earth-Science Reviews*, 58(3–4), 279–310.

Indian Meteorological Department (n.d.). Website: www.imd.gov.in (accessed December 10, 2014).

Indian Standards Institute (1962). *Recommendations for Earthquake Resistance of Structures, IS: 1893–1962.* Indian Standards Institute (ISI), New Delhi.

Indian Standards Institute (1967). *Recommendations for Earthquake Resistance of Structures (First Revision), IS: 1893–1966*, ISI, New Delhi.

Indian Standards Institute (1971). *Criteria for Earthquake Resistance of Structures (Second Revision), IS: 1893–1970*, ISI, New Delhi.

International Association for Earthquake Engineering (IAEE), and National Information Center of Earthquake Engineering (NICEE). *Guidelines for Earthquake Resistant Non-Engineered Construction* (English ed.). Buildings in Fired-Brick and Other Masonry Units, NICEE, Kanpur, India, 2004, 25–42.

Jain, S.K., Lettis, W.R., Murty, C.V.R. et al. (2002). Introduction. *In* Jain, S. K., Lettis, W.R., Murty, C.V.R. et al. (Eds.) *2001 Bhuj, India Earthquake Reconnaissance Report, Earthquake Spectra, Supplement A to Volume 18*, Earthquake Engineering Research Institute, Oakland, 1–4.

Jain, S.K., Murty, C.V.R., Chandak, N. et al. (1994). The September 29, 1993, M6.4 Killari, Maharashtra, Earthquake in Central India. *EERI Newsletter*, 28(1), 1–8.

Kumar, P., Yuan, X., Ravi Kumar, M. et al. (2007). The rapid drift of the Indian tectonic plate. *Nature*, 449, 894–897.

Kumar, V.S., Pathak, K.C., Pednekar, P. et al. (2006). Coastal processes along the Indian coastline. *Current Science*, 91(4), 530–536.

Kynman, Y., Kouppari, N. and Mukhier, N. (Eds). (2007). *World Disasters Report 2007: Focus on Discrimination*. International Federation of Red Cross and Red Crescent Societies (IFRC), Geneva.

Langenbach, R. (2007). *Guidelines for Preserving the Earthquake-Resistant Traditional Construction of Kashmir*, United Nations Educational, Scientific and Cultural Organization (UNESCO), New Delhi.

Mukhopadhyay, P. (2014). Study on Cyclone and Earthquake Resistance of Non-engineered Building Elements. Unpublished Doctoral thesis, Indian Institute of Engineering Science and Technology, Shibpur, India.

Mukhopadhyay, P. and Dutta, S.C. (2008). *A Reconnaissance Based Vulnerability and Damage Survey Report at the Khulna and Barisal Divisions of Bangladesh: Effect of the Cyclone Sidr of 15th November 2007*, Bengal Engineering and Science University, Shibpur, India.

Mukhopadhyay, P. and Dutta, S.C. (2012). Strongest cyclone of the new millennium in the Bay of Bengal: Strategy of RVS for nonengineered structures. *Natural Hazards Review*, 13(2), 97–105.

Mukhopadhyay, P., Dutta, S.C. and Bhattacharya, S. (2009). Lessons learnt from the impact of Sikkim 2006 earthquake on heritage structures. *In* Gazatas, G., Goto, Y., Tazoh, T., (Eds.). *Seismic Design, Observation, Retrofit of Foundations, Proceedings of the 3rd Greece – Japan Workshop (3GJW)*, Santorini, Greece, Sept 22–23, 2009, Laboratory of Soil Mechanics, National Technical University of Athens (NTUA), Athens, 615–628.

Rajendran, K., Rajendran, C.P., Thulasiraman, N. et al. (2011). The 18 September 2011, North Sikkim Earthquake. *Current Science*, 101(11), 1475–1479.

Sheehan, A.F., Wu, F.T., Bilham, R. et al., (2002). Himalayan Nepal Tibet Broadband Seismic Experiment (HIMNT). *American Geophysical Union, Fall Meeting 2002*, abstract #S61D-11.

Simon, F., Prasad, V.S.K.G. and Arora, M. (2001). Performance of built environment in the Oct 1999 Orissa super cyclone. *Proceedings of Conference on Disaster Management*, Birla Institute of Technology & Sciences (BITS), Pilani, India, Mar 5–7, 2001, BITS, Pilani, India.

Wikipedia (2012). Website: http://en.wikipedia.org/wiki/File:India_topo_big.jpg (accessed Dec 2012).

Zetter, R. (Ed.) (2012). *World Disasters Report 2012: Focus on Forced Migration and Displacement*, IFRC, Geneva.

5 Alert Systems in Natural Hazards

Best Practices

*Montserrat Ferrer-Julià, Miguel Mendes,
Joaquín Ramírez and Paul Hirst*

CONTENTS

5.1 INTRODUCTION

Floods, volcanic eruptions and tsunamis are the most often forecasted hazards, and can usually be sensed and tracked with enough time to alert the population of the forthcoming danger ahead of the event, to let them know about the hazard type, where to go and what to do. An alert and warning system is the complete infrastructure system for broadcasting emergency alerts to the public, as well as to personnel who may be required to respond to the emergency.

When an emergency can potentially affect the citizens, the emergency management organizations need to be able to communicate in two directions: one to rapidly inform the population about the impending threat, to allow them to take protective actions; and, in the opposite direction, to spread the same information among personnel intervening in the management of the incident, as well as to the media, to avoid inconsistencies in the sharing of knowledge about the event. Furthermore, once the alert has been issued, communication channels shall be set and used to inform and train the public about how to respond, and how to act in case an emergency takes place. An alert system includes one or more communication networks, protocols to improve their use, procedures for planning the warning, public education and periodical test and evaluation of the system.

The warning planning is a dynamic procedure that should be part of the emergency management planning activities. It should include the establishment of internal and inter-organizational procedures for the parties involved in the warning process, the identification of population groups that need to be warned in case of a crisis event and the identification of the areas that are most vulnerable if a crisis occurs. One of the main aims of warning planning is to allow a common operational picture among the involved authorities, regarding the conveyed warnings to the population during the warning process, so that they will issue consistent and accurate warnings (Reynolds, 2012). Public education also plays a major role in the effectiveness of the warning process (CEMA, 2008). Appropriate response to warnings is more likely to occur when people have been educated about the hazard events that may affect them, and when they have developed a plan of action long before the warning occurs (Reynolds, 2009).

In general terms, when targeted with a warning, the public should be able to comply properly with the requested actions. They also need to know how the warning will reach them, the meaning of the warning and where to find further information regarding the hazard event. The area of public risk education may also include the testing and evaluation of the alerting channels, to enhance the awareness of the population. Hence, as well as allowing the authorities to check the operability of the alerting channels, testing and evaluating these channels increases the citizens' understanding of how they will be reached with a warning, and how the warning process works (Ding, 2009).

In an alert and warning system, different types of messages related to the notification of the population may be distinguished (Figure 5.1):

- *General public education.* This should be carried out regularly during normal circumstances, to prepare the population against any kind of hazard event that may affect them. Thus, these types of messages may be

FIGURE 5.1   Message types in a warning system.

considered preventive information messages (e.g. making regular drills at the local level with the participation of the citizens and/or publishing information about how people can be alerted).

- *Warnings.* These have two main goals: (1) To alert the population, making them aware that something is wrong; and (2) to achieve quick protection in the face of an imminent or an already occurring major hazard event, by spreading crucial information to the population at risk (Ding, 2009).
- *Follow-up information.* This is issued to inform and advise the public during and after the occurrence of short-term and long-term hazard events (Strander, 2011).

Related to the warning effectiveness, some predefined sender and receiver factors are recognized by researchers (Coombs, 2010; Kortom, 2011) as determining the effectiveness of the warning response process. The receiver factors include the characteristics of the population, such as their socioeconomic status and their social contact, whereas the sender factors include hazard-related situational factors and those related to population education, as well as the way the alert message itself is constructed and issued. Of these factors, only the last three can and should be influenced by the warning issuer. If the organization that issues the alert message addresses the sender factors appropriately, the warning response effectiveness among the targeted population can be increased (Coombs, 2010). Through education of the population, the use of the proper channels to reach them and structuring the alert message carefully may increase the effectiveness of the alert and warning system. The knowledge of how to construct an alert message in an effective way should be acquired before the civil protection organization adopts the alert and warning system, whereas the education of the population should take place regularly when no crisis is taking place.

5.2   BEST PRACTICES IN ALERTING THE POPULATION

To implement an effective alert and warning system, the civil protection organization should first determine which types of alerting channels would be more efficient (e.g. TV, radio, sirens, Internet, etc.) according to the characteristics of the population. They should then implement the system and educate the receiving public about how this system works, how they can be reached, and how to behave in case they receive an alert message.

The warning effectiveness may be measured by how appropriately the citizens respond to the recommended protective actions that are advised (Reynolds, 2009). When receiving a warning, people do not act instantly on demand. They usually enter and follow a psychological process from the reception of the warning to the moment they respond (Coombs, 2011; Kortom, 2011), and their reaction depends on different receiver factors such as environmental, social, psychological and physiological aspects. But at the same time, the effectiveness may depend on the types of alerting communication channels used, how many of them are used at the same time, the alert message content and style, the frequency with which the warning is repeated and updated, as well as the source of the message, all of which can also influence the citizens' response. Therefore, the response effectiveness is affected by warning issuer, as well as by receiver factors (Coombs, 2010; Kortom, 2011). Nevertheless, the warning sender can only control the first type of factors. Thus, the following collected best practices are focused on those.

Research on alerting the population has been carried out, since the 1980s, by numerous authors of international reference (e.g. Mileti and Sorensen, 1990), and a compilation of best practices was carried out during a European Union funded research and development project, for the development of a European alert and warning system (http://cordis.europa.eu/project/rcn/98427_en.html). These guidelines are based on an extensive and thorough literature review of existing best practices and recommendations in alerting the population, as well as on the outcomes of interviews with civil protection practitioners. They are built from past experiences where mistakes or errors were detected, corrections were proposed, and effective choices were recognized to improve future actions (Mendes et al., 2012). When manipulated in a proper manner, the factors and attributes that were identified can achieve an optimal outreach of the warning and the resulting response by the population at risk. They have been summarized on four pillars (Figure 5.2): warning variables, alert messages, differentiating between their style and content and the usage of alerting communication channels.

Best Practices			
Warning variables	Alert message style	Alert message content	Usage of alerting communication channels

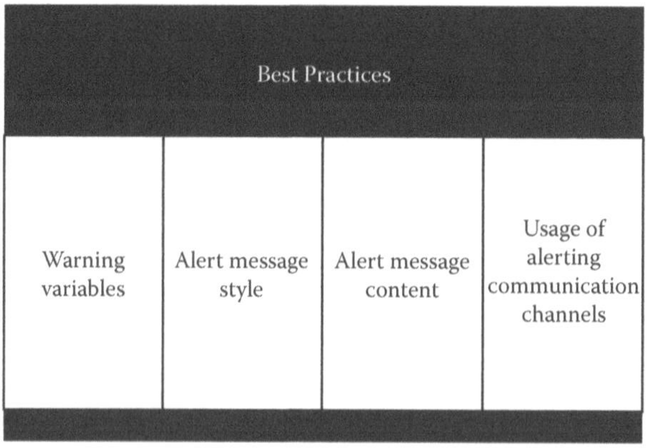

FIGURE 5.2   Best practices pillars to improve effectiveness in alerting the population.

FIGURE 5.3   Warning variables when warning the population.

5.2.1   WARNING VARIABLES

There are five warning variables that should be considered when warning the population (Figure 5.3).

5.2.1.1   Whether to Warn

After the civil protection organization has quantified the risk, and assessed the potential damage and other consequences of the hazard event to the population, if there is a doubt whether to warn, it is better to do so (Kortom, 2011), as it is shown in the case of the Elbe and Danube floods of 2002 (Case 5.1). The consequences of being

---

Case 5.1: During the first two weeks of August 2002, many rain gauges located in parts of Austria, the Czech Republic and Germany in central Europe reached the highest amount of rainfall and intensities ever recorded (Risk Management Solutions, 2003; Ulbrich et al., 2003). This produced several flash floods in the mountain catchments of the River Elbe (Ulbrich et al., 2003) which, together with general flood waves along the river system, led to a high-return period flood in Dresden. Depending on the author, this was either the 400- to 500-year return period flood (Kreibich and Thieken, 2009) or the 100- to 200-year return period flood (Thieken et al., 2007). At the same time, the upper Danube River discharges showed flooding with lower return period (100 years according to Thieken et al., 2007) affecting parts of Austria, Slovakia and Hungary, albeit with minor damages (Risk Management Solutions, 2003). In Germany alone, these events caused damages estimated at around 10 billion euro, as well as killing 21 people (Thieken et al., 2007). This situation could have been different if more inhabitants had been properly warned. Thieken et al. (2007) performed a survey among flood-affected people after the event, based on 1697 interviews. Results showed that more than a quarter of the people did not receive any warning, and only the 40% of the surveyed people received flood warnings disseminated by the authorities. In contrast, Kreibich et al. (2007) performed a survey among 415 German companies affected by the Elbe floods, where damages reached 1.1 million euro. In this sector of the society, only 25% of the companies had been aware of the authorities' warnings, and 45% of the companies did not receive any flood warning and so could not perform any emergency measure. Nevertheless, 49% of the surveyed companies had undertaken precautionary measures before the event, due to their previous flood experience.

wrong are much smaller than if disaster occurs and warning has not been carried out. Sometimes the decision-making process may be improved by a cost–benefit analysis, which sets a threshold on the acceptable probability of false alarms (UNEP, 2012). However, it is always possible that a false alarm may occur. In this situation, the reasons for the issuing of the alarm should be explained by the relevant authorities, and a valid scientific and rational explanation to support the decision taken should be provided to the citizens. Where this is done, affected populations are generally more willing to tolerate false alarms in the future (Kortom, 2011), if this tolerance is not abused (see Case 5.2).

The consequences of false alarms also can be reduced when authorities and the population are educated to fully acknowledge the probabilities corresponding to uncertainties that hazards involve. On the one hand, authorities will better understand the hazard nature and the scientific team warnings before and during an emergency, while, on the other, the population will better accept false alarms and panic

Case 5.2: On July 3, 2006, 190 mm of torrential precipitation fell in the province of Vibo Valentia (Italy), with a maximum intensity of 130 mm/h (Ietto et al., 2009). As a result, a sudden flood laden with high sediment content reached the town of Vibo Valenti. As well as 200 million euro damages (Baglivo, 2016), three people lost their lives (Caloiero et al., 2014). All these damages were the basis for a trial of those civil protection workers who failed to forecast the weather and their legal representatives. The prosecutor wanted to convict them because 'having caused landslides and floods, as officer in duty for the forecast and prevention of natural risks office within the Department of Civil Protection, as negligently failing to comply with the obligations related to that position because despite the existence of a meteorological phenomenon whose severity was predictable and expected, failed to issue an alert of adverse weather conditions' (Vibo Valentia Law Court, Office of the Judge for Preliminary Investigations, Reg. Sent 76/09, 6 September 2009, extracted from Altamura et al., 2011). As it can be observed from these words, the prosecutor didn't consider the uncertainty of weather models and forecasts, and for this reason blamed civil protection operators for negligence in execution of their tasks (Altamura et al., 2014). At the end of the trial, all defendants were acquitted but, afterward, the civil protection increased the number of warnings that finished in false alarms in order to avoid future criminal sentences, which Altamura et al. (2011) describe as a consequence of the over-criminalization of this process. This resulting situation does not provide a useful service to citizens (Cocco et al., 2015), and may lead to a weakening of the trust in civil protection and, hence, in the effective response of the citizens (Reynolds, 2009). Indeed, in 2005 during Hurricane Katrina in the United States, many inhabitants did not want to leave their homes because they had already been evacuated unnecessarily for hurricanes Georges in 1998 and Ivan in 2004 (EUMETSAT, 2014).

situations will be avoided. Where citizens have not received this type of education, special attention should be given to the alert message style, as it is explained later.

This situation has also gotten worse because, in recent years, a public debate about the responsibilities of false or missed alarms has been opened. The major factor in this was the trial of the scientific community after the L'Aquila earthquake in Italy in 2009 (Case 5.3). To avoid this type of situation, a specific and precise agreement about the responsibilities in the case of false or missed alarm should be

Case 5.3: In 2009, an earthquake occurred in Aquila (Italy), causing 29 deaths. Several families of the casualties believed these deaths would have been avoided if the victims had received enough information, so they decided to prosecute several scientists and civil protection workers. Consequently, in the first trial in 2012, seven experts (scientists and civil protection workers) were all sentenced to 6 years in prison, perpetual interdiction from public office, and a common fine of more than €8 million in compensation (Cocco et al., 2015). The judge in the case considered one of the reasons the giving of only an approximate risk assessment was generic and ineffective in relation to the activities and duties of prediction and prevention (Tribunale di l'Aquila, 2012, p. 2 extracted from Benessia and de Marchi, 2017). Later, a second appeal trial in 2014 overturned the sentence and only the former deputy director of the Department of Civil Protection (DPC) was sentenced to 2 years imprisonment, although the sentence was suspended in the same trial (Benessia and de Marchi, 2017).

All these judicial processes led to a big controversy around whether legal trials can understand the uncertainty of natural hazards, or the scientific approach to managing this type of event. It is expected that they have helped to show that society needs to reduce its vulnerability to natural risks, by reinforcing citizen's preparedness, as well as through building performance evaluation and retrofitting in the case of earthquake hazard areas. Another consequence of these trials is that they helped to emphasize the importance of translating the scientific language to citizens. In fact, De la Pomerai and Omer (2012) rightfully pointed out the possibility that there was a misuse of the word 'Prediction' as a 'Real Time Early Warning'. Moving on, Amato et al. (2013a,b) argued that the communication chain showed that, due to misunderstandings, the scientists' message ('Earthquakes are not predictable,' 'A strong earthquake in the next days is unlikely (but not impossible)') was distorted ('The earthquake will not happen,' 'The situation is normal (favorable)'). This, together with contradictory information given by the mass media weeks before the earthquake event, served to increase confusion among citizens (Cocco et al., 2015). Finally, it is important to remark that this type of trial makes scientists and DPC workers feel vulnerable, and it is not difficult to understand why, as was also seen in the case of the Vibo Valentia (Italy) floods described in the previous case, the number of false alarms issued from the DPC increased after the first trial sentence (Amato et al., 2013a).

established (1) between the scientific community and the authorities; and (2) among the different administrative organizations (national, regional and local administrations, depending on the country structure) involved. In this way, if none of the steps defined in the agreement are neglected, no one will be liable to prosecution, and no one will fear making the wrong decisions in future emergency events.

5.2.1.2 When to Warn

Once the need has been identified, many well-recognized best practices stress that the warning about the hazard event should be provided to the public as soon as possible (Partnership for Public Warning, 2004; Perry and Lindell, 2007). Any delay may highly increase the damages (see Case 5.4). In the case of geological hazards such as earthquakes, volcanic eruptions or tsunamis, it may be possible to define where it will take place, but not when, nor the degree or the total affected area. Therefore, it is not always possible to provide recommendations to the population for protective actions due to these uncertainties, although a quick first message can and should be supplied to (Mileti and Sorensen, 1990):

- Draw people's attention to the risk
- Refer to the uncertainty of the event
- Explain what is being done by the authorities to gather more information about the event

When possible, it should be also mentioned when and how updates will be provided. Afterward, when the hazard consequences become clearer, a detailed warning with accurate information about the hazard and protective actions should be supplied (Coombs, 2010; Kortom, 2011).

5.2.1.3 Where and Who to Warn

The warning issuer should consider flexible boundaries when deciding the geographical area(s) to be warned. It is considered wiser to warn a wider area than may be necessary, rather than having to improvise if the hazard impact spreads to other zones (Kortom, 2011). This is especially applicable to geographically random hazards such as forest fires. Furthermore, when issuing a warning it is important to explain, with reasons, those parts of the population who are at risk and those who are not, since the hazard event may not affect all groups of people or areas equally.

5.2.1.4 Warning Frequency

Warning updates should be provided when the hazard conditions change significantly, to allow the population to adapt their actions to the new conditions. This means that the frequency of updates will usually be higher during the initial phase of the hazard, and lower after it (Kortom, 2011) until the end of the event, when the population may be advised to return to their normal activity (Ding, 2009; Nagarajan et al., 2010).

To define an adequate warning frequency strategy, it is necessary to continuously monitor the progress of the hazard event, as well as the reactions and information that is being created by the citizens and the mass media to the warnings that have already been sent (Kortom, 2011).

Case 5.4: Based on World Bank (2010) cost–benefit analysis, Rogers and Tsirkunov (2010) stated that there are three disaster prevention aspects where investments mean benefits when a disaster occurs. They are early warning systems, critical infrastructure and environmental buffers. The second one is the most often used, although early warning systems have already shown their effectiveness (Rogers and Tsirkunov, 2010). For instance, based on the European Flood Alert System, and considering different flood protection scenarios, Thielen-del Pozo et al. (2015) remark that the benefit of flood early warning in Europe has the potential of reducing costs of flood damages by about 25%. This means that Europe could save around 30,000 million euro over the next 20 years, if a proper warning system is used to alert and warn the population.

Standardized methods are used to estimate these benefits, based on the difference between the expected damage with warning and without it. The expected damage is calculated according to the sum of expected annual damage for different probability events (Carsell et al., 2004).

Depending on the type of hazard, the estimation rule of benefits may present some variations. For instance, flood damages are usually estimated according to water depth (USACE, 2000). One of the first researchers to use warning time in damage estimation was Day (1970, extracted from Carsell et al., 2004), who proposed a function curve between the warning time and percent of flood damages, after a survey to Susquehanna River basin property owners (Figure 5.4). One of the limits of this model is that it is assumed that owners' properties will react quickly and efficiently once they receive the notification, knowing what to do at any time, so the graph shows optimistic values.

Carsell et al. (2004) defined a function of floodwater-depth and warning mitigation time to estimate the flood percentage damage, based on the

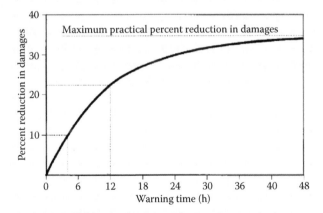

FIGURE 5.4 Day's curve. (Extracted from Carsell, K. M., N. D. Pingel, and D. T. Ford. 2004. Quantifying the benefit of a flood warning system. *Natural Hazards Review*, 5, 131–170.).

responses of a panel of 11 floodplain management and flood damage experts. To apply this value to a realistic citizen response situation, this value is subsequently modified by a warning efficiency factor calculated from (1) the fraction of the public that receives a warning; (2) the fraction of the public that is willing to respond; and (3) the fraction from those who know how to respond effectively, and are able to respond. As a result, this factor value moves from 0.3 to 0.6. The implementation of this method shows that the overall damage of a flood may be reduced by more than 50% when a warning is sent to the population 48 hours before the event.

5.2.1.5 Alert Message Repetition

The more the alert message is repeated and heard, through different warning channels, in the period leading up to a major hazard event, or during it, the better will be people's response to the warning (Coombs, 2010; Strander, 2011; Veil et al., 2011; Hallegate, 2012). Nevertheless, to avoid the population losing interest, when disseminating repeated alert messages this should not be done for too long and should be performed within different time intervals to achieve an effective stimulation of the citizens (Egli, 2002).

5.2.2 ALERT MESSAGE STYLE

There are five stylistic aspects of the alert message (Figure 5.5) that affect its efficiency regarding a citizen's response to it (Kortom, 2011). By following these style

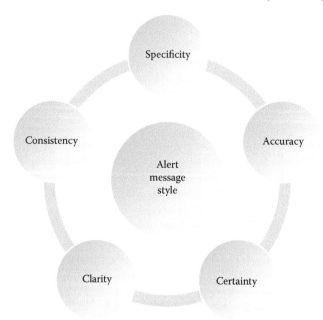

FIGURE 5.5   Stylistic aspects in alert messages.

rules, clear and accurate alert messages may be created to enhance the personalization of its content, and thus increase degree of compliance by the citizens.

5.2.2.1 Specificity

The hazard and protective actions should be described in a precise and non-ambiguous way, avoiding information omission which could lead to wrong interpretation by the citizens (Kortom, 2011).

5.2.2.2 Accuracy

Accuracy in warning may be achieved by being totally open with citizens, providing complete hazard information and avoiding withholding information (Ding, 2009). At the same time, the message needs to be spelled correctly; otherwise, citizens may disregard its content and may not follow the official instructions.

5.2.2.3 Certainty

When transmitting a voice or video warning, trust in the message may be increased when it is transmitted with a confident tone of voice and, in case of video transmission, with appropriate body language (Mulero-Chavez et al., 2011).

5.2.2.4 Clarity

The warning message should be constructed with clear and simple words (see Case 5.5), in a way that any citizen may understand (Kortom, 2011), so codes, jargon, acronyms or scientific expressions such as the 'incident magnitude level' should be avoided. Another way to give clarity to the message is to present the information in sequence, stating the reason for or purpose of the message, its supporting information and conclusion.

5.2.2.5 Consistency

A warning should be consistent with its own content. This means that the message content should be neither contradictory nor ambiguous (Case 5.6). If it is either of these, people may get confused or may distrust the warning content, and may not follow the required instructions (Kortom, 2011; Veil et al., 2011). At the same time, the warning should be consistent with previous alert messages, or with those disseminated by other official sources, to generate belief and avoid confusion within the target population (Kortom, 2011; Veil et al., 2011). This step will be achieved through sharing and coordinating communications in real time with the rest of the parties involved in the warning process.

Case 5.5: In April 1991, the Tropical Cyclone Marian struck Bangladesh, and almost 140,000 people died. According to Bern et al. (1993), 83% of the surveyed coastal population did not follow the warning instructions, 16% of whom said that they did not understand the warning. This means that if the message content had followed the above style rules, at least 22,000 people might have had the opportunity to remain alive.

Case 5.6: The evacuation warning issued after the eruption of the volcano Nevado del Ruiz (Colombia) in 1985 is one of the most famous cases of lack of consistency among messages sent from different authorities. This active volcano is in the Andes Volcanic Belt, where the Nazca tectonic plate sinks under the South American plate. Due to its height (5321 m above sea level), it is covered by glaciers. On November 13, 1985, after several signs detected by scientists during a year's observation, this volcano erupted. Due to the eruption of ejected materials, part of the glacial ice cap was melted, and it generated a lahar reaching speeds of up to 45 km/h (Mileti et al., 1991).

The village of Armero was located a few kilometres below the mountain. For several months the community had pressed the government for deeper studies, and scientists decided to compile a volcanic hazard map. But because the situation seemed to be improving, local financial authorities disputed the economic costs that creating this map involved. It seems that this was an influencing factor in the decision to not warn the population to evacuate until some hours after the eruption occurred, even though the local government was warned by the regional government of Caldas about the high probabilities of the lahars arriving (Carracedo, 2015). At the same time, high-ranking religious representatives warned the people with loudspeakers to stay at home, and not to follow the regional government safety measures, thereby reinforcing the local authority warnings (Carracedo, 2007). This contradictory information persuaded a lot of citizens to remain in their homes, which resulted in at least 23,000 deaths and 5000 injuries (Carsell et al., 2004). If they had left the village when warned by the regional government, although late, most of them could have survived.

5.2.3   ALERT MESSAGE CONTENT

To be able to issue an optimized alert message, with the goal of prompting citizens to follow the required protective actions, the content elements shown in Figure 5.6 should ideally be included. This does not mean that the alert message must be long and vague. Messages should be as brief as possible, but should always include this minimum content, unless the communication channel transmission capability or airtime forces the message to be kept shorter. In these latter cases, it is highly recommended to divide the content into two different messages and send them separately but within a certain time period. The first message would provide information about the hazard and the need to prepare, and the second would focus on safety measures such as ordering an evacuation (Coombs, 2011).

5.2.3.1   Hazard Information, Location, Timing and Duration

The objective of including hazard information, location, timing and duration in the message is to inform the population about the hazard characteristics as well as where, when and for how long the hazard event impact is estimated. The intensity of the hazard event should be transmitted in a way that addresses the health and welfare

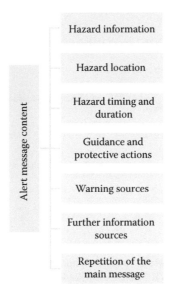

FIGURE 5.6   Alert message content components.

consequences to the citizens, unless the citizens have specific training about the emergency levels or hazard magnitude levels. For instance, the population that lives in an earthquake-prone area is usually familiar with the hazard, and knows which magnitude levels it can reach; however, a foreigner visiting that country or region may not be aware of this. Usually, the population living in a disaster-prone area is more prepared and informed about the hazard than a population that lives in an area with low probability of occurrence of the same type of hazard. It is also important that the hazard location is mentioned in a way the population may be familiar with, describing the affected geographic area(s) with easily recognizable landmarks, and differentiating those specific high-risk populations or assets. Finally, the amount of advance warning that is being given should be stated, to indicate how much time is available for the addressed public to start taking protective actions (e.g. in 2 hours) or otherwise the time when they should start them (e.g. at 15:00h).

5.2.3.2   Guidance and Protective Actions

The alert message should provide specific guidance regarding protective actions and behaviours to be followed, telling people exactly what to do, and when and how to do it depending on their location (Ding, 2009; Reynolds, 2009; Kortom, 2011; Strander, 2011). Otherwise, the alert message simply enhances fear (Strander, 2011).

These safety measures should focus on lifesaving protective actions when the hazard severity is high and the forewarning time is short, but they may also be focused on property protection when more forewarning time is available (Strander, 2011). At the same time, the advice should be accompanied by an explanation of how they provide protection to the citizens, to provide a rationale for the requested actions. In cases where the affected population will be unable, or unlikely, to mitigate

```
Event: Lake Wind Advisory
  Alert:
        ...LAKE WIND ADVISORY REMAINS IN EFFECT UNTIL 9 PM PDT THIS
        EVENING FOR PYRAMID LAKE...

        * Winds: North to northwest winds 15 to 20 mph with gusts to 35
        mph.

        * Wave Heights: 1 to 3 feet with the highest waves from mid
        lake to downwind shores.

        * Small boats, kayaks and paddle boards will be prone to
        capsizing and should remain off lake waters until conditions
        improve.

Instructions: Check lake conditions before heading out and be prepared for a sudden increase in winds and wave
              heights. Consider postponing boating activities on the lake until a day with less wind.

Target Area: Western Nevada Basin and Range including Pyramid Lake

Forecast
  Office: NWS Reno (Western Nevada)
```

FIGURE 5.7   Example of an alert message from the U.S. National Weather Service. (Extracted from https://alerts.weather.gov/cap/wwacapget.php?x=NV12584FDB7234.LakeWindAdvisory .12584FE83500NV.REVNPWREV.e3bffed3632446feb73bf5ddeb25b88f.)

the consequences of the hazard by taking protective actions of their own, then the alert should explain what is being done by the authorities to address the situation (Strander, 2011). An example of these protective action instructions can be found in the guidelines issued by the U.S. National Weather Service, as shown in Figure 5.7.

5.2.3.3   Repetition

It may be helpful to end the warning message with repeated information about to whom it is addressed, and a summary of what actions should be taken and by when.

5.2.3.4   Sources of Information

In any alert message, it is important to state the source of the message, since this may improve its credibility to the population (Reynolds, 2009; Strander, 2011). As shown in the cases presented in Cases 5.7 and 5.8, the effectiveness of the warning will improve

Case 5.7: During Hurricane Michelle in November 2001, 700,000 Cuban people were evacuated. Although it was a Category 4 hurricane on the Saffir-Simpson Scale, and wind speeds reached up to 220 km/h, the human losses (5 people) were low compared with the 36 people who died a few days earlier in Central America when Michelle was just a tropical depression (Fitzpatrick and Mileti, 1991). There were many reasons for the great effectiveness of the warning given (Thompson and Gaviria, 2004), such as the adequate planning, and effective implementation of the contingency plan in evacuating 712,000 people. However, this alone would not have been sufficient if the citizens had not trusted the Cuban authorities (Direction de la Défense et de la Sécurité Civiles, 2005).

Case 5.8: During the cyclone that struck Bangladesh in April 1991, at least 45% of the population did not look for shelter before the storm, even though they had received warning messages (Bern et al., 1993), although Haque (1995) points to other numbers: 49% in urban/mainland and 71% of the rural/offshore island. The main reported reason for this was that they did not believe that an event of the predicted characteristics was going to happen (Bern et al., 1993). In other words, they did not trust the source of the warnings. This situation was repeated in the subsequent Sidr (2007) and Mahasen (2013) cyclones. After they happened, a study was carried on two coastal zones (Bagerhat and Patuakhali) and a survey showed that the proportion of evacuated population was 28% and 43%, respectively. Again, among the key reasons for non-evacuation appeared to be mistrust of the warning messages (Roy and Kovordany, 2015).

if the message is enforced by different sources (such as experts, organizations and other authorities) since no single source is credible to everyone (Veil et al., 2011).

Further information sources, where the recipients may find additional information about the incident, should also be mentioned in the alert message where available (Reynolds, 2009).

5.2.4   ALERTING COMMUNICATION CHANNELS USAGE

Another important factor that influences the effectiveness of an alert is the use and choice of the alerting channels for the dissemination of the warning. The choice of channels used to warn the population during an emergency should be based on the type of message (i.e. whether it is just to let people know that something is wrong, or to transmit specific information about the event and instructions for protective actions to be taken), the hazard event characteristics (i.e. if it is during day or night, long or short time event, geographically localized or a large-scale event, predictability, etc.), its onset speed, the desired reach and frequency of the message, as well as the specific characteristics of the different population segments to be warned (Kortom, 2011; Mulero-Chaves et al., 2011; Strander, 2011). The high number of possible combinations among these factors is translated into an impediment to completely automate the warning process.

No single alerting channel can ensure reception of the warning by all intended recipients in an infallible way (Ding, 2009). Therefore, when issuing a warning, it is recommended to use a combination of several alerting channels, to maximize the warning audience coverage in the minimum time possible (Kortom, 2011; Strander, 2011), and to ensure that the affected population receive the alert message through at least one of these channels (Perry and Lindell, 2007).

The most effective alerting channels are those that simultaneously include a wake up and an information transmission component. This type of alerting channel catches peoples' attention, and also provides them hazard-related information at the same time (CEMA, 2008). Nevertheless, since each an event scenario is likely to be quite different, as a rule alerting channels that are capable of targeting members of the population individually with a warning (e.g. cell broadcast, SMS) are especially suitable for rapid-onset or short-term hazard events, due to the rapid dissemination of the message. Meanwhile, alerting channels with the capability of issuing a mass warning to the population at large (e.g. TV and radio) are especially suitable for hazard events where a longer alert and preparation time is forecast, or during longer duration events, due to the common delay in delivering the alert message to the affected population. Table 5.1 summarizes the suitability of a selection of possible alerting channels, and offers recommendations for their appropriate use, based on their characteristics and limitations.

In most developed countries, the mass media is still currently the main tool and best ally of the civil authorities, when disseminating warnings to the citizens at the time a hazard event occurs. At the same time, social media networks offer a new communication channel that is increasingly gaining adoption among first responders, as a means of disseminating warnings. Each of these is discussed in greater detail below.

5.2.4.1   Mass Media (Television and Radio)

In many countries, the mass media is considered to be the main alerting channel for hazard events. It provides an efficient way of transmitting simple information quickly and to a large sector of the public (Perry and Lindell, 2007). Television has the particular advantage of allowing the transmission of hazard event information through use of diagrams and graphics, which enhances greatly the understanding of the alert message by the public. Also, the resilience of satellite TV during and after the occurrence of crisis situations turns it into a reliable communication channel to transmit alert messages to the citizens in such situations.

However, it should be borne in mind that it may not reach the entire population, due to dependency on the TV channel's programming schedule, and the fact that there may be a large percentage of people who, at the time the alert message is broadcast, may not be watching TV at all, or may not be watching the channel where the message is being transmitted. Nevertheless, transmission of the warnings through 24/7 news channels reduces the time required for dissemination of the message due to speed of delivery and frequent repetition of last-minute (breaking) news. Also, the possibility of issuing the message through the insertion of a banner during a TV broadcast, or even by interrupting the broadcast, can also accelerate dissemination of the alert.

Donnelley (2008) and Perry and Lindell (2007) draw attention to the fact that broadcast media will be a more effective alerting channel during the day than during the night. For instance, when dealing with a short-term hazard event during the night, the use of TV and radio as alerting channels are not effective because they do not have the ability to wake up the population at risk to get their attention. In these cases, direct alerting channels such as telephone/cell phone calls, cell broadcasting/SMS messaging, public address systems, sirens or even door-to-door warnings are considered to be more effective, due to their higher penetration effect (Perry and Lindell, 2007).

When broadcasting alert messages through the television or radio, it should be considered that these are very imprecise alerting channels. The warning may reach

TABLE 5.1

Alerting Channels Usage Recommendations and Suitability of Use

Communication Channel	Description	Suitability	Caution and Tips for Use
Auto dialer	System that calls a number of predefined listed phone numbers with the transmission of a pre-recorded message to mobile phones and/or fixed phones.	Can be used as mass alerting channel but due to high delivery delay it is recommended more to use it as targeted alerting channel. Mostly suitable for internal communication and to warn other organizations (Mileti and Sorensen, 1990) (e.g. schools, prisons, hospitals, etc.). Very suitable to be used during sleeping hours due to the 'wake up' capability of this channel (Perry and Lindell, 2007). May potentially reach visually impaired people.	If used as mass alerting channel, it should be used in rather small areas and in the cases that the hazard events have medium or high forecast due to the high delivery delay of the alert messages (an average of 1500–15,000 messages in one hour in best cases). When using this alerting channel to send alert messages exclusively to fixed landline phones it should be taken into account that not everyone owns such a device anymore (Párraga et al., 2011). Thus, in this case it is recommended to use it as a supporting means of other alerting channels. It should also be considered that recipients are not always near the phone and that busy phones will inhibit the delivery of the alert message (Mileti and Sorensen, 1990). In this case registration is needed, telephone contact lists should be kept up to date. This should be done with the acknowledgment and acceptance of the contacts (Mulero-Chaves et al., 2011).
Mobile phones SMS sending	Conveys text messages through SMS sending.	Mostly recommended to be used as a targeted alerting channel.	Only individuals who have subscribed to this service may be targeted with an alert message. When using this alerting channel it should be taken into account that people will only receive the alert message if the device is on (Párraga et al., 2011). Recipients may not realize about the received alert message (e.g. when they are not close to it) (Wiersma et al., 2008). This is especially applicable to the hard of hearing or deaf community.

(Continued)

TABLE 5.1 (CONTINUED)

Alerting Channels Usage Recommendations and Suitability of Use

Communication Channel	Description	Suitability	Caution and Tips for Use
Mobile phones Cell broadcast	Alert message broadcasting by targeting all mobile phones on specific cell sites with a text message.	It has the capability of targeting the message to specific geographical locations; especially suitable for its use in imminent or short-term hazard events due to rapid dissemination. Potential channel to reach deaf or hard of hearing people.	When using this alerting channel it should be taken into account that people will only receive the alert message if the device is on (Párraga et al., 2011). Recipients may not realize about the received alert message (e.g. if they are not close to it) (Wiersma et al., 2008). This is especially applicable to the hard of hearing or deaf community.
Fax messaging	Usage of conventional fax system to issue an alert message.	Mostly recommended as targeted alerting channel. Usually used for internal or inter-organizational communication, or to warn vulnerable organisations (e.g. schools, prisons, hospitals, nursing homes, etc.). May potentially reach subscribed deaf or hard of hearing citizens.	Rules for good writing should be followed. A cover sheet should be included to ensure reception. The number of pages should be included on the cover sheet. Confirmation of reception should be requested. It should be taken into account that the speed of transmission does not negate the need for good writing and good manners (FEMA, 2005).
Internet subscription service (pop-up message or e-mail)	'Push' system that will send out a pop-up message or e-mail to the subscribed persons at risk.	Most suitable to reach younger generations. Most suitable to be used during working hours. May potentially reach deaf or hard of hearing people.	It should be taken into consideration that there is the possibility that the content of the sent e-mails may be tampered by the recipients in case these change and forward the message to others. It should be considered that the reception of the alert message depends on when the recipients access the Internet or check their e-mail. Also, the e-mail may be forwarded by the recipients to other audiences that are not intended to be warned (Mulero-Chaves et al., 2011).

(Continued)

TABLE 5.1 (CONTINUED)

TABLE 5.1 (CONTINUED)
Alerting Channels Usage Recommendations and Suitability of Use

Communication Channel	Description	Suitability	Caution and Tips for Use
Internet web portals Online news	Internet websites, web portals or online news web displaying alert messages and hazard event-related information.	Mass alerting channel (reaches anyone who reads it). Most suitable in reaching urban areas and mainly the younger generations. Most effective when used during the day. May potentially reach hard of hearing or deaf communities (Stout et al., 2004).	The Internet should be considered as a supporting alerting channel, and not as the primary source of emergency warning for life-threatening events (Web City of Hobbs, 2009). As during an impending disaster usually there is no time to enter warning information into websites (Perry and Lindell, 2007; Párraga et al., 2011), this communication channel should be considered to be used for high time forecasted or long-term hazard events. When inserting warning videos into the Internet sites, these should be captioned to make it accessible to the hard of hearing and deaf community.
Pagers and similar devices	Hand-held text message receiving units.	Most suitable to be used as targeted alerting channel or to warn subscribed users. Usually used for internal or inter-organizational communication. May potentially reach subscribed people including deaf or hard of hearing individuals.	Two-way messaging pagers can receive full text messages up to 20,000 characters (American Messaging Service, 2006). Text pagers can receive full text messages usually up to 100 characters (some pager models and service areas may support traditional text paging up to 250 characters) (American Messaging Service, 2006). Numeric pagers can only receive the numbers '0' through '9' (American Messaging Service, 2006).

(Continued)

TABLE 5.1 (CONTINUED)

Alerting Channels Usage Recommendations and Suitability of Use

Communication Channel	Description	Suitability	Caution and Tips for Use
Public address systems	Mobile and fixed loudhailers transmitting voice messages to specific areas as necessary.	Mainly recommended to be used as targeted alerting channel (FEMA, 2005). Most effective when used in small areas. Should be used as supporting alerting channel as it has the capability to override misinformation transmitted through other channels (e.g. TV, radio, social media, SMS broadcast, etc.) (Mileti and Sorensen, 1990; FEMA, 2005). Suitable alerting channel for its use during the night when people are asleep due to high penetration level (Mileti and Sorensen, 1990) and when other forms of communication are not available or working (FEMA, 2005).	Effective during the night due to its high penetration level (Mileti and Sorensen, 1990; FEMA, 2005), although in some areas it can be difficult to hear it because of air-conditioner noise or the existence of insulated windows. It is important to enunciate the message clearly. Alert messages should be kept simple but accurate. Key information should be repeated often enough to account for changes in audience and considering that the message may be not understood clearly without repetition (Mileti and Sorensen, 1990).
Electronic billboards	Electronic displays in public spaces, including gantry signs on motorways and railways.	Mostly suitable to be used as targeted alerting channel. To be used as auxiliary and preventive information dissemination rather than to disseminate alert messages. Useful for the management of traffic flows during a hazard event.	It should be considered that through the use of this communication mean it is difficult to convey different messages to different segments of the population. It is difficult to assess the demographics or special communication needs of a mobile group (FEMA, 2005). These types of public signs address a very limited portion of the population (FEMA, 2005).

(Continued)

TABLE 5.1 (CONTINUED)

Alerting Channels Usage Recommendations and Suitability of Use

Communication Channel	Description	Suitability	Caution and Tips for Use
Knock on doors approach	The most basic and personal warning method—Civil protection officials go door to door warning the population.	Suitable to be used: • As targeted warning and when no other means are available, as mass notification channel • In sparsely populated areas • Whenever enough personnel is available to perform this warning method • In areas with a large seasonal or daytime active population • In areas with scarce or without any alternative alerting channels (e.g. rural areas)	A major advantage of this warning method is that people are more willing to respond to an alert message delivered personally because they are more likely to believe that the risk exists (Mileti and Sorensen, 1990). When using this warning method it is recommended to drive emergency vehicles, wear organisational uniforms and protective gear in order to increase seriousness and belief to the affected population (Perry and Lindell, 2007).
Sirens/klaxons	Loud device disseminating a single or multitone sound.	Depending on the demographics, number and dispersion rate, in some cases may be used as mass alerting channel and in others as targeted alerting channel. Suitable to be used as an alerting channel by requesting the public to seek further information related to the emergency in other alerting channels. Ideally, it should be used as an alerting complement of other alerting channels (e.g. TV, radio, social media, SMS broadcast etc.).	Possible usage in all types of risks and especially those with rapid onset. Effective during the night due to its high penetration level. A siren has 62% probability of wakening someone up when sounding 10 decibels during 3 minutes during sleeping hours. When using this alerting channel it should be considered that usually people do not pay much attention to it (Sorensen, 1993). Usually, the general public does not understand the meaning of different sound signals unless they are provided with specialized training (Sorensen, 1993). This alerting channel may not reach the deaf or hard of hearing community (FEMA, 2005).

(Continued)

TABLE 5.1 (CONTINUED)

Alerting Channels Usage Recommendations and Suitability of Use

Communication Channel	Description	Suitability	Caution and Tips for Use
Aircraft with banner or helicopters dropping leaflets	Use of planes with banners or helicopters dropping leaflets to warn the population.	Mass alerting channel May be useful for remote and rural areas where electronic alerting and warning devices are very scarce or inexistent.	May be an expensive method to warn the population. Risk of incident may exist in difficult conditions (Mileti and Sorensen, 1990). This alerting method presents a major disadvantage with regard to the common large delay in disseminating the information to the population.
Radio	Broadcasting of alert messages through the news editions or special broadcasts.	Suitable to be used as a mass alerting channel One of the most resilient mass alerting channels. Even when power and phone lines are down, alert messages may be received via battery, windup portable radio (FEMA, 2015), car radios or even mobile phone or Internet radios. Has the potential to reach blind individuals.	Efficient during the day, weak during sleeping hours. The use of amateur radio operators during hazard events is especially useful to access remote and rural areas. When issuing an alert message through this communication channel: • The use of jargon and codes should be avoided. • It is important to be brief, communicating essential information only. • Protocols for identification, communication and signoff should be followed. • It should be taken into account that other people may overhear the alert message in addition to the intended audience.

(Continued)

TABLE 5.1 (CONTINUED)

Alerting Channels Usage Recommendations and Suitability of Use

Communication Channel	Description	Suitability	Caution and Tips for Use
Television (terrestrial or satellite TV)	Broadcasting of alert messages through news programs, special broadcasts or specific news channels.	Works as a mass alerting channel. More suitable to be used during high-forecasted or long-term events than imminent or short-term events due to the long time it usually takes to issue warnings.	Efficient during the day, weak during the sleeping hours (Perry and Lindell, 2007). When pre-prepared messages exist, the delivery time of the alert message through this mean may be diminished. When broadcasting hazard event related news, these should be captioned to allow accessibility to the hard of hearing and deaf community (Stout et al., 2004).
Television banners (terrestrial or satellite TV)	Dissemination of alert messages through the displaying of a text banner during TV broadcast.	Works as a mass alerting channel.	Efficient during the day, weak during the sleeping hours (Perry and Lindell, 2007). When pre-prepared messages exist, the delivery time of the alert message through this mean may be diminished.
Paper media	Published documents containing information related to the hazard event.	Mass alerting channel. Suitable as supporting alerting channel for high time forecasted or long-term hazard events. Not suitable for short-term hazard events due to its low and slow dissemination levels.	Useful mean to be used for public risk education and awareness.
Authority contact number	To provide to citizens special emergency call numbers to obtain further information about a hazard event.	Mass alerting channel.	Requires prior divulgation through other alerting channels.

everyone in the broadcast area, which may include population segments that are not at risk, and therefore not intended to be addressed by the warning. This fact may be mitigated if the broadcasters provide specific information regarding who is and is not at risk (Perry and Lindell, 2007). Also, to emphasize the seriousness of the message to its recipients, when making use of these channels it is recommended to compare the existing hazard event with similar past events, mention vulnerability assessment of the affected jurisdictions and enforce the message by citing external hazard experts (Perry and Lindell, 2007). There may be the possibility that broadcasters cut part of the official message, indeed sometimes the most important part of the message is missing (information obtained from field practitioners' interviews during the research). This happens because journalists may not always correctly understand the message to be transmitted. To avoid this situation, during normal situations drills should be carried out or, at least, agreed templates should be developed to differentiate and highlight the essential fields of information that must be communicated to the population.

5.2.4.2   Social Media Networks

Social media networks offer an alternative group of alerting channels. Their appearance has changed communication characteristics before, during and after an event (Nagarajan et al., 2010; Mendes et al., 2012), to such an extent that, if an emergency event is not mentioned on social media networks, people may ignore it (Birgfeld, 2010).

Accordingly, there is no doubt about the need to use this type of channel, although it is important to keep in mind that inappropriate use of it may lead to crisis response impediments such as rumours, or access to fragmented or unreliable information, and hence, citizens making decisions that may not protect them or that put them in greater danger. Due to the specific characteristics of these social media, besides the above-mentioned best practices, other aspects should be considered to improve their results as part of the warning process. Because of the comparative novelty of these media, it is difficult to as yet determine fully their suitability for alert dissemination, or the relevance of the best practices compiled in the previous sections for their use. Nevertheless, experiences reported thus far suggest that social media show a high potential (Case 5.9). Therefore, their introduction into crisis communication plans is highly recommended, although guidelines for their use will have to be frequently updated due to the arrival of new platforms which outdate the current theory and information (Strander, 2011). This means that crisis communication plans should be flexible enough to adapt to future social media changes (Wang et al., 2003).

Case 5.9: During the Japanese earthquake of March 2011, due to the nature of the hazard it was impossible to send any warning message before the event took place. Nevertheless, the Facebook social network constituted an important post-disaster communication channel, by allowing the affected population to post information about their situation to inform their concerned friends and relatives (FEMA, 2005).

5.3 CONCLUSION

Human populations are growing rapidly, and are occupying new places where natural hazards can hit strongly. This situation is expected to increase further into the foreseeable future, due to improvements in living conditions as well as the anticipated consequences of climate changes. To protect the populations that are more vulnerable to hazards, effective alerting systems should be implemented. Over the last few decades, experience and research into the use of such systems has enabled recognition of the best-practice approaches to be adopted when seeking to warn affected populations. These best practices include choice of warning variables, alert messages (style and content) and alerting communication channels usage, but the number of variables involved and their many potential combinations make it currently impossible to completely automate the warning process. Nevertheless, the identification and application of these best practices should speed up and improve the effectiveness of the warning process, and thereby help achieve a comprehensive risk reduction for the population likely to be affected.

It must also be taken into account that there will always be sudden-onset natural hazards that arrive without warning, or that cannot be detected with enough time to alert the population or evacuate them. In these cases, the only way to reduce risk is to have an alert system that considers the importance of educating the general public regarding existing risks and how to effectively comply with warnings in case a hazard event takes place.

ACKNOWLEDGEMENTS

This research work has received funding from the European Union Seventh Framework Program (FP7/2007-2013) under Grant Agreement Number [261732].

REFERENCES

Altamura, M., L. Ferraris, D. Miozzo, L. Musso, and F. Siccardi. 2011. The legal status of uncertainty. *Natural Hazards and Earth System Sciences*, 11, 797–806. http://www.nat-hazards-earth-syst-sci.net/11/797/2011/nhess-11-797-2011.pdf.

Altamura, M., D. Miozzo, and L. Ferraris. 2014. The role and responsibility of scientists in the Italian Civil Protection. Presentation CIMA Research Foundation, Berlin, February 23, 2014. http://www.cimafoundation.org/wp-content/uploads/News/OECD_presentation.pdf (accessed May, 9, 2017).

Amato, A., M. Cocco, G. Cultrera, F. Galadini, L. Margheriti, C. Nostro, and D. Pantosti. 2013a. The impact of the L'Aquila trial on the scientific community. Seismological Society of America, annual meeting, Salt Lake City, UT. http://processoaquila.files.wordpress.com/2013/04/amato-ssa2013.pdf (accessed on April 28, 2017).

Amato, A., M. Cocco, G. Cultrera, F. Galadini, L. Margheriti, C. Nostro, and D. Pantosti. 2013b. The l'Aquila trial. *Geophysical Research Abstracts*, vol. 15, EGU2013-12140. EGU General Assembly 2013. http://meetingorganizer.copernicus.org/EGU2013/EGU2013-12140.pdf (accessed May, 9, 2017).

Baglivo, G. 2016. Vibo e l'alluvione del 2006: A 10 anni dal disastro si attende ancora giustizia. Zoom 24 dentro la Calabria. Available at http://www.zoom24.it/2016/07/03/processo-alluvione-vibo-10-anni-disastro-attende-giustizia-pm-callea-23913/ (accessed on April, 28, 2017).

Benessia, A. and B. de Marchi. 2017. When the earth shakes… and science with it. The management and communication of uncertainty in the L'Aquila earthquake. *Futures.* http://www.sciencedirect.com/science/article/pii/S0016328717300460 (accessed April, 28, 2017).

Bern, C., J. Sniezek, G. M. Mathbor et al. 1993. Risk factors for mortality in the Bangladesh cyclone of 1991. *Bulletin of the World Health Organization*, 71(1), 73–78.

Birgfeld, R. 2010. Why crisis management and social media must co-exist. Smartblog on Social media, [blog] October 27, 2010. http://smartblogs.com/social-media/2010/10/27/why-social-media-crisis-management-must-co-exist/ (accessed on May 9, 2017).

Caloiero, T., A. A. Pasqua, and O. Petrucci. 2014. Damaging hydrogeological events: A procedure for the assessment of severity levels and an application to Calabria (Southern Italy). *Water*, 6, 3652–3670.

Carracedo, J. C. 2007. La erupción del Nevado del Ruiz y el lahar catastrófico del 13 de noviembre de 1985. In *Riesgos naturales,* F.J. Ayala-Carcedo and J. Olcina, Eds., 295–306, Ariel, Colección Ciencia, Barcelona.

Carracedo, J. C. 2015. Peligros asociados a megadeslizamientos y lahares. *Enseñanza de las Ciencias de la Tierra*, 23.1, 66–72.

Carsell, K. M., N. D. Pingel, and D. T. Ford. 2004. Quantifying the benefit of a flood warning system. *Natural Hazards Review*, 5, 131–170.

CEMA (California Emergency Management Agency). 2008. Alert and warning report to the California State Legislature. http://www.csus.edu/ccp/documents/publications/alert_warning_report_final_calema.pdf (accessed on May 9, 2017).

Cocco, M., G. Cultrera, A. Amat et al. 2015. The L'Aquila trial. In *Geoethics: The Role and Responsibility of Geoscientists*, S. Peppoloni and G. di Capua, Eds., 43–56. Geological Society, London, Special Publications.

Coombs, W. T. 2010. Parameters for crisis communication. In *The Handbook of Crisis Communication*, W. T. Coombs and S. J. Holladay, Eds., 17–53, Wiley-Blackwell, Oxford, UK.

Coombs, W. T. 2011. *Ongoing Crisis Communication: Planning, Managing and Responding,* 3rd ed. SAGE Publications, Thousand Oaks, CA.

Day, H. J. 1970. *Flood Warning Benefit Evaluation – Susquehanna River Basin – Urban Residences.* ESSA Technical Memorandum WBTM Hydro-10, National Weather Service, Silver Spring, MD.

De la Pomerai, G. and K. Omer. 2012. Seismic prediction and real time early warning make a perfect combination. In *Integrative Risk Management in a Changing World – Pathways to a Resilient Society,* Global Risk Forum (GRF), 34–36, Davos Poster Collection, IDRC Davos, Ref. 571.

Ding, A. W. (2009) *Social Computing in Homeland Security: Disaster Promulgation and Response.* Hershey, PA: IGI Global. doi:10.4018/978-1-60566-228-2.

Direction de la Défense et de la Sécurité civiles (2005). *Plan Communal de Sauvegarde. Guide Pratique d'élaboration.* Ministère de l'Intérieur et de l'Aménagement du Territoire (República Française). www.interieur.gouv.fr/content/download/73159/535169/file/guide%20PCS.pdf (accessed on May 9, 2017).

Donnelley, R. R. 2008. Warning and informing Scotland, communicating with the public. Safer Scotland, Scottish Government. http://www.scotland.gov.uk/Resource/0038/00388646.pdf (accessed on May 9, 2017).

Egli, T. 2002. Non-structural flood plain management – measures and their effectiveness, International Commission for the Protection of the Rhine (ICPR), Koblenz. http://www.iksr.org/index.php?id=266&L=1 (accessed on May 9, 2017).

EUMETSAT. 2014. The case for EPS/METOP Second-Generation: Cost Benefit Analysis. Full Report. EUMETSAT, Darmstadt. http://www.eumetsat.int/website/home/Satellites/FutureSatellites/EUMETSATPolarSystemSecondGeneration/index.html (accessed on May 9, 2017).

FEMA (Federal Emergency Management Agency). 2005. Effective communication, independent study. FEMA Manual. http://training.fema.gov/emiweb/downloads/is242.pdf (accessed on May 9, 2017).

FEMA (Federal Emergency Management Agency). 2015. An emergency alert system best practices guide, version 1.0. https://www.fema.gov/media-library-data/20130726-1839-25045 -9302/eas_best_practices_guide.pdf (accessed on December 1, 2017).

Fitzpatrick, C. and D. S. Mileti. 1991. Motivating public evacuation. *International Journal of Mass Emergencies and Disasters,* August 1991. http://ijmed.org/articles/274/download/ (accessed on May 9, 2017).

Hallegate, S. 2012. A cost-effective solution to reduce disaster losses in developing countries. Hydro-meteorological services, early warning and evacuation. World Bank, Policy Research Working Paper, n. 6058. http://elibrary.worldbank.org/content/workingpaper /10.1596/1813-9450-6058 (accessed on May 9, 2017).

Haque, C. E. 1995. Climatic hazards warning process in Bangladesh: Experience of, and lessons from, the 1991 April cyclone. *Environmental Management*, 19(5), 719–734.

Ietto, F., F. Talarico, and S. Francolino. 2009. Cause dell'alluvione di Vibo Valentia del 2006 e caratteristiche dell'ambiente costiero dopo l'evento. *Biologia Ambientale*, 23(1), 3–12.

Kortom (2011). Use of social media in crisis communication. Kortom (Belgium), http://www .kortom.be/file_uploads/5352.pdf (accessed on May 9, 2017).

Kreibich, H., M. Müller, A. H. Thieken and B. Merz. 2007. Flood precaution of companies and their ability to cope with the flood in August 2002 in Saxony, Germany, *Water Resources Research*, 43, W03408.

Kreibich, H. and A. H. Thieken. 2009. Coping with floods in the city of Dresden, Germany. *Natural Hazards*, 51(3), 423–436.

Mendes, M., M. Ferrer-Julià, J. Ramírez et al. 2012. Best practices manual to alert population. Alert4All Internal Report D.3.2.

Mileti, D. S., P. A. Bolton, G. Fernandez, and R. G. Updike. 1991. *The Eruption of Nevado Del Ruiz Volcano Colombia, South America November 13, 1985.* National Academy Press, Washington, DC, Natural Disaster Studies, Vol, Four. http://www.nap.edu/catalog /1784.html (accessed on May 9, 2017).

Mileti, D. S. and J. H. Sorensen.1990. Communication of emergency public warning, a social science perspective and state-of-the-art assessment. Oak Ridge National Laboratory, Report ORNL-6609, August 1990.

Mulero-Chaves, J., C. Párraga, I. Sivarajah, C. Garcia-Monteiro, and E. Barbosa. 2011. Analysis on information and communication technologies: State of the art report. Alert4All Internal Report D.2.4.

Nagarajan, M., D. Shaw, and P. Albores. 2010. Informal dissemination scenarios and the effectiveness of evacuation warning dissemination of households – A simulation study. *Procedia Engineering*, 3, 139–152.

Partnership for Public Warning. 2004. Protecting America's communities. An introduction to public alert and warning. Report 2004-2 (June). http://tap.gallaudet.edu/emergency /nov05conference/EmergencyReports/handbook.pdf (accessed on May 9, 2017).

Párraga, C., Hirst, P., Mendes, M., Sigmund, K., Johannes, S., Maribel, N., Matthias, M., Joel, B. 2011. Communication plans and impact workshop. Alert4All Internal Report, D.2.5.

Perry, R. W. and M. K. Lindell. 2007. *Emergency Planning.* John Wiley & Sons, Hoboken, NJ.

Reynolds, B. 2009. Crisis and emergency risk communications: Best practices. CDC (Centers for Disease Control and Prevention), Atlanta. http://www2c.cdc.gov/podcasts/player .asp?f=11509 (Podcast accessed on May 9, 2017).

Reynolds, B. 2012. Crisis and emergency risk communications – By leaders for leaders. CDC (Centers for Disease Control and Prevention), Atlanta. https://emergency.cdc.gov/cerc /resources/pdf/leaders.pdf (accessed on May 9, 2017).

Risk Management Solutions. 2003. Central Europe flooding, August 2002. Event Report. Risk Management Solutions. http://forms2.rms.com/rs/729-DJX-565/images/fl_2002 _central_europe_flooding.pdf (accessed on May 9, 2017).

Rogers, D. and V. Tsirkunov. 2010. Costs and benefits of early warning systems. ISDR-World Bank. Global Assessment Report on Disaster Risk Reduction, 69358.

Roy, C. and R. Kovordanyi. 2015. The current cyclone early warning system in Bangladesh: Providers' and receivers' views. *International Journal of Disaster Risk Reduction*, (12), 285–299. http://dx.doi.org/10.1016/j.ijdrr.2015.02.004.

Sorensen, J. H. 1993. Warning systems and public warning response, Workshop Socioeconomic Aspects of Disaster in Latin America, Costa Rica, January, 21–23, 1993. Available at: http:// desastres.usac.edu.gt/documentos/pdf/eng/doc6361/doc6361-contenido.pdf (accessed on November 2012).

Stout, C., Brick, K. and Heppner, C. A. 2004. Emergency preparedness and emergency communication access: Lessons learned since 9/11 and recommendations. Report, Deaf and Hard of Hearing Consumer Advocacy Network (DHHCAN). Available at: http://www.prevention web.net/files/9232_DHHCANEmergencyReport.pdf (accessed on September 2013).

Strander, I. 2011. Effective use of social media in crisis communication: Recommendations for Norwegian organization. University of Leeds. http://www.kommunikationsforum .dk/log/multimedia/PDF%20og%20andre%20dokumenter/Specialer/Effective_Use _of_Social%20Media_in_Crisis_Communication.pdf (accessed on May 9, 2017).

Thieken, A., H. Kreibich, M. Müller, and B. Merz. 2007. Coping with floods: Preparedness, response and recovery of flood-affected residents in Germany in 2002, *Hydrological Sciences Journal*, 52(5), 1016–1037.

Thielen-del Pozo, J., V. Thiemig, F. Pappenberger et al. 2015. The benefit of continental flood early warning systems to reduce the impact of flood disasters. EUR 27533 EN, http://publications.jrc.ec.europa.eu/repository/bitstream/JRC97266/lbna27533enn.pdf (accessed on May 9, 2017).

Thompson, M. and I. Gaviria. 2004. Cuba. Weathering the storm: Lessons in risk reduction from Cuba. Oxfam America Report. http://www.oxfamamerica.org/files/OA-Cuba _Weathering_the_Storm-2004.pdf (accessed on May 9, 2017).

Tribunale di L'Aquila and Sezione penale. 2012. Motivazione Sentenza n. 380 del 22/10/2012, Depositata il 19/01/2013. See http://www.magistraturademocratica.it/mdem/qg/doc /Tribunale_di_LAquila_sentenza_condanna_Grandi_Rischi_terremoto.pdf (accessed January 5, 2016).

Ulbrich, U., T. Brücher, A. H. Fink, G. C. Leckebusch, A. Drüger, and J. G. Pinto. 2003. The central European floods of August 2002: Part 1: Rainfall periods and flood development. *Weather*, 58, 371–377, http://onlinelibrary.wiley.com/doi/10.1256/wea.61.03A /pdf (accessed on April 28, 2017).

UNEP. 2012. Early warning systems. A state of the art analysis and future directions. UNEP, http://na.unep.net/siouxfalls/publications/Early_Warning.pdf (accessed on May 9, 2017).

U.S. Army Corps of Engineers (USACE). 2000. Generic depth damage relationships. Economic guidance memorandum (EGM) 01-03. USACE, Office of Chief of Planning and Policy, Washington, DC.

Veil, S. R., T. Buehner, and M. J. Palenchar. 2011. A work-in-process literature review: Incorporating social media in risk and crisis communication. *Journal of Contingencies and Crisis Management*, 19(2), 110–122.

Wang, Z. Y., Y.-C. Zheng, and J. X. Li. 2003. Early warning systems for the reduction of natural disasters in China. In *Early Warning Systems for Natural Disaster Reduction*, J. Zschau and A. N. Küppers, Eds., Chapter 1.2, 15–17, Springer-Verlag, Berlin, Heidelberg.

Web American Messaging Service. 2006. Available at: http://www.americanmessaging.net
/he1b_emailfaq.asp#5 (Accessed on September 2013).

Web City of Hobbs. 2009. Alerting and warning system. Available at: http://www.hobbsnm
.org/em_warning.html (Accessed on September 2013).

Wiersma, J. W. F., Jagtman H. M. and Ale B. J. M. 2008. Report on the use of cell broadcast as
a citizen alert system, lessons from a two-year study in the Netherlands (2005–2007),
commissioned by the Expertise Centre for Risk and Crisis Communication, Ministry of
the Interior and Kingdom Relations. Available at: http://www.ceasa-int.eu/wp-content
/uploads/2010/04/Cell-Broadcast-Trials-University-Delft.pdf (accessed on September
2013).

World Bank. 2010. Natural hazards, unnatural disasters: Effective prevention through an economic lens. World Bank Report, 231 pp.

6 Tropical Cyclone Activities

Asia-Pacific Region

Lindsey M. Harriman

CONTENTS

6.1 INTRODUCTION

The Asia-Pacific region is known to experience many tropical cyclones during the summer and autumn months with two primary basins of tropical cyclone origin and activity: the western North Pacific Ocean (WNP) and the North Indian Ocean (NIO). Subregions of these basins in which there is also frequent tropical cyclone activity include the South China Sea and the Bay of Bengal. The WNP basin experiences more tropical cyclones, or typhoons as they are known in this region, per year than any other region worldwide, affecting island nations such as the Philippines and vulnerable coastlines of Vietnam. Tropical cyclones in the NIO affect countries such as India when a tropical cyclone forms over the Bay of Bengal. This chapter explains the characteristics of tropical cyclones, the impacts these systems can have on humans and the environment and provides insight on the recent trends of activity in the Asia-Pacific region with primary focus on the WNP and NIO basins. Two of the most intense storms in history, Very Severe Cyclonic Storm Phailin and Super Typhoon Haiyan, both of which struck land in 2013, are profiled.

6.2   CYCLONES AND TYPHOONS

6.2.1   What Is a Cyclone or Typhoon?

The terms cyclones and typhoons, and even the term hurricanes, all refer to the same general natural phenomenon: a tropical cyclone. A tropical cyclone is an organized, rotating storm system that forms over tropical or sub-tropical waters (National Hurricane Center, n.d.(a)). Tropical cyclones are warm-core, non-frontal, low-pressure systems that pull their energy from the ocean surface, as opposed to being a cold-core system that forms along fronts in higher latitudes (Hobgood, 2005). The vernacular distinction between a cyclone, typhoon and hurricane lies in regional application and severity of the storm, the latter being based upon sustained wind speed. Over the Atlantic, these storms are referred to as hurricanes, in the WNP, they are typhoons, and in the NIO they are referred to as cyclones (National Hurricane Center, n.d.(a)). In its weakest form, a tropical cyclone is termed a 'tropical depression', and is distinguished by sustained winds of less than 60 kilometres per hour (km/h) (34 knots). When winds surpass 60 km/h, the disturbance is then referred to as a tropical storm in the WNP and a cyclonic storm in the NIO. Continued increases in wind speed define the severity of the storm and regional vernacular influences what the storm is called (Table 6.1).

Cyclogenesis is the term used to describe the process of cyclone formation from an existing disturbance. The existing disturbance needs to be over warm ocean surface temperatures ($\geq 26°C$) with enough moisture above the surface, and unstable atmospheric conditions so that the storm can intensify, ultimately becoming a tropical cyclone (Hobgood, 2005). A tropical cyclone is composed of three main parts: the eye, the eyewall and rain bands (National Weather Service, n.d.(a)). A well-formed eye is distinct in the centre of the storm (Figure 6.1) and is usually an area of weak horizontal temperature and pressure gradients, which strongly contrast with the high-speed winds and thick clouds that surround it (Jordan, 1961). A well-defined eye is indicative of a very strong storm – a Category 4 or 5 hurricane or a super typhoon. Another indicator of the storm's intensity is when there are changes to the structure of the eye and eyewall, which can lead to changes in wind speed (National Weather Service, n.d.(a)). The eyewall is made up of a ring of tall

---

TABLE 6.1

Tropical Cyclone Classifications

	Maximum Sustained Winds	
Basin	119–220 km/h (64–119 knots)	≥220 km/h (≥119 knots)
WNP	Typhoon	Super typhoon[a]
NIO	Very severe cyclonic storm	Super cyclonic storm

Source: National Weather Service (n.d.(b)). Tropical Cyclone Classification. http://www.srh.noaa.gov /jetstream/tropics/tc_classification.htm (accessed 28 March, 2017).

[a]   When winds reach ≥ 241 km/h.

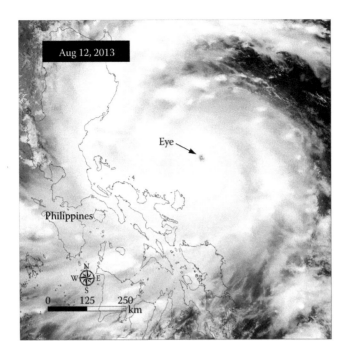

FIGURE 6.1   In this satellite image of Super Typhoon Utor captured by Terra Moderate Resolution Imaging Spectroradiometer (MODIS) on August 12, 2013, the eye of the storm is clearly visible. (From NASA Earth Observatory, http://earthobservatory.nasa.gov/Natural Hazards/view.php?id=81837, accessed January 31, 2014.)

thunderstorms and usually this is where the strongest winds can be found. The rain bands extend out from the eyewall, bending in a spiral manner (Barcikowska et al., 2013). The tropical cyclone itself can extend for hundreds of kilometres, typically around 480 km, but size is not necessarily an indicator of intensity (National Weather Service, n.d.(b)).

The warm sea surface temperatures of the Asia-Pacific region, such as in the Bay of Bengal in the NIO basin and the South China Sea of the WNP basin, are common areas for cyclogenesis and tropical cyclone activity. In the WNP, cyclogenesis primarily occurs between latitudes 10° and 30°N and longitudes 120° and 150°E (Barcikowska et al., 2013) (Figure 6.2). Tropical cyclones cannot form within 500 km of the equator (Hobgood, 2005). It is very possible for a tropical cyclone to form in the WNP, but migrate westward toward the NIO basin, like Very Severe Cyclonic Storm Phailin did in October 2013 (CWD-IMD, 2013).

As a cyclone leaves these warmer latitudes and areas of converging winds, the storm typically loses strength. For example, Super Typhoon Guchol began as a disturbance off the southeast coast of Guam in June 2012 and rapidly intensified to a super typhoon when it moved north of 10°N latitude. When it surpassed 30°N latitude, cooler sea surface temperatures, among other factors, weakened the typhoon before it made landfall in Japan (Evans and Falvey, 2012).

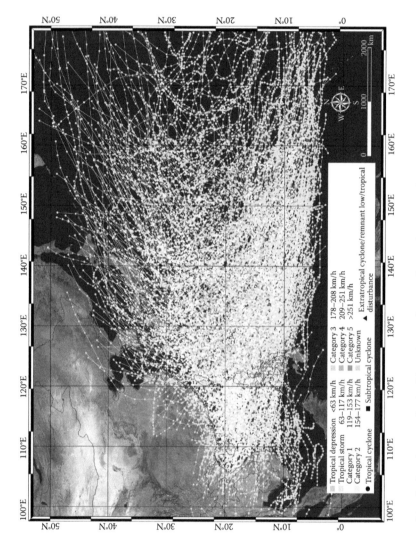

FIGURE 6.2   A map of tropical cyclone tracks, in 6-hour intervals, over the northwest Pacific Ocean from 1980 to 2005 based on data from the Joint Typhoon Warning Center and laid over a Terra MODIS image from the National Aeronautics and Space Administration (NASA). (http://en.wikipedia .org/wiki/File:Pacific_typhoon_tracks_1980-2005.jpg via Wikimedia Commons.)

6.3 METEOROLOGICAL INFLUENCES ON TROPICAL CYCLONES

Meteorological patterns such as the Asia–Pacific Oscillation (APO) and the El Niño–Southern Oscillation (ENSO) can have an impact on tropical cyclone formation and intensity in the Asia–Pacific region. A strong APO is indicated by warmer upper-tropospheric eddy air temperatures over Asia and cooler temperatures over the North Pacific. According to Zhou et al. (2008), tropical cyclone frequency tends to increase if the APO is stronger than normal in the summer, and decreases in response to a weaker APO than normal. For example, prior to 1975, the APO was stronger than normal and after 1975, the APO weakened. Following this trend, the frequency of tropical cyclones in the WNP was higher before the mid-1970s when compared to later in the decade (Zhou et al., 2008).

The terms El Niño and La Niña refer to phenomena surrounding the equatorial Pacific, where years with sea surface temperatures warmer than normal are considered to be El Niño years and, alternatively, when sea surface temperatures are cooler than normal, are considered to be La Niña years (Philander, 1985). On average, an El Niño period can last 12 to 18 months at a time and occur every 3 or 4 years. ENSO refers to the influence of the Southern Oscillation (SO), which induces variability in tropical sea level pressure between the eastern and Western Hemispheres. In El Niño years, this influence typically results in high pressures in the western tropical Pacific and Indian Ocean basins and lower pressures in the southeast tropical Pacific Ocean (Pacific Marine Environmental Laboratory, n.d.). Studies have found that in El Niño years, the average tropical storm and typhoon genesis region has a southward displacement in the WNP (Carmargo and Sobel, 2005) and other studies have found that the frequency of tropical cyclones with varying intensity (such as typhoons and super typhoons) changes based on influences from El Niño or La Niña (Li and Zhou, 2012).

Li and Zhou (2012) use upper-ocean heat content as a proxy for ENSO, to determine the influences on three different types of tropical cyclone frequency: (1) tropical storms and depressions, (2) typhoons and (3) super typhoons. They conclude that the changes incurred in upper ocean heat content vary according to the phase of the ENSO cycle. For example, the frequency of super typhoons increases during the mature phase of El Niño and the frequency of typhoons increases when La Niña transitions to El Niño.

6.4 IMPACTS

No matter the basin of origin, tropical cyclones are known for heavy rainfall, high winds and ensuing floods when and if they hit land (Nguyen-Thi et al., 2012). The World Meteorological Organization (WMO) has established a network of Regional Specialized Meteorological Centres (RSMCs), specifically dedicated to understanding and managing the impacts of tropical cyclones (Obasi, 1974). Of these, the RSMC located in Tokyo, Japan, has the regional responsibility for monitoring tropical cyclones and providing forecasts and warnings in the WNP and the South China Sea, and the RSMC in New Delhi, India, is responsible for the NIO, which includes the Bay of Bengal (WMO, n.d.(b); Tyagi et al., 2010). The Saffir–Simpson hurricane

TABLE 6.2

Saffir–Simpson Hurricane Wind Scale Categories and Potential Damage That Could Occur Because of Winds

Category	Sustained Winds	Types of Damage due to Winds
1	119–153 km/h (64–82 knots)	Well-constructed frame homes could have damage to roof, shingles, vinyl siding and gutters. Large branches of trees will snap, and shallowly rooted trees may be toppled. Extensive damage to power lines and poles likely will result in power outages that could last a few to several days.
2	154–177 km/h (83–95 knots)	Well-constructed frame homes could sustain major roof and siding damage. Many shallowly rooted trees will be snapped or uprooted and block numerous roads. Near-total power loss is expected with outages that could last from several days to weeks.
3	178–208 km/h (96–112 knots)	Well-built framed homes may incur major damage or removal of roof decking and gable ends. Many trees will be snapped or uprooted, blocking numerous roads. Electricity and water will be unavailable for several days to weeks after the storm passes.
4	209–251 km/h (113–136 knots)	Well-built framed homes can sustain severe damage with loss of most of the roof structure and/or some exterior walls. Most trees will be snapped or uprooted and power poles downed. Fallen trees and power poles will isolate residential areas. Power outages will last weeks to possibly months. Most of the area will be uninhabitable for weeks or months.
5	>252 km/h (> 137 knots)	A high percentage of framed homes will be destroyed, with total roof failure and wall collapse. Fallen trees and power poles will isolate residential areas. Power outages will last for weeks to possibly months. Most of the area will be uninhabitable for weeks or months.

Source: Adapted from National Hurricane Center (n.d,(b)). Saffir–Simpson Hurricane Wind Scale. http://www.nhc.noaa.gov/aboutsshws.php (accessed 28 March, 2017).

wind scale (Table 6.2) breaks storms down into categories based upon wind speed and the estimated property damage they could cause (Schott et al., 2012; National Hurricane Center, n.d.(b)).

According to the Emergency Events Database (EM-DAT) and the Centre for Research on the Epidemiology of Disasters (CRED) International Disaster Database, both based in Belgium (www.cred.be), and the U.S. Office of Foreign Disaster Assistance (OFDA), some of the deadliest storms that have occurred worldwide since 1900 have all been of tropical cyclone origin and have occurred in the Asia-Pacific region (Table 6.3). The Asia-Pacific region is vulnerable to some of the most severe impacts because of the extent of coastline, the number of island nations present and the heavily populated coastal areas. The mere infrastructure of many communities, such as buildings made of clay or mud, unpaved roads and weak power and water networks can exacerbate any immediate effects of a tropical cyclone (Dube et al.,

TABLE 6.3

Top 10 Most Important Tropical Cyclones for 1900–2014, Based on Fatalities

Country	Date (mm/dd/yyyy)	Number of Fatalities
Bangladesh	11/12/1970	300,000
Bangladesh	4/29/1991	138,866
Myanmar	5/2/2008	138,366
China	7/27/1922	100,000
Bangladesh	10/1942	61,000
India	1935	60,000
China	8/1912	50,000
India	10/14/1942	40,000
Bangladesh	5/11/1965	36,000
Bangladesh	5/28/1963	22,000

Source: International Disaster Database EM-DAT: The Emergency Events Database – Université Catholique de Louvain (UCL) – CRED, D. Guha-Sapir - www.emdat.be, Brussels, Belgium.

2006; UN-ESCAP and UNISDR, 2012), especially in rural areas (Peduzzi et al., 2012). Furthermore, the prevalence of poverty can be an inhibiting factor to financing for both preparedness and hazard mitigation and this creates economic vulnerability (UN-ESCAP and UNISDR, 2012).

The total population of the Asia-Pacific region increased from 2.2 billion in 1970 to 4.2 billion in 2010 and during this time, the number of people living in cyclone-prone areas increased approximately 68% from 71.8 million to 120.7 million (UN-ESCAP and UNISDR, 2012). Some impacts of tropical cyclones can include flooding, erosion and destruction of dykes and sand dunes, harm to crops and saltwater intrusion of agricultural land (Nguyen et al., 2012). Coastal populations are especially vulnerable as they bear the brunt of the initial landfall of a tropical cyclone. Tropical cyclones can have direct impact on daily activities such as fishing, transportation and tourism (Kotal et al., 2014) in addition to destroying homes, roads and utility infrastructures. These impacts can lead to lasting damage on the environment, economies and livelihoods; it is estimated that more than 85% of global economic exposure to tropical cyclones occurs in the Asia-Pacific region (UN-ESCAP and UNISDR, 2012). Future forecasting of tropical cyclone activity in response to climate change could be of practical value (Stowasser et al., 2006) and help to contribute to already improving forecasting abilities of current tropical cyclone events (WMO, 2014).

Storm surge, especially for island nations or nations with exposed and heavily developed coastlines, is a considerable concern when anticipating a tropical cyclone. The magnitude of storm surge depends on factors such as speed, angle of approach toward the coast, size and spread of the tropical cyclone. Geomorphology of the land, including positioning and structure of bays and estuaries, near-shore bathymetry and width and slope of the continental shelf can also be determining factors for magnitude of storm surge (Lin et al., 2013). In many cases, natural barriers such as wetlands and mangroves can help to mitigate some impacts of storm surge, breaking the waves and absorbing much of the rush of water (McIvor et al., 2012).

The strong winds of a tropical cyclone can stir up nutrients in the ocean and when these nutrients reach the surface and become exposed to sunlight, a massive phytoplankton bloom can result (Lin et al., 2003). Phytoplankton are tiny marine plants that contain chlorophyll, requiring sunlight for survival, much like plants on land (National Ocean Service, 2014). The size, speed of movement and preconditions determining where cold, nutrient-rich water lies are also factors in whether a plankton bloom can be induced by a tropical cyclone (Lin, 2012). For example, in the WNP basin, slow moving, large-diameter tropical cyclones with intense winds moving over cooler waters can cause a plankton bloom to occur (Lin, 2012). In other cases, such as in 1999 over what is now the state of Odisha in India, intense rains from a super cyclone caused rivers and streams to flood, dumping a considerable amount of nutrients into the ocean, resulting in a plankton bloom (Kundu et al., 2001). Plankton blooms can remove greenhouse gases from the atmosphere and provide an ample feeding ground for fish and other marine life (National Ocean Service, 2014). However, very thick blooms can also act as a barrier to sunlight, which is necessary to most marine life below the ocean's surface. Fish mortality and negative impacts to human health, among other issues, can result from excessive plankton blooms.

In 2000, off the coast of Taiwan, Japan and the Philippines, a major plankton bloom occurred when Cyclone Kai-Tak stirred up nutrients from below the surface of the South China Sea; the plankton accounted for 2–4% of the total new phytoplankton production in the South China Sea that year (Lin et al., 2003). Plankton blooms can have long-lasting residual effects such as the two plankton blooms that lingered for about 2 weeks following two of the eleven typhoons the WNP experienced in 2003 (Lin, 2012).

There is also evidence of the influence of cyclones on dissolved oxygen (DO) levels of the open oceans and estuaries. DO levels were measured by Lin et al., (2014) after the passage of Super Typhoon Nanmadol over the western Pacific Ocean and the South China Sea in August 2011. They found that DO concentrations increased between the surface and 40-metre (m) depth, an indication that oxygen from the air was entrained after the typhoon passed. The maximum DO concentration was found at 5–50 m higher than the average summer DO depth in the South China Sea – providing further evidence that the strong typhoon winds affected DO concentrations by uplifting the maximum level. An increase of chlorophyll levels was also observed during this study. Estuaries are vulnerable to intrusion of ocean water due to high winds, altering the chemical composition of the water and harming coastal resources. Mitra et al., (2011) monitored conditions of the Hooghly–Matla estuarine system after Severe Cyclonic Storm Aila passed over the Bay of Bengal in May 2009. They observed an intrusion of saline water into the estuarine system from the Bay of Bengal, which resulted in an increase of salinity levels in surface water. Increased salinity levels can lead to harmful algae blooms, which can also lead to fish kills.

6.5 RECENT TRENDS OF TROPICAL CYCLONE ACTIVITY

6.5.1 TROPICAL CYCLONE ACTIVITY IN THE WNP BASIN

As previously stated, the WNP is one of the primary basins of tropical cyclone development and experiences the most tropical cyclone activity in the world (D'Asaro et al., 2011; Lin, 2012; Barcikowska et al., 2013). Tropical cyclones in the WNP most frequently form between July and October and although tropical cyclones can form over the southern Indian Ocean and the southwestern Pacific Ocean at any time, most tropical cyclones form in the summer months (Hobgood, 2005). Over the central North Pacific Ocean, the average number of tropical cyclone occurrence is low, only 6 per year, but the inter-annual variability is much higher, with some years experiencing upward of 10 tropical cyclones (Clark and Chu, 2002).

Trends in changes of frequency and intensity of tropical cyclone activity in the WNP have been ambiguous over the past three decades (Barcikowska et al., 2013). A review of literature in Liu and Chan (2012) indicates a decrease in tropical cyclone activity between 1998 and 2011 in the WNP based upon reports of tropical cyclone occurrence of at least a tropical storm intensity (where winds are greater than or equal to 60 km/h) from the Joint Typhoon Warning Center (JTWC). Cycles of active and inactive years are typical, with a normal year consisting of between 25 and 29 tropical cyclones (Liu and Chan, 2012). The inactive period of 1998 to 2011 had an average of 23.2 tropical cyclones per year, making it the lowest average of noted cycle periods since 1960 (active periods = 1960–1974; 1989–1997; inactive periods = 1975–1988, 1998–2011) and two of the years, 1998 (UN-ESCAP and UNISDR, 2012) and 2010 (WMO, n.d.(b)), had the least number of tropical cyclones since 1960 (Liu and Chan, 2012). Liu and Chan (2012) attribute the decrease in observed tropical cyclone activity to a decrease in instances of formation in the southeastern WNP (0°N–20°N, 140°E–180°E) and an overall northward shift in formation between longitudes 120°E and 160°E.

In 2012, the JTWC issued warnings on 27 tropical cyclones in the WNP basin, four of which intensified to super typhoons (Evans and Falvey, 2012). The number of total tropical cyclone occurrences for 2012 was below the long-term average of 31 (Evans and Falvey, 2012), but can be considered within the normal range of occurrences (Liu and Chan, 2012). However, the Philippines experienced its second deadliest typhoon in history when Super Typhoon Bopha struck the island of Mindanao on December 3, 2012, and then continued onto central Visayas and Palawan the next day, leaving a total of 1901 people dead (Masters, 2013). Super Typhoon Bopha exhibited maximum wind speeds of 278 km/h and sustained winds of at least 213 km/h when it first made landfall in the Philippines (Evans and Falvey, 2012).

For the WNP basin, 2013 was the most active tropical cyclone season since 2004 (WMO, 2014), with the area experiencing not only more tropical cyclones than average, but also the most intense tropical cyclone in history, Super Typhoon Haiyan. Super Typhoon Haiyan brought 315 km/h winds to the Philippines, resulting in over 6000 deaths and impacting 16 million people (NDRRMC, 2014; NOAA National Centres for Environmental Information, 2014); the storm also had residual effects on southern China and Vietnam. The following year, the World Meteorological

Organization (WMO) reported 31 storms in the WNP, which was also well above the average of 26, with 13 of these storms intensifying to typhoon strength (WMO, 2014).

6.5.1.1 Philippines, Vietnam and South China: Super Typhoon Haiyan in 2013

On average, the Philippines experience 20 tropical cyclones that form or pass through the area surrounding the country each year (PAGASA, 2011). Super Typhoon Haiyan (or 'Yolanda' as it is better known in the Philippines) is the most powerful storm to ever hit land worldwide, based upon sustained wind speed (WMO, 2014). The first signs of Haiyan were thunderstorms that occurred in Micronesia on November 4, 2013, and warnings of intensification were issued throughout November 5, with indications that the storm would transform from a tropical storm to a typhoon headed for the Philippines by November 8 (NASA, 2013). Maximum sustained winds prior to reaching the Philippines were recorded at almost 315 km/h, classifying Haiyan as a super typhoon. Winds slightly subsided to 269 km/h when the storm first made landfall (NASA, 2013). Haiyan initially made landfall early on November 8 in the Philippines, not once, but six times by the end of the day and at super typhoon strength (NDRRMC, 2014). The dense populations of low-lying Tacloban City saw storm surges of up to 7.5 m, swallowing the city as most of it lies below 5 m (NASA, 2013). The island of Leyte received as much as 685 mm of rain over the course of the storm (NASA, 2013). Haiyan continued past the islands of the Philippines, weakening as it approached the West Philippine Sea. The system left the Philippines Area of Responsibility in the afternoon of November 9 (NDRRMC, 2014) and continued toward Vietnam (Figure 6.3).

Haiyan made landfall over Vietnam in the early morning hours of November 11, gusts of wind up to 157 km/h near Ha Long Bay (BBC, 2013). Although the storm had been downgraded to a tropical depression by this time (Haeseler and Lefebvre, 2013), Haiyan was still the most severe storm many coastal cities of Vietnam had experienced in years, causing power outages, uprooting large trees and destroying roofs (ABC, 2013). Storm surge estimates were placed at 3–5 m and measurements of up to 100 mm of rain in 24 hours were reported (WMO, n.d.(a)). As Haiyan progressed inland, it changed course and headed toward China with winds exceeding 70 km/h; the system finally weakened to a tropical storm by November 11 (WMO, n.d.(a)).

The Philippines National Disaster Risk Reduction and Management Council (NDRRMC) recorded a total cost of damages in the Philippines at more than US$840 million (PhP36.7 billion) with upward of 1.1 million homes damaged. Although the super typhoon resulted in more than 6000 deaths, more than 4 million people were able to be sheltered in evacuation centres (NDRRMC, 2014).

6.5.2 Tropical Cyclone Activity in the North Indian Ocean Basin

In the North Indian Ocean (NIO) basin in 2012, the JTWC recorded four tropical cyclones, with two having formed in the Persian Sea and two in the Bay of Bengal, and none of which had peak winds that exceeded 93 km/h (CWD-IMD, 2013). Only two of

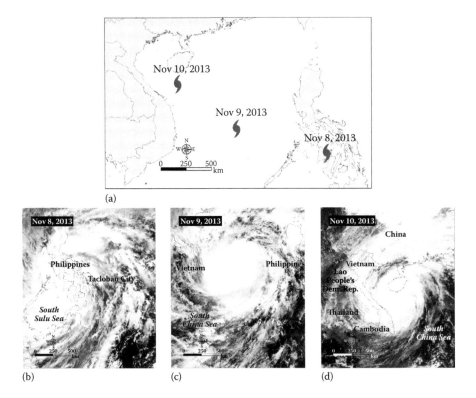

FIGURE 6.3   Path of Super Typhoon Haiyan. (a) Map of the relative location of the eye of the storm as it corresponds with the MODIS images in b, c, d. (b) Aqua MODIS image from November 8, 2013, of Super Typhoon Haiyan as it made landfall over the Philippines. (c) Aqua MODIS image from November 9, 2013, as Typhoon Haiyan passed over the Philippines, weakening slightly and continuing over the South China Sea toward Vietnam. (d) Terra MODIS image from November 10, 2013, as the eye of the typhoon approached Vietnam and southern China. (MODIS imagery retrieved from NASA EOSDIS Worldview, https://worldview.earthdata.nasa.gov/.)

these cyclonic storms made landfall; this number was three tropical cyclones below the normal range of occurrences in the NIO (Kotal et al., 2014).

The NIO experienced five tropical cyclones with an intensity of a tropical storm or greater in 2013, and three of the five storms were categorized as very severe cyclonic storms; no super cyclones were reported (Kotal et al., 2014). The most significant of these, Very Severe Cyclonic Storm (VSCS) Phailin, is profiled in detail below.

6.5.2.1   Bay of Bengal, North Indian Ocean: Very Severe Cyclonic Storm Phailin in 2013

VSCS Phailin was the worst storm to strike India since October 1999, when a super cyclone over Orissa (now Odisha) had left more than 10,000 people dead and caused US$2.5 billion in damage (Kalsi, 2006). When VSCS Phailin intensified to similar strength of this super cyclone, worries spread that it might have the same impact. However, the 1999 super cyclone had reaffirmed the value of knowing what effects a tropical cyclone can have on inland areas as well as at the coast, as it had maintained its intensity as a cyclonic storm for 30 hours after reaching land (Kotal et al., 2014); and it also reaffirmed the need for early warning and forecasting to prevent such losses in the future, a lesson that was heeded during preparations for VSCS Phailin (Mühr, 2013; Ghosh, 2014; WMO, 2014).

According to a comprehensive report from the Cyclone Warning Division of the India Meteorological Division (CWD-IMD, 2013), VSCS Phailin formed from remnants of a cyclonic circulation over the South China Sea on 6 October 2013. The system intensified over the North Andaman Sea, becoming a depression by 8 October and a cyclonic storm by the evening of 9 October. Continuing to progress northwestward over the Bay of Bengal, Phailin evolved into a very severe cyclonic storm less than 24 hours later by the afternoon of 10 October. VSCS Phailin made landfall on 12 October, first passing over the Indian states of Odisha (Ganjam and Puri Districts) and then Andhra Pradesh with maximum sustained winds of 210 km/h and gusts up to 220 km/h (CWD-IMD, 2013; Mühr et al., 2013).

Due to early warnings, approximately 1 million people from Puri and Ganjam Districts were able to evacuate to safe shelters within 3 days before VSCS Phailin made landfall (ADB et al., 2013). After the storm passed, 21 deaths were attributed to the cyclone and 17 caused by flooding were recorded (CWD-IMD, 2013).

Regardless, extensive damage was caused to agricultural land, resulting in a 50% loss to nearly 1.3 million hectares of crops due to impacts from the storm. Approximately 90,000 homes in Ganjam District were damaged to some extent, if not lost completely. The estimated cost of reconstruction for all sectors including infrastructure, housing and agriculture and others, totalled US$1.45 billion (ADB et al., 2013).

The fragile coastline also sustained damage because of VSCS Phailin. The coastal Chilika Lake flooded and experienced sedimentation (Figure 6.4) (Hindustan Times, 2013). Excessive winds that accompanied VSCS Phailin wreaked havoc on the mangroves around the lake, which serve as a coastal barrier, protecting against erosion and minimizing intrusion of saltwater from the Bay of Bengal (Hindustan Times, 2013). Impacts from VSCS Phailin also affected wildlife in the area. Chilika Lake is a favoured location for many species of migratory birds and the aftermath of VSCS Phailin has inhibited the return of some of these migratory birds. In November 2013, the Bombay Natural History Society (BNHS) reported that it would begin a study on the impact of VSCS Phailin on the migratory bird populations and migration patterns (ENVIS Centre, 2013). In early 2014, approximately 158,000 fewer birds than normal were observed as reported by local ornithologists, experts and organizations such as the BNHS, possibly because the storm had removed much of the vegetation that the birds tend to feed on (Zee News, 2014).

6.6 FUTURE OUTLOOK

Several studies have investigated the potential future impact of various climate change implications on storm surge, tropical cyclone frequency and tropical cyclone intensity in the Asia-Pacific region (Chan and Liu, 2004; Stowasser et al., 2006; Mase et al., 2013). Some studies assessing the impacts of warming climate scenarios indicate that the intensity and frequency of storms could increase in parts of the

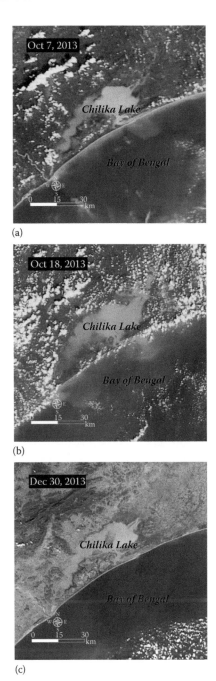

FIGURE 6.4   Status of Chilika Lake and the coastline (a) on October 7, 2013, before VSCS Phailin struck; (b) 6 days after VSCS Phailin made landfall; (c) approximately 2 months after VSCS Phailin made landfall. (MODIS imagery retrieved from NASA EOSDIS Worldview, https://worldview.earthdata.nasa.gov/.)

Asia-Pacific region (Stowasser et al., 2006). Likewise, the Intergovernmental Panel on Climate Change (IPCC) finds that climate variability, including patterns such as ENSO, is 'very likely' to continue to influence tropical cyclone frequency, intensity and global distribution (Kirtman et al., 2013). Furthermore, the IPCC finds that tropical cyclones will continue to vary on an annual and decadal scale (Kirtman et al., 2013) with the general thought that though frequency may decrease or remain the same, intensity of individual storms, in terms of wind speed and precipitation, may increase (Christensen et al., 2013). An increase in intensity could lead to an increase in damage to the rapidly growing Asia-Pacific region if exposure to risks and vulnerability to tropical cyclone impacts is not reduced (Peduzzi et al., 2012; UN-ESCAP and UNISDR, 2012).

6.7   CONCLUSION

Favourable conditions for tropical cyclogenesis exist in the Asia-Pacific region. Variations in intensity and frequency are typical, as has been exhibited recently with years of below average activity, which have been observed in the past decade. However, the intense tropical cyclones that the Asia-Pacific region experienced in 2013 are a reminder that continued monitoring of oceanic and atmospheric conditions remains necessary to mitigate impacts to the more than 120 million people vulnerable in the region, and to increase the certainty of climate change projections.

ACKNOWLEDGEMENTS

Many thanks to Alyson Kauffman, Sean Miller and James Rowland for their insight, review and comments.

REFERENCES

ABC. (2013). Weakened Typhoon Haiyan sweeps over Vietnam and into China. http://www.abc .net.au/news/2013-11-11/typhoon-haiyan-sweeps-over-vietnam-and-china/5084232 (accessed 12 July, 2017).

ADB, GoO, WB. (2013). *India: Cyclone Phailin in Odisha. Rapid Damage and Needs Assessment Report*, Asian Development Bank (ADB), Government of Odisha (GoO) and the World Bank (WB), Odisha.

Barcikowska, M., Feser, F. and von Storch, H. (2013). Changes in Tropical Cyclone activity for the western North Pacific during the last decades, derived from a regional model simulation. *27th Conference on Hydrology American Meteorology Society.* Austin, 2013.

BBC. (2013). Severe Tropical Storm Haiyan makes landfall in Vietnam. http://www.bbc.com /news/world-asia-24890114 (accessed 10 January, 2014).

Camargo, S.J. and Sobel, A.H. (2005). Western North Pacific tropical cyclone intensity and ENSO. *Journal of Climate.* 18, 2996–3006.

Chan, J.C.L. and Liu, K.S. (2004). Global warming and Western North Pacific typhoon activity from an observational perspective. *Journal of Climate.* 17, 4590–4602.

Christensen, J.H., Krishna Kumar, K., Aldrian, E., An, S.-I., Cavalcanti, I.F.A., de Castro, M., Dong, W. et al. (2013). Climate phenomena and their relevance for future regional climate change. In *Climate Change 2013: The Physical Science Basis. Contribution of Working Group I to the Fifth Assessment Report of the Intergovernmental Panel on Climate Change*; Stocker, T.F., Qin, D., Plattner, G.-K., Tignor, M., Allen, S.K., Boschung, J., Nauels, A. et al., Eds.; Cambridge University Press, Cambridge and New York, 1217–1308.

Clark, J.D. and Chu, P.-S. (2002). Interannual variation of tropical cyclone activity over the Central North Pacific. *Journal of the Meteorological Society of Japan.* 80(3), 403–418.

CWD-IMD. (2013). *Very Severe Cyclonic Storm Phailin over the Bay of Bengal (08-14 October 2013): A Report.* Cyclone Warning Division India Meteorological Department, New Delhi.

D'Asaro, E., Black, P., Centurioni, L., Harr, P., Jayne, S., Lin, I.-I., Lee, C. et al. (2011). Typhoon-ocean interaction in the western North Pacific: Part 1. *Oceanography.* 24(4), 24–31.

Dube, S.K., Mazumder, T. and Das, A. (2006). An approach to vulnerability assessment for tropical cyclones: A case study of a coastal district in West Bengal. *ITPI Journal.* 3(4), 15–27.

ENVIS Centre. (2013). http://www.orienvis.nic.in/searchdisplay.aspx?st=1&lid=384&langid =1&mid=4 (accessed 28 March, 2014).

Evans, A.D. and Falvey, R.J. (2012). *Annual Tropical Cyclone Report 2012.* Annual Report, Joint Typhoon Warning Center (JTWC), Pearl Harbor, HI.

Ghosh, S., Vidyasagaran, V. and Sandeep, S. (2014). Smart cyclone alerts over the Indian subcontinent. *Atmospheric Science Letters* 15, 157–158. (Published online in Wiley Online Library). DOI: 10.1002/asl2.486.

Haeseler, S. and Lefebvre C. (2013). Super typhoon HAIYAN crossed the Philippines with high intensity in November 2013. Deutscher Wetterdienst (Published online: https://www .dwd.de/EN/ourservices/specialevents/storms/20131213_taifun_haiyan_philippinen _en.pdf?__blob=publicationFile&v=3).

Hindustan Times. (2013). http://www.hindustantimes.com/india/odisha-phailin-hit-chilika -may-take-years-to-recover/story-aEN6BWnRAXvFVEjYTmgYMN.html (accessed 28 March, 2014).

Hobgood, J. (2005). Tropical cyclones. In *Encyclopedia of World Climatology*, Oliver, J.E., Ed., Springer, the Netherlands, 750–756.

EM-DAT: The Emergency Events Database. (n.d.). Université Catholique de Louvain (UCL) – CRED, D. Guha-Sapir, www.emdat.be, Brussels, Belgium (accessed 3 March, 2014).

http://www.rsmcnewdelhi.imd.gov.in/images/pdf/cyclone-awareness/terminology/faq.pdf (accessed 12 July, 2017).

Jordan, C.L. (1961). Marked changes in the characteristics of the eye of intense typhoons between the deepening and filling stages. *J. Meteor.* 18, 779–789.

Kalsi, S.R. (2006). Orissa super cyclone – A Synopsis. *Mausam.* 57, 1–20.

Kirtman, B., Power, S.B., Adedoyin, J.A., Boer, G.J., Bojariu, R., Camilloni, I., Doblas-Reyes, F.J. et al. (2013). Near-term climate change: Projections and predictability. In *Contribution of Working Group I to the Fifth Assessment Report of the Intergovernmental Panel on Climate Change 2013: The Physical Science Basis,* Stocker, T.F., Qin, D., Plattner, G.-K., Tignor, M., Allen, S.K., Boschung, J., Nauels, A. et al., Eds., Cambridge University Press, Cambridge and New York, 953–1028.

Kotal, S.D., Bhattacharya, S.K., Bhowmik, S.K.R., Rao, Y.V.R. and Sharma, A. (2014). *NWP Report on Cyclonic Storms Over the North Indian Ocean During 2013.* India Meteorological Department Numerical Weather Prediction Division, New Delhi.

Kundu, S.N., Sahoo, A.K., Mohapatra, S. and Singh, R.P. (2001). Change analysis using IRS-P4 OCM data after the Orissa super cyclone. *International Journal of Remote Sensing*. 22(7), 1383–1389.

Li, R.C.Y. and Zhou, W. (2012). Changes in Western Pacific tropical cyclones associated with the El Nino-southern oscillation cycle. *Journal of Climate*. 25, 5864–5878.

Lin, I., Liu, W.T., Wu, C.-C., Wong, G.T.F., Hu, C., Chen, Z., Liang, W. et al. (2003). New evidence for enhanced ocean primary production triggered by tropical cyclone. *Geophysical Research Letters*. 30(13), 51–54.

Lin, I.I. (2012). Typhoon-induced phytoplankton blooms and primary productivity increase in the western North Pacific subtropical ocean. *Journal of Geophysical Research*. 117, C03039.

Lin, I.-I., Goni, G.J., Knaff, J.A., Forbes, C., and Ali, M.M. (2013). Ocean heat content for tropical cyclone intensity forecasting and its impact on storm surge. *Natural Hazards*. 66(3), 1481–1500.

Lin, J., Tang, D., Alpers, W. and Wang, S. (2014). Response of dissolved oxygen and related marine ecological parameters to a tropical cyclone in the South China Sea. *Advances in Space Research*. 53, 1081–1091.

Liu, K.S. and Chan, J.C.L. (2012). Inactive period of Western North Pacific tropical cyclone activity in 1998–2011. *Journal of Climate*. 26, 2614–2630.

Mase, H., Mori, N. and Yasuda, T. (2013). Climate change effects on waves, typhoons and storm surges. *Journal of Disaster Research*. 8, 145–146.

Masters, J. (2013). Category 5 Super Typhoon Haiyan heads towards the Philippines (Blog entry). http://www.wunderground.com/blog/JeffMasters/category-5-super-typhoon-haiyan-headed-towards-the-philippines (accessed 29 March, 2017).

McIvor, A., Spencer, T., Möller, I. and Spalding, M. (2012). Natural Coastal Protection Series: Report 2. *Cambridge Coastal Research Unit Working Paper 41*, The Nature Conservancy and the Department of Zoology, University of Cambridge, Cambridge.

Mitra, A., Halder, P. and Banerjee, K. (2011). Changes of selected hydrological parameters in Hooghly estuary in response to a severe tropical cyclone (Aila). *Indian Journal of Geo-Marine Sciences*. 40(1), 32–36.

Mühr, B., Köbele, D., Bessel, T., Fohringer, J., Lucas, C. and Girard, T. (2013). *2nd Report on Super Cyclonic Storm "Phailin"*. Centre for Disaster Management and Risk Reduction Technology (CEDIM). (Available online at http://www.cedim.de/download/CEDIM-Phailin_Report2.pdf; accessed 28 March, 2017.)

NASA. (2013). Haiyan (Northwestern Pacific Ocean) http://www.nasa.gov/content/goddard/haiyan-northwestern-pacific-ocean/#.UzrHzPldXA0 (accessed 29 March, 2017).

National Hurricane Center. (n.d.(a)). Tropical cyclone climatology. http://www.nhc.noaa.gov/climo/(accessed 29 March, 2017).

National Hurricane Center. (n.d.(b)). Saffir–Simpson hurricane wind scale. http://www.nhc.noaa.gov/aboutsshws.php (accessed 28 March, 2017).

National Ocean Service. (2014). http://oceanservice.noaa.gov/facts/phyto.html (accessed 29 March, 2017).

National Weather Service. (n.d.(a)). Tropical cyclone structure. http://www.srh.noaa.gov/jetstream/tropics/tc_structure.htm (accessed 28 March, 2017).

National Weather Service. (n.d.(b)). Tropical cyclone classification. http://www.srh.noaa.gov/jetstream/tropics/tc_classification.htm (accessed 28 March, 2017).

NDRRMC. (2014). *SitRep No. 90 Effects of Typhoon Yolanda (Haiyan)*. Update, Quezon City, Republic of the Philippines National Disaster Risk Reduction and Management Council.

Nguyen-Thi, H.A., Matsumoto, J., Ngo-Duc, T. and Endo, N. (2012). Long-term trends in tropical cyclone rainfall in Vietnam. *J. Agrofor. Environ.* 6(2), 89–92.

NOAA National Centers for Environmental Information. (2014). *State of the Climate: Hurricanes and Tropical Storms for Annual 2013* (accessed 29 March, 2017 from https://www.ncdc.noaa.gov/sotc/tropical-cyclones/201313).

Obasi, G.O.P. (1994). WMO's role in the international decade for natural disaster reduction. *Bulletin of the American Meteorological Society.* 75(9), 1655–1661. DOI: http://dx.doi.org/10.1175/1520-0477(1994)075<1655:WRITID>2.0.CO;2.

Pacific Marine Environmental Laboratory. (n.d.). https://www.pmel.noaa.gov/elnino/faq#what (accessed July 12, 2017).

PAGASA. (2011). *Climate Change in the Philippines.* Philippine Atmospheric, Geophysical and Astronomical Services Administration, Quezon City.

Peduzzi, P., Chatenoux, B., Dao, H., De Bono, A., Herold, C., Kossin, J., Mouton, F. and Nordbeck, O. (2012). Global trends in tropical cyclone risk. *Nature Climate Change.* 2, 289–294.

Philander, S.G.H. (1985). El Niño and La Niña. *Journal of the Atmospheric Sciences* 42(23), 2652–2662. DOI: http://dx.doi.org/10.1175/1520-0469(1985)042<2652:ENALN>2.0.CO;2.

RMSC. (n.d.). Regional Specialized Meteorological Centre (RSMC) for Tropical Cyclones over North Indian Ocean (http://www.rsmcnewdelhi.imd.gov.in/images/pdf/cyclone-awareness/terminology/faq.pdf).

Schott, T., Landsea, C., Hafele, G., Lorens, J., Taylor, A., Thurm, H., Ward, B., Willis, M. and Zaleski, W. 2012. The Saffir–Simpson hurricane wind scale. *National Weather Services, National Hurricane Centre, National Oceanic and Atmospheric Administration (NOAA) factsheet.* http://www.nhc.noaa.gov/pdf/sshws.pdf.

Stowasser, M., Wang, Y. and Hamilton, K. (2006). Tropical cyclone changes in the Western North Pacific in a global warming scenario. *Journal of Climate.* 20, 2378–2396.

Tyagi, A., Bandyopadhyay, B.K. and Mohapatra, M. (2010). Monitoring and prediction of cyclonic disturbances over North Indian Ocean by Regional Specialised Meteorological Centre, New Delhi (India): Problems and prospective. In Y. Charabi (Ed.), *Indian Ocean Tropical Cyclones and Climate Change.* pp. 93–103. Springer Science + Business Media BV. DOI 10.1007/978-90-481-3109-9_13.

UN-ESCAP and UNISDR. (2012). *Reducing Vulnerability and Exposure to Disasters: The Asia-Pacific Disaster Report.* United Nations Economic and Social Commission for Asia and the Pacific (UN-ESCAP) and the United Nations Office for Disaster Risk Reduction (UNISDR), Bangkok.

WMO. (2014). WMO statement on the status of the global climate in 2013. World Meteorological Organization, Geneva.

WMO. (n.d.(a)). https://public.wmo.int/en/media/news/wmo-community-mobilizes-haiyan-yolanda (accessed July 12, 2017).

WMO. (n.d.(b)). Latest advisories on current tropical cyclones, hurricanes, typhoons. http://www.wmo.int/pages/prog/www/tcp/Advisories-RSMCs.html (accessed 29 March, 2017).

Zee News. (2014). Drop in the number of migratory birds visiting Chilika Lake. Online at http://zeenews.india.com/news/eco-news/drop-in-the-number-of-migratory-birds-visiting-chilika-lake_903951.html (accessed 29 March, 2017).

Zhou, B., Cui, X. and Zhao, P. (2008). Relationship between the Asian-Pacific oscillation and the tropical cyclone frequency in the western North Pacific. *Science in China Series D: Earth Sciences.* 51(3), 300–385.

7 Desertification
Causes and Effects

Amal Kar

CONTENTS

7.1 INTRODUCTION

The word 'desertification' brings immediately to the mind the images of an expanding desert, advancing sand dunes and an overall barren landscape. Desertification is not, however, about expanding boundaries of an existing desert, but about land degradation in the arid, semi-arid and dry sub-humid regions, that results from various factors, including climatic variations and human activities (UN, 1992; UNCCD, 1995). It is a slow and less perceptible process that is expected to gradually lead to the loss of land's biological and economic productivity and complexity (UNCCD, 1995), and thus affect the socioeconomic conditions of people whose livelihoods depend on that land. The three climatic regions mentioned above, together with the hyper-arid regions, form the drylands, which cover about 41% of Earth's land surface and are inhabited by more than 2 billion people (Safriel and Adeel, 2005). Figure 7.1 provides the broad global extent of drylands. The areas that are considered to be more vulnerable to land degradation in the drylands are typically those having less plant cover and low soil moisture, factors which discourage anchorage of the soil. The arid lands (covering about 42% of the global drylands) and their semi-arid fringes are thus most vulnerable to desertification.

Typically the biophysical processes identified for assessment and monitoring of desertification are wind erosion (mostly as deflation hollows, and formation of sand sheets and sand dunes), water erosion (mostly expressed as sheet-wash and gully erosion), waterlogging and salinization (especially due to excessive use of irrigation water) and other forms of chemical degradation of the land (e.g. pollution of surface water and groundwater due to industrial wastes, leading to soil quality degradation, including acidification), degradation of soil physical condition (e.g. hard-setting caused by mechanization of tillage in heavy-textured soils, resulting in compaction, crusting or slumping of soils, and consequent problems in seed germination), degradation of natural vegetation (especially through overgrazing and overuse of biomass for fuel, fodder, etc. in the rangelands and encroachment of forests), degradation due to mine spoils, desiccation of the land (e.g., through long drought), and so on. The acceleration of these processes beyond certain long-term threshold values can lead to excessive loss of soil cover and soil fertility, changes in vegetation ecology and a host of other changes that result in a decline in land production and productivity.

There are also some typical desertification processes that are specifically related to mountainous terrain and cold arid regions. These include mass movement (e.g. landslide and debris flow in mountainous region), frost shattering and frost heaving due to freeze-thaw effects in glacial and periglacial regions that lead to expansion of gravel spread and patterned ground formation and overall expansion of stony/gravelly deserts (e.g. the Gobi Desert). Expansion of stony areas in the karstic terrain has been considered as a notable process in some parts of China. Most of these processes are natural, but in some places the rate at which they operate is accelerating due to increased human pressures from activities such as forest clearance, encroachment onto hill slopes, overcultivation, and overgrazing, often with disastrous consequences. A recent example is the extensive damage caused in the semi-arid upper catchments of the Ganga River in the Himalayas in India after a one-day rainfall of ~300 mm on 16 June 2013. It resulted in numerous landslides, debris flows and a

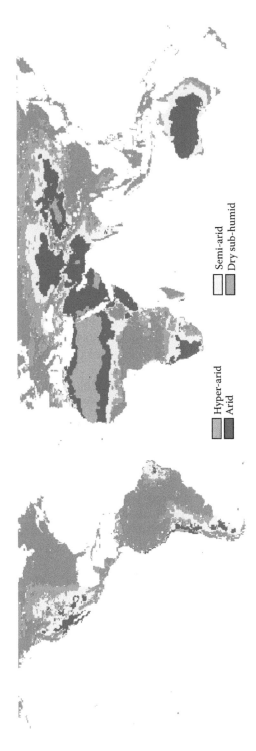

FIGURE 7.1 Approximate global distribution of the four climatic zones under drylands at 0.5° spatial resolution. The zones are based on mean annual Thornthwaite's moisture deficiency index (MDI) [calculated as ((P-PET/PET) × 100)], for the period 1951–2000. (Original values of P [precipitation] and PET [potential evapotranspiration] for the period were sourced from the Terrestrial Water Budget Data Archive: Monthly Time Series (1900–2010) [Version 2.01] of C.J. Willmott and K. Matsuura, Department of Geography, University of Delaware, http://climate.geog.udel.edu/~climate/html_pages /Global2011; downloaded on 25 September 2012.)

Hyper-arid
Arid
Semi-arid
Dry sub-humid

major flood in the Kedarnath–Badrinath–Rudraprayag area, with consequent very high loss of life and property (Sundriyal et al., 2013). Cloud burst is not uncommon in the Himalayas, but one of the major reasons for extensive damage this time was the high and non-regulated human encroachments along the narrow valleys in recent decades. This led to numerous impacts, including unscientific modification of the hill slopes, widespread mining of river sand for building construction, rerouting of channel flow for hydroelectric power without safeguarding the stream banks, dumping of rubble in channel beds, large-scale clearance of the climax oak forest which is a good soil binder and replantation of some cleared areas with more remunerative but less protective pine trees, and so on.

Indeed, one of the major factors responsible for acceleration of the natural processes is human activities through land use decisions, while another is abnormal changes in atmospheric conditions, loosely called climatic variation. Since the end-result of desertification is considered to be a decline in land production and productivity, a serious analysis of its causes and consequences should involve quantification of the rates of processes, magnitude of loss/gain in the land resources and a host of other factors involved in assessing the nature of the change, including the role of historical land uses. Measurement of the biophysical aspect of desertification is not an easy task, since the list of variables to be monitored is usually long and in some cases nested, the interaction between them may be complex and sometimes poorly understood in the local context, and the outcomes are often multicausal. Added to this is the problem of identification of the appropriate socioeconomic variables that trigger and/or assist in the degradation, and their analysis conjointly with the biophysical variables. Viewed from this angle, understanding desertification and its causes through measurement of forms and processes is far more complex than the subject is treated normally.

The prescriptive suggestions for 'amelioration' of desertification, based on insufficient or questionable understanding of the interlinked factors, may lead to wastage of resources. For example, during the UNCOD (1977), one of the proposals to tackle desertification in North Africa was to initiate two transnational projects on afforestation: a North Saharan Green Belt and a Sahel Green Belt. Although the projects failed due to lack of funding (Odingo, 1992), the wisdom of planting a belt of trees across countries without adequate consideration of the different climatic, edaphic, terrain, as well as socioeconomic conditions and needs, has been questioned (CSFD, 2011). Historically, another such experiment, the execution of a 'Great Green Wall of China', involving plantation of a few barely adapted tree species along the southern border of the Gobi Desert, failed through large-scale die-back of trees, and hence the programme was oriented toward overall greening, rather than on forest plantation (CSFD, 2011). Despite the above experiences, another 'Great Green Wall Strategic Action Plan' across the Sahel for 7775 km, from Dakar in the west to Djibouti in the east, was proposed in 2005. Fortunately, after deliberations its objectives were modified to include activities related to sustainable land management, rather than forest plantation (Anon., 2012).

In India, a proposal was mooted in the mid-1980s to construct a high stone wall across some of the 'wind gaps' in the Aravalli Hill ranges along the eastern margin of Thar Desert in Rajasthan state. The state government thought that the problem of sand reactivation to the east of the Aravalli Hills was caused by sand invading from across the hills, and hence the proposal. Fortunately, they decided to discuss the proposal with a team of researchers (the author being a part of it) before its formalization. It was explained to the bureaucrats how such walls might attract more sand deposition along them and eventually make them ineffective through burial. The real cause of the problem was explained as localized sand reactivation in the vast aeolian landscape to the east of the Aravalli Hills. The sand was largely inherited from a past harsher climate when the desert had spread far to the east, but got naturally stabilized by a subsequent wetter climate. Recent overexploitation of the natural resources of this aeolian landscape, especially through intensification of cropping under well irrigation and clearance of natural plant cover for expansion of the croplands, led to localized sand reactivation, and provided a false impression of sand invasion from the Thar Desert. Adequate land and water management practices on both sides of the Aravalli Hills were suggested as a more reliable step to minimize the problem.

Similarly, in Morocco, Davis (2005) reported how the change of narrative on range management practices, from the traditional concept of open grazing by the pastoralists, who moved with their stock following a knowledge of the local grass ecology and availability, to the one where such mobility was considered as causing overgrazing and land degradation, led to a prescription of destocking, privatization and enclosure. This led to accentuation of land degradation.

Before further discussing the related issues of causes and effects of desertification, it may be worthwhile to look into the changing paradigm of how the term has been defined.

7.2   DEFINING DESERTIFICATION

The term 'desertification' was first coined by Lavauden (1927) to describe severely overgrazed rangelands in Tunisia (Dregne, 2000). Subsequently, the term was used to refer to excessive soil erosion due to deforestation in French West Africa (Aubreville, 1949). Worldwide interest in the term surfaced in the context of the first UN Conference on Desertification (UNCOD) in Nairobi in 1977, which was held against the background of a long drought in sub-Saharan Africa during the early 1970s. At that time, the term was defined as diminution of the biological potential of land in any ecosystem (UNCOD, 1977). UNCOD, however, decided to exclude the hyper-arid regions from the areas to be considered for desertification assessment, based on the argument that these areas were already 'desertified'. Many experts (see, e.g. Bauer and Stringer, 2008; DSE, 2009) felt that this definition betrayed a focus on the degradation problems in the drylands, and more specifically in Africa, for which the conference was held. The 1992 UN Conference on Environment and Development (UNCED) negotiated a definition of desertification as 'land degradation in arid, semi-arid, and dry sub-humid areas, resulting from various factors, including climatic variation as well as human activities' (UN, 1992). The UN Convention to Combat Desertification (UNCCD) accepted this definition for all practical purposes (UNCCD, 1995). In Africa the emphasis is more on drought, and so 'drought and desertification' have become familiar terms in that continent. Elsewhere, land degradation is a major focus, but the climatic domain of its operation is fixed for UN-related activity and funding.

It has since been felt (Safriel, 2006) that the exclusion of the hyper-arid areas from even the new definition is a disservice to the population residing in nations within that harsh climatic zone (e.g. in the extremely dry areas of Israel and Saudi Arabia). Similarly, some developing nations along the more-vegetated wetter fringes of the drylands like Myanmar, Sri Lanka, and so on, which are dominated by a humid climate but have a small core of sub-humid to semi-arid climate (or, the 'dry zone'), found themselves left out, although they had terrain vulnerable to land degradation. To accommodate the former group, a proposal has been made to redefine the hyper-arid areas as those having values for the ratio of annual precipitation (P) and annual potential evapotranspiration (PET) of 0.03 and less, instead of the earlier 0.05 and less (CGIAR in their website http://csi.cgiar.org/aridity/Global_Aridity_PET _Methodolgy.asp, downloaded on 12 April 2010). For the countries on the wetter margin of the drylands, some concession was made to include the developing countries that have a 'dry zone' (e.g. Bangladesh, Indonesia, Nepal, Sri Lanka, Vietnam) and, possibly to accommodate the different perspectives, a new term, 'desertification, land degradation and drought' (DLDD) is now often seen in UNCCD literature. The wider latitude given to demarcate the areas under UNCCD's monitoring system after the term 'desertification' was defined, however, left scope for some confusion, especially when the need was felt for developing a robust system of measurements to quantify the desertification status and prioritisation of ameliorative steps.

7.3   RECENT PARADIGMS OF DESERTIFICATION

Defining desertification is becoming increasingly difficult with time. As we have seen, the term has undergone a number of etymological shifts and intended meanings, but a universally acceptable definition is still eluding the scientific community. By the early 1980s, more than 100 definitions of the term existed (Glantz and Orlovsky, 1983). Since then, despite attempts to standardize its definition and meaning at the international level, debate on what desertification should (or should not) mean still continues. Summaries of the changing contexts of the debate are compiled and discussed in Mainguet (1994), Thomas (1997), Warren (2002), Warren and Olsson (2003), Herrmann and Hutchinson (2005) and Veron et al. (2006). Some of the conclusions from these studies are that assessing desertification is still an unsolved issue, and that the coexistence of conflicting definitions and divergent estimates negatively affects the societal perception, leading to scepticism and delay of solutions to the problems (Mainguet, 1994; Thomas, 1997; Warren, 2002; Warren and Olsson, 2003; Herrmann and Hutchinson, 2005; Veron et al. 2006). Common to these perspectives, however, is recognition that the prominent core variables that drive degradation are considered to be climatic, economic, institutional, national policies, population growth and remote influences (Geist and Lambin, 2004).

Broadly, the concept of desertification that prevails at any given time, also called the 'desertification paradigm' (Safriel and Adeel, 2005), suggests that apart from drought, the nonsustainable management and land uses within the drylands often lead to overexploitation of the natural resources and a decline in productivity, the consequent desertification, followed by poverty and migration. Thus, the roles of humans in land degradation, and in decline in land productivity, get a close scrutiny by the experts, involving either a detailed analysis of the physical aspects of degradation with presumed links with prevailing land use practices (Dregne, 1983), or an analysis of the socioeconomic styles of the primary stakeholders (i.e. those engaged in farming, animal husbandry, etc. [Salvati et al., 2011]), or a mix of both (Schlesinger et al., 1990; Seaquist et al., 2009), but the results sometimes become too judgemental against those land users, and subject to criticism. A typical example of such judgemental notices is to blame the centuries-old and traditional practices of sheep and goat rearing, and of livestock migration, as being responsible for degradation through 'overgrazing', surface instability, and so on, especially in the context of the Sahel in Africa, and to then suggest prescriptions for a cure that are alien to the prevailing socioeconomic ethos of the region, and hence unwanted (Charney et al., 1975; WCED, 1987). This prompted the counter-suggestion that the typical land uses of deserts and drylands provided some of the best possible strategies to cope with the uncertainties present in these environments, and that desertification did not automatically follow from those land uses (Mortimore, 1998; Mortimore and Turner, 2005). According to this paradigm, dryland ecosystems are inherently non-equilibrium systems where the ecosystem dynamics are usually event-triggered, and therefore, an ecosystem approach is necessary in understanding the desertification problems (Puigdefabregas, 1995, 1998). The relationship between the biophysical variables and the socioeconomic ones is more complex than is thought to be, and the resilience of the landscape needs to be factored properly (Warren and Khogali, 1992; Bharara, 1993; Reynolds and Stafford Smith, 2002; Warren, 2002).

The dryland development paradigm (DDP) (Reynolds et al., 2007) offers a different perspective on the problem, based on the interaction between the demographic and sociopolitical drivers on the one hand, and the inherently low productivity of the drylands on the other. It recognizes the close dependence of human livelihoods on the environmental conditions in the drylands, and suggests the construction of an integrated research framework for measuring and understanding the dynamics of coupled human (H) – environmental (E) systems. The evolved framework is expected to internalize the heterogeneity, variability, self-organization and nonlinearity of the H–E systems (Reynolds et al., 2011). Although complex, this concept recognizes the need for understanding the many different variables involved in desertification, as well as to integrate them quantitatively through appropriate measurements for monitoring and development purposes.

A third paradigm, called the 'counter-paradigm', was then proposed (Safriel and Adeel, 2008), which suggested that to meet the challenges of inherently low productivity and other constraints, the land users would apply innovative means to surmount the problems of desertification, poverty and other consequences, and the efforts will be supported and sustained by policies and infrastructural facilities, and thereby make the system more sustainable than before. The concept appears to be too much driven by technological inputs for maximizing production for poverty reduction, and so on, but is not very clear on how the risks involved are sought to be understood and countered. The question assumes more importance in the resource-poor developing nations where, once the 'innovative means' are introduced, but the traditional wisdom and the natural resource base get eroded in its wake, a failure of the system may lead to abysmal problems for the poor inhabitants. It tends to work better, however, in the developed land-based economies like that of Israel or Saudi Arabia (e.g. in Israel, Kolkovsky et al., 2003; Adeel and Safriel, 2008).

Yet another 'paradigm' was proposed by Stringer et al. (2017), who, while elaborating on the concept of DDP (Reynolds et al., 2007), argued that the major environmental challenges in the drylands still remained attached to improvement in the dominantly occurring smallholder-agriculture, especially to address the issues of low agricultural productivity, low investment, poverty, livelihoods and land degradation. They felt that the framework for development pathways for DDP provided in Reynold et al. (2007) needed some updating and refinement, despite being conceptually sound, to make it more actionable and relevant to the evolving global context. One of the issues, which Stringer et al. (2017) felt was missing from Reynold et al.'s (2007) work, was a focus on understanding and analysis of interactions between the vast and diverse networks of value chains of a produce.

In the context of DDP, the 'value chain' refers to agricultural products, and means a series of activities through which a primary produce from the crop field is processed and passed on to the next processing step, so that the product gradually becomes more attractive and marketable, that is, the agro-product gets more value-added, which may potentially benefit all those involved in the production process from crop field to the finished product.

Called the 'new dryland development paradigm', Stringer et al. (2017) proposed adherence to three integrative principles of 'unpack', 'traverse' and 'share' to effectively focus on the farming systems and livelihoods through multidisciplinary approaches. The 'unpacking' of relationships and interactions in the system was suggested to facilitate proper structuring of the problems, identification of the linkages and feedbacks and understanding of the roles and relations between stakeholders. The 'traversing' of scales, sectors and stakeholders was proposed to improve the creation, availability and access to the assets for a more contextual, sustainable and efficient people-centred networking and facilitation. The 'sharing' of knowledge and experiences of all the dryland stakeholders, including the land users, other actors, researchers, policy makers, and so on, was suggested to reduce the trade-offs, avoid externalities and help in upscaling.

The above discourses emphasize the need for identification and cataloguing of the pathways and processes of desertification, as a primary step toward a holistic understanding. Some notable examples of the constructed pathways of changes toward desertification syndrome are provided in Lambin et al. (2001), Geist and Lambin (2004) and Hellden (2008), while reviews of the recent holistic literature on research into desertification processes and their assessment are provided in Baartman et al. (2007) and Nkonya et al. (2011), and a recent review of the desertification research in India is available in Kar et al. (2009). It is apparent from the above discussion that the meaning of the term desertification is changing with time, such that the focus of desertification paradigms is gradually shifting from degradation alone to sustainability, from monitoring of the biophysical processes alone to giving more emphasis on the socioeconomic factors, and from a top-down science-led programme for amelioration to a programme involving people's participation.

7.4   DESERTIFICATION CONCEPT AT THE UN CONVENTION TO COMBAT DESERTIFICATION (UNCCD)

At UNCCD today, desertification is viewed as a societal problem that creates conditions favourable for land degradation, and hence needs to be addressed through overall improvement in the living conditions and economic well-being of the people. Under the Convention, the affected developing country parties are obliged to:

- Give due priority to combating desertification, and mitigating the effects of drought
- Establish strategies and priorities within the framework of sustainable development
- Address underlying causes of desertification, particularly to the socio-economic factors contributing to the desertification process
- Promote awareness and facilitate the participation of local populations, particularly the women and youth nongovernmental organizations, in efforts to combat desertification and mitigate the effects of drought
- Provide an enabling environment by strengthening the relevant existing legislation, enacting new laws, where they do not exist, and establish long-term policies and action programs (UNCCD, 1995)

Consequently, all affected parties are to prepare a suitable National Action Programme and monitoring mechanism, through appropriate framework for policy, legal, strategy, programme, institutional development and R&D (see http://www2.unccd.int/convention/action-programmes). Themes like poverty alleviation, land reform, education, infrastructural development, and so on, are therefore, now as important an element of the countermeasures in the affected countries, as are soil and water conservation, wind erosion control, afforestation, and so on (e.g. India [Anonymous, 2001]). In India, for example, a 'Mahatma Gandhi National Rural Employment Generation Scheme' (MGNREGS) was introduced by the Government of India to attempt to address the related issues through providing at least 100 days' employment in a year to a rural family (150 days proposed for the 'calamity-stricken areas') for the creation of natural resource assets as well as for poverty alleviation and stabilization of rural income (Anonymous, 2011; Ministry of Rural Development, 2014). The scheme also aims to develop synergies between the country's programmes on natural resources conservation, as well as the other development and management programmes in rain-fed drylands, through linkages between watershed development, farming systems development, and other natural resources management activities outside the watershed programme areas, drinking water mission, decentralized food security through dryland crops, and so on.

In 2007, the UNCCD adopted a 10-year (2008–2018) strategic plan and framework (UNCCD, 2007) to attain the objectives of improving the living conditions of affected populations, improving the condition of the ecosystems, and generating global benefits through the above implementations. A set of 12 impact indicators were proposed for the purpose, at national and global levels: water availability per capita, change in land use, proportion of population above the relative poverty line, food consumption per capita, capacity of soils to sustain agro-pastoral use, degree of land degradation, plant and animal biodiversity, drought index, land cover status, carbon stocks above and below ground and land under sustainable land management (SLM). A White Paper on review of the impact indicators clarified that 'desertification, land degradation and drought as defined by the United Nations Convention to Combat Desertification results from dynamic, interconnected, human-environment interactions in land systems, where land includes water, soil, vegetation and humans – requiring a rigorous scientific framework for monitoring and assessment, which has heretofore been lacking... ... desertification, land degradation and drought (DLDD) is caused by complex interactions among physical, biological, political, social, cultural and economic factors, and is interrelated with social problems such as poverty, poor health and nutrition, lack of food security and other factors' (Orr, 2011: 15). Thus, at the UNCCD level, the focus is now more on monitoring the human-environment interactions, and on addressing the social problems while trying to improve the ecosystem services in the drylands.

7.5   DESERTIFICATION PROCESSES

In so far as desertification is thought to mean 'land degradation', it is commonly believed that the term refers essentially to a problem of faster-than-normal degradation of the land, which takes place when the land surface processes found under natural conditions are accelerated due to climatic aberrations or increased human pressures (the driving forces that ultimately trigger degradation). While these processes are clubbed together under the head 'biophysical', there are also the 'socioeconomic' processes that aid in the acceleration of normal rates of degradation. If this simplistic framework will continue to draw the attention of researchers, as a first step toward understanding the subject, a historical perspective will be in order.

7.5.1   BIOPHYSICAL PROCESSES

Commonly the proximal biophysical processes that cause desertification over large areas are wind erosion/deposition, water erosion, waterlogging and salinization and degradation of vegetation. It is also recognized that acceleration of these processes is important, rather than the normal rates of their functioning, and Dregne (1991) recommends that their status should be monitored in rain-fed croplands, irrigated croplands and rangelands. While mapping the distribution of the areas affected by the different processes, it is usual to tag one dominant process to one affected area. However, as discussed in Kar and Takeuchi (2003), mapping of desertification for UNCCD by several countries, especially in Africa, has revealed some confusion on what features (termed the indicators) are to be measured to recognize a certain category of degradation, as well as the responses of the society to control it. They suggest that the basic reasons for the difference lie in the complexities or ambiguities in evaluation of the indicators, wide differences in the nature of problems in different regions, as well as large variations in the database collection system and availability of necessary infrastructures (Kar and Takeuchi, 2003).

Since it is recognized that desertification can result from both climatic and human-induced causes, attempts have been made to differentiate between the signatures of human-induced degradation from the natural ones, using modern tools. A notable study in the semi-arid areas of North China revealed that rapid desertification during the 1970s and early 1980s was more a result of higher wind speed and sand drift potential than human action (Wang et al., 2006). In the Sahel region, a study on the relationship between potential vegetation, satellite-derived vegetation greenness and human pressures during 1982–2002 showed that grazing might not have appreciable impact on vegetation dynamics (Seaquist et al., 2009); while in South Africa a comparative analysis of rain-use efficiency of the NDVI and the residual trends in NDVI predicted by rainfall suggested that the latter could identify the potential problem areas of community land degradation at the regional scale, but it needs to be applied with caution (Wessels et al., 2007). Considering the complexities involved in the degradation processes, more definitive studies are required to segregate the signatures of human-induced degradation from the natural ones. For example, a sudden increase in bare sand thickness may not always be a result of faulty land use; it can form even independently of any land use change, especially during severe sandstorms, say within a prolonged drought period when most annual and much perennial natural vegetation is dead, but the land use may not have changed much. Similarly, the appearance of a field of barchans in an area may not necessarily mean that land use has been alarmingly different; it is also possible that the marked barchans were some distance upwind previously, that these were moving at a constant rate, and that no relationship existed between their occurrence at a particular site and land use. These features could be expected to travel further downwind with time.

In the semi-arid areas, rill and gully erosion usually takes place in pulses of fast advancement, separated by years with almost no advancement (Wolman and Gerson, 1978; Poesen et al., 1996). The processes are activated during periods of high-intensity above-normal rainfall. A good vegetation cover can minimize the rate of advancement of rill and gully heads, but under a constantly degrading environment it is difficult to say if land use during a particular period can be held responsible for the measured advancement. It is most likely that rainfall amount and rainfall intensity will determine the rate of rill and gully formation, while the sites of their occurrences may be linked to fractures and other weaker zones within the sediments, as well as geologically recent mild tectonic events (Kar, 1993a). Considering that many biophysical indicators can develop from a variety of causes, it is difficult to prejudge the features and link them to a particular pressure indicator.

Notwithstanding the debate on the responsibilities of humans and/or nature in causing land degradation, many land features are good indicators of degradation. Stocking and Murnaghan (2000) have listed many field indicators of degradation, and simple methods for measuring them, especially for the benefit of small farmers and other land users. Some field indicators of wind erosion and water erosion were also catalogued for mapping degradation in the Thar Desert (Singh et al., 1992). Surface manifestation of some physical and chemical processes of degradation, such as waterlogging and secondary salinity due to overuse of irrigation, is now easy from remote sensing, but estimation of their likely future trends needs field monitoring of the rates of water table rise, subsurface salinity level, and so on (Mainguet, 1994; Kar et al., 2009, 2016).

Several attempts have been made to model degradation scenarios. Many of the interesting advancements have been made in understanding the vegetation response to pressure, and application of the knowledge to modelling the degradation patterns. For example, studies in natural rangelands of the United States showed that a thinning of the plant stands in an arid area gradually leads to a more heterogeneous distribution of soil fertility and soil moisture, which provokes further degradation (Schlesinger et al., 1990). Using the knowledge on soil erosion and grazing dynamics, and utilizing satellite sensor data, the effects of degradation on rangeland productivity was estimated in Australia (Pickup, 1996). Analysis of the major land use/land cover signatures from low-resolution AVHRR data has helped several researchers to estimate the vegetation dynamics and their significance or otherwise for desertification, in Africa and elsewhere (Tucker et al., 1991; Prince, 2002; Eklundh and Olsson, 2003; Hellden and Tottrup, 2008; Fensholt and Rasmussen, 2011).

7.5.2 SOCIOECONOMIC PROCESSES

Compared to the biophysical processes, the socioeconomic factors have received less attention, especially because the analysis involves a very large size of sample population and, while the information can be gathered mostly through inter-personal dialogues using some standard questionnaire, the process is time-consuming, needs large human-hours to gather the data at sample household to village levels and is often costly. Even then, the level of quantification that can be achieved with conviction for biophysical variables, in a shorter period of time and at pixel level, can hardly be achieved at any comparable scale or resolution for socioeconomic variables.

Some socioeconomic indicators, such as changes in land use and herd composition, child health, utilization of fuel wood, local perceptions, and so on, were provided by Dregne (1983). Analysis of the data for a degradation perspective is, however, subjective. Some notable studies from Africa have shown that land users generally have greater wisdom about the vulnerability and resilience of their land, and can judge better about its degradation status (Tiffen et al., 1994; Swift, 1996; Mortimore, 1998).

In a study at the global level, Lambin et al. have shown that neither population growth nor poverty alone constitute the sole or major causes of land-cover change globally, but people's response to economic opportunities created at local to regional levels (that are getting increasingly triggered by global factors) drive these changes (Lambin et al., 2001). Different human-environment conditions react to the impacts of the drivers of change differently, which need to be understood and considered while explaining the land-use changes. Lambin et al. called for developing systems that capture 'the generic qualities of both socioeconomic and biophysical drivers as well as place-based, human-environment conditions that direct land-use and land-cover change' (Lambin et al., 2001: 267). Based on these understandings, Hellden (2008) developed a model using long-term (~150 years) data for Kordofan, Sudan, that amalgamated both human and environmental factors of desertification, to simulate the Sahelian desertification syndrome suggested by Downing and Ludeke (2002). Through use of this model, Hellden concluded that generating irreversible desertification for Kordofan was difficult (Hellden, 2008) and that this would also apply to any area with open market and free population mobility, unless some extraneous forces like those arising from climate change intervene.

Meta-analysis of 132 case studies on desertification (Geist and Lambin, 2004) revealed that it is difficult to support some of the earlier beliefs that desertification is caused by a single factor (e.g. increasing number of poor land users, or sedentarization of the nomads), or that it leads to 'irreducible' complexity. Multiple agents and multiple responses to social, climatic and ecological changes at multiple spatial and temporal scales are usually the rules, of which the more dominant drivers are the climatic factors, economic factors, institutions, national policies, population growth and remote influences (Geist and Lambin, 2004). These factors trigger cropland expansion, overgrazing and infrastructure extension. Therefore, no universal policy for mitigating desertification could be conceived for all the drylands, and a detailed understanding of the complex set of proximate causes and underlying driving forces in a given location would be required for any effective assessment and policy intervention (Geist and Lambin, 2004).

International politics and macro-economic policies have also been found to contribute to desertification (Danish, 1995). Introduction of cash crops like groundnut and cotton for international markets in the Sahel region of Africa has been linked to widespread degradation due to intensive use of soil and water, and replacement of traditional sustainable farming practices; while the dumping of subsidized European beef in the Sahelian markets, in the wake of a major drought from 1984, left the African pastoralists unable to compete with their own animal products, and as a result their grazing lands saw higher pressure due to the mounting numbers of unsold cattle (Danish, 1995).

Studies in Inner Mongolia of China (Hoffmann et al., 2008) have shown how some of the state land use policies can impact the rangeland conditions, the social fabric of people dependent on rangeland resources, and emission and long-distance transport of desert dust. For example, in the Xilingol Prefecture (Hoffmann et al., 2008), which was known previously for its vast areas under good grasslands, that were maintained by an ecologically sustainable traditional nomadic pastoralism by the Mongols, large-scale introduction of crop cultivation in the post-independence era created huge problems of land degradation and livelihood support. The policy of bringing every possible piece of land in the country under the plough resulted in very high influx of farming communities to the grasslands (Dahl and McKell, 1986; Armstrong, 2001). Livestock rearing, which was earlier almost the sole occupation, became an additional source of income. Meanwhile, due to the inherent terrain and climatic constraints, this newly introduced agricultural system soon became less sustainable as recurring droughts led to crop failure, and sandy terrain without adequate grass-shrub cover became more vulnerable to wind erosion. Xilingol Prefecture became a major additional source of the dust that gets transported almost every year from the Gobi Desert and surrounding areas, by strong wind during spring and summer months, to affect a large part of northeast China, Korean Peninsula and Japan (Kar and Takeuchi, 2004; Hoffman et al., 2008). Steps to boost the pastoral economy through privatization of the rangelands for maximizing production exacerbated the problems of overgrazing and desertification in the pastoral areas, especially due to intensification of grazing and hypercritical stocking ratios on the vulnerable rangelands (Williams, 1997; Ho, 2000). Fortunately, learning from previous mistakes, the country has launched major programmes to improve the condition of natural grasslands and livelihoods of people (CCICCD, 2006).

Desertification in some areas has a strong historical narrative, especially in old civilizations or flourishing pre-twentieth century agricultural systems that have perished due to faulty land management. Typical examples include waterlogging and salinization due to overirrigation in the alluvial plains of Egypt and Mesopotamia. In China, ruins of many ancient towns have been found in oases along the margins of the Taklamakan Desert or in Hexi Corridor, where a system of irrigated farming existed previously, based principally on the seasonal stream flow. It was postulated by early researchers (e.g. Stein, 1921; Zhu and Liu, 1983) that, as dryness increased with time, the water availability declined, aeolian sand invasion increased and the flourishing agriculture collapsed, leading to abandonment of the settlements. Since many of those ancient towns were along the Old Silk Road to Europe, they were also the victims of periodic wars. Mainguet (1994) has discussed a number of these narratives.

A review by Bolwig et al. (2009) of the sustainability and management of natural resources in the Sahel, especially in the context of four current policy-relevant drivers (markets and value chains, property rights, decentralized governance and climate change), showed that the agricultural markets in the Sahel are very imperfect, and that the rural poor are highly exposed to market risks, with adverse impact on their livelihoods and income. These, it was suggested, need to be addressed through infrastructure and institution building, like 'value-chain' management and 'public governance'. The latter term has been used to mean an administrative mechanism through which an effective dialogue and power sharing can take place between the government bureaucrats, the civil societies, the market players, and so on, who are involved in delivery of a social good. The territorial domain of such a system has been considered to be larger than an administrative district.

The relationship between natural resource management (NRM), land tenure security and property right regimes is not straightforward, and obtaining a formal right to use land and water in a village does not necessarily translate into actual uses by the poor, as the prevailing social structures and hierarchies often intervene. Although decentralization of power and democratic reforms are taking place slowly and with periodic setbacks and reversals in some of the Sahelian countries like Ethiopia, Ghana (Ribot, 2002), Mali (Umutoni et al., 2016) and Burkina Faso (Diep et al., 2016), the democratic rights of most inhabitants to NRM are still poorly realized, and act against the stated objectives of any development assistance, especially from the international agencies. Bolwig et al. (2009) argue that the assistance policy for ameliorating the much-hyped deteriorated conditions of the Sahel environment, as well as for adaptation to a worsening climate in the future, needs to be viewed against the real problems of the region that are social in nature.

7.6   PROXIMAL CAUSES AND EFFECTS
OF DESERTIFICATION IN THE THAR DESERT

There is much confusion over the proximal causes of a certain type of degradation. One suggestion, though one that is not universally accepted, is that for effective control measures the types of degradation faced in the drylands need to be identified, mapped and understood in the light of the changing land uses and stakeholder interests. The following narrative of the degradation paradigm in the Thar Desert, based on research by Kar (2011a), may clarify the issues.

7.6.1 TRADITIONAL AGRICULTURE AND LAND MANAGEMENT PRACTICES

Within the Thar Desert, a distinctive set of traditional wisdom prevails, at the core of which lie the themes of water conservation, mixed farming of crops and livestock, agro-forestry and land care for best utilization of the capricious rainfall during the summer monsoon. This is unlike the history of settlement in many of the world deserts, where animal husbandry and migration were the core concepts. The mixed cropping helped to take care of grain production in years of small monsoon aberrations, while animal husbandry helped the most during droughts, especially through sales proceeds of the live animals, but also through animal products, especially milk. Practices such as keeping the land fallow for some seasons (long fallow for 2–5 years; short fallow for a year) to regain the soil nutrients that are lost due to cropping; erecting of fences around fields during summer to trap the suspended silt that blows in from the fertile plains during sandstorms in the hot premonsoon months, or to prevent soil from blowing away; lopping of trees (rather than felling) for fuel and fodder; management of permanent pastures for grazing animals; rotational grazing practices; and a host of other practices of land care and water conservation, are still in-built in the traditional customs and agricultural practices of the rural population.

7.6.2 EFFECTS OF POPULATION PRESSURE ON PASTURE CONDITIONS

The situation in the Thar Desert started changing with population growth, especially after the independence in 1947 when in-migration due to partition of the desert territory between India and Pakistan, and changes in administrative mechanisms and land governance at the grassroots level in its wake, led to a very high pressure on the region's land and water resources. The system of land fallowing became almost the first casualty of the need to feed a large population. Permanent pastures (a form of common property resources, CPR) around the village settlements soon became almost bereft of ground flora, and shrubs for animals to browse became fewer, as the system of controlled grazing through a traditional village monitoring mechanism collapsed under the mounting pressure of increasing population, combined with land reforms that liberally distributed the many submarginal CPR lands (held earlier by the large owners/feudal heads) to common people at a nominal price. In addition to the consequent privatization and shrinkage of the CPRs, the advent of modern agricultural practices put more emphasis on cropping, even on the submarginal lands (Jodha, 1985). Sparse natural woody vegetation on the sand dunes and low sandy hummocks then gradually became the targets of fuel wood collectors, which started loosening the structure of sand, making the dunes more vulnerable to wind erosion during dry summer months (Kar, 1986; Singh et al., 1994).

7.6.3 Ushering in the Green Revolution through Groundwater Irrigation and Tractor Use

The major changes in agricultural land use in the Thar Desert started in the wake of the 'Green Revolution' in the neighbouring fertile alluvial plains of Punjab and Haryana states of India. 'Green Revolution' (GR) refers to an agricultural technology that was developed through research and extension from the 1960s onward, especially encouraged by the stupendous success of a dwarf wheat variety that was developed by Norman Borlaugh of the United States. GR enhanced food crop production by many times over the conventional system, especially using high-yielding crop varieties, irrigation water, chemical fertilizers and pesticides. Adoption of this technology helped many countries, including India, to increase their crop yields to a level where they became net exporters of food grains rather than net importers (Brown, 1977). However, it also gradually revealed its limitations and some new challenges of land management, crop yield, socioeconomic stability, and so on (Chambers, 1984; Chand, 2008; Hazell, 2009; Pingali, 2012). We provide here a summary of the changes brought about by GR in the Thar Desert within Rajasthan state of India. As the Green Revolution started to show results, farmers in the desert margins first opted for diesel pump sets to energize their wells, especially for winter cropping. With time, as rural electrification progressed, and the state Ground Water Department moved in to sink tube wells for drinking purposes, the farmers followed in their footsteps for sinking their own wells for irrigation. The total sown area increased from 7.8 m ha in 1950–1951 to 10.09 m ha in 1980 and to 10.94 m ha by 2005. At the same time irrigated land in the desert tract of Rajasthan increased from 363 thousand ha in 1950–1951 (4.7% of sown area) to 1.39 m ha in 1980 (13.8%) and to 2.77 m ha in 2005 (25.3%), where canal networks (essentially the Indira Gandhi Canal system) accounted for 43% of the irrigated area, and electrified wells the remaining 57%. Of the total groundwater utilized, less than 15% was for drinking purpose, and about 85% for irrigation. The pipeline grids for drinking water helped people to avoid the drudgery of fetching water from long distances, but this also started a neglect of the traditional water harvesting infrastructures, many of which became silted up and their catchments disturbed and encroached upon (Kar, 2011b).

As irrigated cropping with groundwater showed promise, tractors swelled in number from 14,500 thousand in 1980 to about 200,000 by 2005, which was justified by the need for quick tillage and sowing operations after the rains in a sandy terrain, where farmers have to complete the sowing operation within 2 days of a 30-mm rainfall event at the break of monsoon (usually early July). Otherwise the strong sun evaporates the soil moisture and the opportunity is lost. Tractor operation, however, is antagonistic to the random distribution of trees and shrubs in a field. The easiest choice was, therefore, to uproot these natural trees and shrubs in farmers' fields, which formerly acted as models of traditional agro-forestry. With increasing demand for irrigated croplands, the tractors gradually began to climb the high, naturally stabilized sand dunes that earlier used to be under rain-fed crops only during good rains, but mostly served as natural rangelands. Gradually, much of the sandy tract in the desert became deep-ploughed by tractors, which meant destabilization of sand over a large area (Kar, 1996). By 2005 many sand dunes in the eastern half of the desert were under crops where tractors ploughed the land. Co-evaluating the TOMS-derived aerosol index (AI) over the region, with annual rainfall, wind speed for June and number of tractors for the period 1980–2002, a steep increase in AI was noticed from the second half of the 1990s despite a drop in wind strength and variable rainfall, which possibly reflected a greater control of critical changes in land use in the sandy tract brought about with large-scale use of tractors, although other factors might also be involved (Kar, 2011a). The trajectory of the emitted dust is toward the Indo-Gangetic Plains, with high potential impacts on the Indian summer monsoon (Niyogi et al., 2007; Bollasina et al., 2011; Dey et al., 2014).

GCM simulation studies by Bollasina et al. (2011), supported by meteorological data for the second half of the twentieth century, have shown that the atmospheric dust and aerosol loads over north India reduce the rain-cloud formation, and also lead to anomalous cooling of the land and sea surfaces, which in turn slows down the tropical meridional overturning circulation between the Indian Ocean and the Pacific Ocean. It also restricts the local Hadley circulation, but strengthens the zonal circulation over the eastern Indian Ocean. Such behaviour impacts the monsoon strength and distribution, leading especially to the drying of the central-northern India, as has been observed during the late-twentieth century.

A previous simulation study by Patra et al. (2005), backed by observation data of the first decade of the current century, had confirmed that an increase in atmospheric aerosol load through dust emission and biomass burning in the Indian subcontinent inhibited the cloud droplet growth during summer monsoon and cooled the Arabian Sea surface, which reduced the rainfall.

Another study (Dipu et al., 2013), involving simultaneous observation and analysis of meteorological parameters from ground to the upper atmosphere, has found that the dust emission from the Thar Desert results in an increased radiative heating and a decreased cloud water-ice content over the peninsular India, but an increase in cloud-water content over the Bay of Bengal, thus influencing the spatial distribution of rainfall in the region. Desert dust is also known to cause some health hazards, especially respiratory problems like asthma, tracheitis and silicosis, as well as skin irritations, conjunctivitis, and so on (Goudie, 2014).

The promises of irrigated cropping and almost free availability of groundwater resulted in overirrigation of fields. As pumping of groundwater increased, the discharge from many wells began to dwindle, and affected farmers started going deeper for water, which not only increased the cost of lifting water, but in many cases, yielded poor-quality water, especially waters high in residual sodium carbonate. Irrigation with such water turned the medium-to-heavy textured soils, sodic and reduced crop yield drastically. In many areas, the wells became dry. It forced many affected farmers to shift back from the irrigated winter cropping to the former practices of rain-fed farming during the monsoon, leading to new sociocultural problems for the affected families. Since the farmers had previously resorted to large-scale levelling of sandy undulations, and uprooting of natural vegetation to introduce irrigated crops, reverting to rain-fed cropping also increased the vulnerability of the land to wind erosion, a problem that becomes more severe during the drought years, due to reduced plant cover and drier soils. In the canal command areas (i.e. a large area served or 'commanded' by a canal network, where many individual farmers access the water, and where the regulation and distribution of the water to individual plots is managed by an authority, the command authority. In the case of well irrigation, the area being served by a well is limited to a few hectares, which is managed by an individual farmer, and hence no such commanding is needed), overuse of water by the farmers, and diversion of excess water to dune-covered areas by the canal engineers (in the hope that aeolian sand thickness would absorb the excess water, but not realizing that underneath the sand was an impervious gypsum bed) led to waterlogging and salinity in many areas (Figure 7.2). In fact, irrigated farming was a new introduction in this millennia-old rain-fed agriculture region, and the farmers became so overwhelmed with its success that they paid no heed to the experts' suggestions to use the water optimally. Consequently, the boons of higher production, crop diversification and soil nutrient gains gradually became a bane, as rising water level and higher salinity led to abandonment of farm lands. The land use changes brought about by the increasing population, and the transition from traditional subsistence agriculture to a modern production-intensive agriculture, especially overenthusiastically using technologies in a fragile, drought-prone land, created such severe problems of land conservation in some places that the researchers increasingly felt the need to implement viable land management practices for the long-term well-being of the land resources (Ghose et al., 1977; Dhir, 2003).

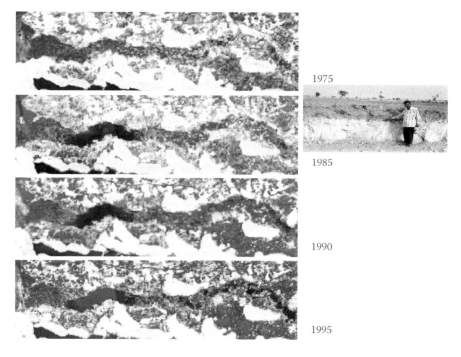

FIGURE 7.2 Sequential satellite images (Landsat FCC) of a part of the canal-irrigated northern part of Rajasthan near Suratgarh, showing development of waterlogging (light to dark blue) along a paleo-channel that used to support good croplands (seen as a winding strip of red through patches of bluish white that represents sand dunes). The field photograph shows a massive gypsum bed at ~1 m depth, which hinders subsurface drainage.

7.6.4 LEARNING FROM PAST MISTAKES AND ADOPTION OF MANAGEMENT PRACTICES

Gradually, farmers started learning their lessons from the past mistakes. There is now a growing awareness that keeping the land in good shape is the key to sustained production from it. In the groundwater-irrigated areas farmers in large numbers have started to purchase sprinkler systems, of which they were advised long ago, and are taking soft loans from the state. Small windbreaks and shelterbelts are coming up along many field boundaries, especially on dune slopes and sandy undulations, mostly in the western part where the summer wind speed is much higher. In the canal-command areas, farmers now use less water for irrigation, and consult the agricultural experts on choice of crops and soil drainage to avoid waterlogging in the root zone. To keep a minimum plant cover on the sandy surface, large and medium farmers having irrigation facilities are trying to keep a part of their land under seasonal fallow, such that the land parcels are alternately used for rain-fed summer cropping and irrigated winter cropping. Winter crops that can continue until March or early April are getting preference along the wetter desert fringes. Choice of crops, however, depends heavily on the market price, and a calculated risk is taken by the farmers on the permissible trade-offs between farm income, likely groundwater depletion, land quality degradation and the next year's yield potentials under the prevailing input strategies, factoring the likely soil nutrient depletion with continuous cultivation in the same parcel of land.

It is perhaps a lingering faith in the values of traditional wisdom of keeping domestic animals, especially to counter the threats of droughts and crop variability (now due to the rising winter temperature, when a number of high-value temperature-sensitive crops like wheat, mustard, etc., become vulnerable), that the farmers continue to keep their stocks of sheep, goats and livestock despite the increasingly degraded state of the permanent pastures and rangelands, and despite a decline in per capita land (current average ~7.5 ha). The increasing cropland productivity due to technological interventions has allowed them to generate part of the fodder needs of their livestock from the crop wastes (Kar, 2014). In earlier times, the low-input and low-returns croplands could hardly supply the required feed for the animals, and the permanent pastures used to supply the bulk of the feed. As the croplands gradually became more productive to sustain their animals to a large extent, the permanent pastures and rangelands became less useful for them, except for fuel wood collection. Most of the farmers, therefore, felt less urge to maintain those community grazing lands although, when in healthy condition, these grazing lands provide many other ecosystem services, like yielding more water than sediments in the ponds from the catchments, wind and water erosion control, fuel wood on sustainable basis, and so on.

The future patterns of wind and water erosion in the Thar Desert will perhaps depend more on how the croplands will be used. The growing awareness of the farmers about caring for the land, when under technology-mediated uses, infuses some hopes of land conservation. This is despite the greater fears of escalating land degradation, due especially to deep ploughing by tractors and constantly falling groundwater reserves that are being encouraged by the choices now available for growing high-value cash crops.

7.6.5 Other Emerging Factors of Degradation

Among the other emerging degradation processes observed in the Thar Desert, industrial effluence needs a mention. As infrastructural facilities in the desert are increasing, it is also attracting industries and urbanization. Several towns in the southeastern part of the desert have become the hub of textile dyeing and printing due to potable water along some of the ephemeral stream valleys. The author's discussions with some of the old textile printers in the region revealed that traditionally the textile dyeing and printing works in the region used natural dye extracted from plant materials, the 'fixation' of which on cotton clothes required water of neutral to slightly alkaline nature (pH of ~7–8), and less contaminated by iron and other impurities, so that there was no fudging of the colour. Potable water in the wells along the ephemeral streams provided such 'good' quality water, and attracted the industry. These industrial establishments now use chemical dye, and release a huge quantity of untreated effluents into the ephemeral streams, while the municipalities also channel the sewages there. As a result, the aquifers have become contaminated to a distance of >50 km downstream, and the farmers downstream who formerly used to tap the potable groundwater for irrigated crops during winter receive highly contaminated groundwater instead. On an average, the effluent waters have pH >9, residual sodium carbonate (RSC) of ~35 me L^{-1}, and SAR (sodium adsorption ratio) exceeding 150. The concentrations of arsenic, copper, lead and zinc in the groundwater in village wells downstream are alarmingly high and pose health hazards (CSE, 2007). The problem has become a rallying point for different water users, and has led to the commissioning of new effluent treatment plants in selected towns, but the problem still persists because the production of effluents is much higher than the capability of the treatment plants. Frustrated, the farmers were compelled to petition the court, which led to closure of several noncompliant plants, followed by worker unrest.

7.6.6 Lessons Learnt and the Way Forward in View of Global Warming

The above narratives show how land use decisions in a vulnerable dryland can lead to land degradation, but also that the land quality can be partly regained through ameliorative measures. With greater availability of improved management and technological options with time there is now a diminishing chance of turning a land irrevocably bad. As the stakeholders become aware of the worsening land condition, they try to adopt new technologies or shift to time-tested traditional ones that would improve their land condition for a sustainable livelihood. The success rate depends on the awareness of stakeholders, access to infrastructures, knowledge and technology, policy support for stakeholder economy through sustainable land management and alternate livelihood options, especially in view of the inherent limitations of the land and a greater threat of drought. A good example of meeting successfully the threats of drought is the response system implemented by the state and the society in the wake of a century-scale 'disastrous' drought in 2002, when the region received a meagre 86 mm rain during the summer monsoon months of June–September as compared to the mean rainfall of 276 mm for the period 1961–2010. Although, in previous times, the region used to suffer immensely in the wake of even moderate to severe droughts, with consequent large-scale human and livestock migration, livestock casualties and high debt of the rural population, the sufferings during the 2002 drought were much less, due to better relief work and investments in land management activities (Narain and Kar, 2005). Viewed from this perspective, the changing perception of the research community on desertification as a people-centric one, with an aim of uncertainty management, is a welcome step forward because the scale of uncertainty that the region is likely to face in the near-future due to global warming may require a healthier asset of natural resources and a stronger infrastructural and management base. Wind erosion and drought are the two major threats that would need the most attention.

Overall, the threats of wind erosion now loom large due to large areas of the sandy desert coming under deep tillage through tractors, with destruction of natural vegetation, and also negative balance of groundwater in the aquifers due to overuse. Kar (2012) carried out a study on the likely future impacts of global warming in the Thar Desert, using the IPCC AR-4 simulation data for the A2 scenario from the following 15 GCMs at half-degree grids:

BCCR_BCM-2.0 (Norway); CCCMA_CGCM-3.1 (Canada); CNRM_CM-3.1 (France); CSIRO_MK-3.0 (Australia); MPI_ECHAM-5.1 (Germany); ECHO_G-1.0 (Germany/S. Korea); GFDL-2.01 (USA); GFDL-2.11 (USA); NASA_GISS_ER-1.0 (USA); UKMO_HADCM-3.0 (UK); INM_CM-3.01 (Russia); IPSL_CM-4.0 (France); MIROC-3.2 (Japan); NCAR_CCSM-3.01 (USA); NCAR_PCM-1.1 (USA); and the Ensemble runs of the above 15 GCMs.

The basic data were first tested for their deviation from the observed values of precipitation, temperature and wind speed during 2001–2005. Simulation data from GFDL 2.01 and ECHAM 5.01 were found to be nearer to the observed values, and therefore the likely moisture availability index and wind erosion index in the future, calculated from these two models, was considered to be more reliable and was reported (Kar, 2012). The results suggested that the soil moisture loss, the drying of the top soil and the wind erosivity are likely to gradually increase over the next few decades. Spatially, the very high wind erosivity in the Thar Desert, as determined through a wind erosion index (WEI) value of 119 and above, is presently confined to the westernmost part of the desert (Kar, 1993b), but calculations based on GCM-derived future atmospheric properties suggest that WEI will most likely increase many-fold from the end of the second decade of this century, and high wind erosivity areas may cover much larger space than at present. When seen in the context of changes made in land tillage and the impacts of near-empty aquifers, there is a greater likelihood that wind-blown sand will start spreading beyond the wetter eastern margin of the desert, where the process will be assisted by the cultivation pressures similar to that within the desert on the more stabilized paleo-bedforms of aeolian sand (Kar, 2011b). Since this may lead to higher atmospheric dust load, one may expect not only a greater export of soil nutrients with the wind beyond the desert boundaries, but also rainfall modification over northern India (Gautam et al., 2009; Bollasina and Nigam, 2011). Global warming-related increases in wind erosion and dust emission have been predicted for other arid areas also (Tegen et al.,

2004; Thomas et al., 2005; Warren, 2010; Munson et al., 2011). The need is not only for a proper understanding of the issues involved, and continuous monitoring of the human-environment system with a set of reliable indicators, but also to develop a good response system to mellow the impacts.

7.7 LINKING THE BIOPHYSICAL AND THE SOCIOECONOMIC PROCESSES

There is a growing feeling among researchers that, in order to construct the different pathways leading to desertification scenarios, it is necessary to critically examine the causes and effects of desertification, under a system that accommodates and analyses all the variables in the chain from the factors that trigger processes of desertification to the society's response to the state of desertification. In other words, modelling desertification over space and time must be carried out with quantified datasets for convincing the scientific communities and for effective policy interventions.

A well-structured analytical approach to assess the type and degree of the land degradation processes was first suggested in the 1980s (Dregne, 1983; FAO, 1984). While most of the biophysical variables are amenable to direct measurement using field methods or aerial and satellite-sensor methods, analyses of the socioeconomic variables are largely based on informed opinions. This is one reason why a socioeconomic database did not figure much in the desertification modelling efforts until very recently, save and except for some weightage provided to the concerned factors (e.g. Kharin et al., 1985; Grunblatt et al., 1992; Babaev et al., 1993; Kust and Andreeva, 1998). These and other early efforts, however, helped researchers to develop a concept of linking the causes and effects of desertification, as well as the societal response to desertification, for proper understanding of the facets of the problem and effective implementation of the countermeasures. This conceptual model, consisting of driving force (D), pressure (P), state (S), impact (I) and response (R), is described in the next section.

7.7.1 THE D-P-S-I-R CONCEPT

The D-P-S-I-R concept was first proposed in 1979 (Anonymous, 1979), and was found useful in assessment of the problem of desertification, especially in the drylands of Europe (Enne and Zucca, 2000). In the model, D represents the climatic and demographic/socioeconomic conditions, changes in which are broadly responsible for the degradation processes. Because of the changes, pressure (P) might build up on the land, and the existing land use systems might change or intensify. The pressure on the land resources triggers acceleration of the land surface processes which lead to changes in the state (S) of the land cover, manifested as different types of degradation. As the forms of degradation engulf newer areas, and more specifically the economically more productive areas, they start impacting (I) the socioeconomic fabric of the local stakeholders in terms of loss of production and income, and so on. The stakeholders/society, therefore, start responding (R) to the situation by either moving out of the area, or by taking ameliorative steps so that they can avoid future risks. This is in agreement with the current concept that (1) the dryland human-environment systems are coupled, dynamic and co-adapting, so that their structure, function and inter-relationships change over time; (2) the systems, including the scales of processes, occur in nested, hierarchical planes, necessitating their operations over multiple scales; (3) often the critical variables that determine the system dynamics are much slower than the processes that lead to geological 'disasters'; and (4) the thresholds of the variables are not static but can change with time, which may lead to some unexpected outcomes like faster degradation or slower recovery (Reynolds and Stafford Smith, 2002; Reynolds et al., 2007). Indicators, benchmarks, assessment, monitoring and early warning are the integral parts of a holistic approach to understand the causal factors and spatio-temporal characteristics of desertification processes, which can be used to inform the society about possible actions for counter-measures and find out future trends.

7.7.2 MODELING OF D-P-S-I-R

Although conceptually mature, implementation of the D-P-S-I-R model is not an easy task, given the complex relationships between the indicators involved and the quantitative nature of the input data needed to implement the model (Kar and Takeuchi, 2003). The basic data requirements for these activities are beyond the reach of many dryland-dominated countries. One of the dangers in blindly following models without verification is gross miscalculation of the patterns (e.g. in case of sediment yield and soil erosion). Many soil erosion models can work better in small watersheds with relatively homogeneous features, but as the system becomes larger and complex the calibration of input parameters becomes increasingly less precise, and with time/length of run the systematic errors become so large that the results tend to become unrealistic (Pickup, 1988).

Modelling of the biophysical parameters with remote sensing and geographic information systems (GIS) has progressed better than that with the socioeconomic parameters. Despite criticism that soil erosion rates and their implications suggested by the models are questionable (Trimble and Crosson, 2000), soil erosion modelling through fluvial processes has advanced significantly, and now enables the modeller to quantify rates to a reasonably tolerable degree of accuracy. Notable among these models are RUSLE (Renard et al., 1997), MMF (Van Rompaey et al., 2003; Morgan, 2005; Morgan and Duzant, 2008) and PESERA (PESERA, 2007), each of which have been applied in many countries. Quantification of wind erosion at local to regional scales through models, including sand/dust emission to the atmosphere and sand budgeting, still has a lower confidence level due to paucity of field data and somewhat complex relationship between the terrain and the atmospheric variables, but many useful methods and results are available. For example, the revised wind erosion equation (RWEQ) model (Fryrear et al., 1998) was used on a GIS database on soil and land uses to find out the potentials of wind erosion (Zobeck et al., 2000), but the difficulty of modelling wind erosion from a mosaic of shrub, grass and bare ground, which is common in most deserts, has also been highlighted to emphasize the need for a model that correctly accounts for vegetation structural diversity (Okin et al., 2006). For Australia, an integrated computational environmental management system (CEMSYS) was developed to predict wind erosion at national and regional scales (Butler et al., 2007). The system comprised of linked models to describe atmosphere, land surface, wind erosion, transport and deposition of sediments, as well as the required land surface database from field to satellite sensor observations. Dust emission from different landscapes in the central and western Sahara was simulated (Callot et al., 2000) for 1990–1992 using a model by Marticorena and Bergametti (1997), for which the inputs on various geomorphic properties were derived from a GIS database and from the interpretation of SPOT and Landsat images. The results from this study agreed much better with the observed data than was achieved in an earlier study by Marticorena et al. (1995), where no such detailed geomorphic database was used.

As satellite sensing technology and systems for monitoring land and atmospheric parameters are improving (some parameters are now being monitored at daily to hourly intervals, and at tolerably good spatial resolutions), a huge near real-time database is becoming available to assess the driving forces, the processes and the degradation scenarios for meaningful modelling of the desertification trends, both present and the future. Improvements in the socioeconomic indicators, their availability and linking those indicators with appropriate biophysical indicators are getting due attention for modelling under a coupled GIS-remote sensing environment (e.g. Mouat et al., 1997; Symeonakis and Drake, 2004; Santini et al., 2010; Molinari, 2014; Al-Bakri et al., 2016; Fleskens et al., 2016; Tombolini et al., 2016).

Remote sensing itself is proving to be a highly reliable tool for quantification and modelling of many biophysical parameters for desertification (e.g. Coppin et al., 2004; Roder and Hill, 2005, 2009; Vreiling, 2006; Kaneko et al., 2010; Lam et al., 2011) and drought (e.g. Singh et al., 2003; Boer and Puigdefabregas, 2005; Bayarjargal et al., 2006; Murthy et al., 2006; Rhee et al., 2010). Increased spatial resolution of the sensors, choice of object-specific spectral bands (Gibbes et al., 2010; Shruthi et al., 2011; Blaschke et al., 2014), and better revisit facilities of the satellites have now helped researchers to quantify the patterns of land surface variability and their changes more confidently, including the analysis of wind and water erosion, salinization, changes in net primary productivity, changes in albedo, and so on. New results from ground penetrating radars (GPRs) have opened the opportunities of monitoring the status of subsurface resources. For example, micro-gravity changes recorded by the Gravity Recovery and Climate Experiment (GRACE) satellites, first launched in 2002 (NASA, 2012), have helped to monitor the groundwater exploitation along the eastern fringe of the Thar Desert from 2002 to 2008 (Rodell et al., 2009). Precipitation measurements from TRMM, dust and cloud measurements from TOMS, OMI, CALIPSO, AIRS, MODIS and MISR, and the results from AMSRE and other upcoming satellites on soil moisture changes are also vastly changing our perspectives on the capabilities of remote sensing to measure environmental parameters.

Methods for linking biophysical and the socioeconomic variables within a GIS framework have also seen recent advances. For example, the application by Ibanez et al. (2008) of an eight-equation dynamic model of a generic human-resource system to a rain-fed cropping system with high soil erosion risk, an irrigated cropping system and commercial rangelands in Spain, revealed that high-profit scenarios were a major cause for desertification (Ibanez et al., 2008). Attempts have also been made to link the land use changes with the landscape processes in southern Spain, through coupling a land use change model (CLUE) with a landscape process model, simulating water and tillage erosion and sedimentation (LAPSUS) (Claessens et al., 2009). The MEDALUS model of the European Commission was applied in the western Nile Delta of Egypt by Rasmy et al. (2010), who found that urban expansion, salinization and lack of policy enforcement were the most dominant variables provoking desertification. Recently, a GIS-based software tool was developed by Santini et al. (2010) to link the status of overgrazing, vegetation productivity, soil fertility, water erosion, wind erosion and seawater intrusion, through a suite of models, to produce desertification maps of the Italian island of Sardinia. Significantly, this tool gives more importance to simulating the actual states of degradation, rather than the risk factors involved (Santini et al., 2010). Another study, in the south Churu area of Rajasthan (India) within the Thar Desert, used digital remote sensing methods to classify sand colour brightness as surrogates of sand-reactivated areas (Figure 7.3). This was followed by calculation of cropland use intensity index from land revenue data, grazing pressure index from carrying capacity of grazing areas and livestock census, and their relationship with sand reactivation pattern. It revealed that despite the poor condition of grazing areas, it was the cultivation pressure that was more responsible for sand reactivation than the grazing by animals (Kar, 2011a).

Although such modelling leads to improved understanding of the desertification conundrum, its execution needs a large database of several variables, in different time steps, which is difficult to get in many of the data-poor arid regions, especially in the developing countries. Fortunately, the biophysical data are now largely amenable to pixel-based satellite sensing, and are being increasingly derived globally through standardized procedures and therefore comparable over time and space, but the socioeconomic data are still collected through ground survey and usually by means of a set of questionnaires that suit a country's requirements and resource availability. Often the data are inadequate in terms of spatial resolution, repetitiveness or even methodologies, making it difficult to use them in spatio-temporal modelling. Large-area mapping of the state of desertification through modelling, therefore, becomes problematic. Any plan to combat desertification, however, requires first an assessment of the status of desertification, including the type and degree of degradation. Faced with such compulsion, the UNCCD decided to develop a minimum set of indicators of desertification for assessing the status of desertification at local to regional levels.

7.8 TOWARDS A COMMON MONITORING AND ASSESSMENT METHOD

Even though UNCCD is now more than two decades old, and the control measures are continuing in many countries, there is still no consensus on the methodologies for assessment of desertification at regional, sub-regional or local levels. Different regions have developed their own methods of assessment and indicators, but unless these are brought to a common platform any reliable global picture is difficult to get. Concerned about the urgent need for an improved global monitoring and assessment (M&A) of dryland degradation to support decision-making in land and water management, UNCCD commissioned a Scientific Conference in 2009, which dealt with the relevant issues and resulted in some recommendations being issued (Winslow et al., 2011).

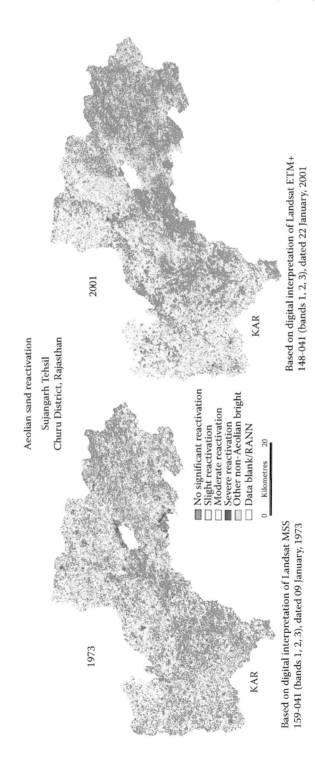

FIGURE 7.3 Separation of aeolian sand reactivation units on satellite images from other similarly bright areas in a part of the eastern Thar, using digital remote sensing methods, helped to find out the temporal variation in sand reactivation patterns and link it with the driving forces.

The Conference argued for a more holistic, harmonized and integrated approach to dryland M&A, with a scientific and institutional approach. Out of its 11 recommendations, notable were that a global M&A regime should be established to gather and analyse relevant data on a routine basis, allowing locally relevant indicators to be aggregated into meaningful classes appropriate at different decision-making levels. Since many public land use and land management decisions are taken at national and subnational levels, the UNCCD may design a strategy of data assimilation and use that are compatible with the data generated at those national and subnational levels. Since there is a need to establish synergies between UNCCD, UN Framework Convention on Climate Change and the UN Convention on Biological Diversity, and since sustainable land management (SLM) is a core issue in all the above three conventions, SLM should be fully integrated into the M&A methodologies. Additionally, while gathering data on degradation and SLM, information that relates them with the issues on climate change and biodiversity, and to other land-related issues, needs to be gathered. To aid the decision makers in setting priorities, information needs to be collected on the economic, social and environmental costs of degradation, and the benefits of SLM also (Winslow et al., 2011). The conference also suggested the use of a flexible framework of analytical approaches in identifying key indicators that are adapted to specific objectives or desertification issues (Sommer et al., 2011).

7.8.1 Global Estimates of Desertification

The first major attempt for global estimate of desertification was made for UNCOD in 1977, which suggested that at least 3.97 billion ha of drylands (or 75.1% of the drylands) were affected moderately by desertification (UNCOD, 1977). This was followed by another exercise during 1983–1984 for UNEP, which estimated that at least 3.47 billion ha (70% of drylands) were moderately affected (Dregn et al., 1991). In the beginning of the 1990s a Global Assessment of Human-Induced Soil Degradation (GLASOD), commissioned by the UNEP, was carried out by the International Soil Reference and Information Centre (ISRIC) at 1:10 M scale, mostly through harmonization of the soil datasets available at the national level, and opinions of 'experts' on the distribution maps in different regions, but it followed well-structured and consistent parameters for evaluation, involving water erosion, wind erosion, physical deterioration (i.e. crusting, soil compaction, waterlogging, etc.) and chemical degradation (i.e. salinization, loss of nutrient and organic matter, acidification and pollution). Degree of degradation was mapped as light, moderate, strong and extreme. According to it, out of 5.17 billion ha area of the global drylands, about 1.14 billion ha area was under degradation, of which 648 million ha (m ha) was moderate to severely degraded. About 478 m ha was estimated to be affected by water erosion and 513 m ha by wind erosion (Oldeman and Van Lynden, 1996; UNEP, 1997). Vegetation degradation without recording soil degradation was not considered.

At the same time the International Center for Arid and Semi-Arid Land Studies (USA) carried out a global land degradation mapping, in which the degraded rangelands were also mapped. When harmonized with the GLASOD map for the dryland areas, the degraded drylands (excluding the hyper-arid areas) were calculated as 3.59 billion ha (Dregne et al., 1991). A sequel to the GLASOD was the Assessment of Human-Induced Soil Degradation in South and Southeast Asia (ASSOD) at 1:5 M scale, which estimated that out of 958 m ha area affected in the region, moderate to strong degradation occupied 355 m ha. In India, 153 m ha area of drylands was estimated to be under degradation, out of which 60 m ha was moderate to strongly affected (UNEP, 1997). Since these assessments did not show the actual degradation status but a speculation of the potential risks based on vulnerability of the soil properties, much caution is required in using the data (Thomas and Middleton, 1994; Reynolds and Stafford Smith, 2002).

Economic evaluation of land degradation in terms of monetary loss of the stakeholders due to degradation, or for cost of rehabilitation of the land, has received some attention. It was estimated that approximately 3% area of the drylands were irrigated croplands, 9% were rain-fed croplands, and 88% rangelands, and that about 30% of the irrigated croplands, 47% of the rain-fed croplands and 73% of the rangelands were degraded, out of which the most affected 18.60 million km^2 desertified area would cost at least $213 thousand million for rehabilitation to make them profitable, at a cost-benefit ratio of 1:2.5 (Dregne, 1991; Dregne and Cho, 1992). Studies in the Caspian Sea and Aral Sea regions showed how restricting the water flow into the seas and overenthusiasm with irrigation water led to gradual drying of the two seas, increased land salinity and a huge loss of agricultural production that impacted the economy (Mainguet, 1994; Zonn, 1995; Alibekov and Alibekov, 2008). Several other assessments have also been made, especially for the funding agencies who would like to know about the size of investment and the fate of it, while a training manual for evaluating the cost of environmental degradation was prepared for the World Bank by Bolt et al. (2005).

7.8.2 TWENTY-FIRST CENTURY ASSESSMENTS

In the early 2000s, the Millennium Ecosystem Assessment (MA) commissioned its own land degradation assessment in the drylands, which showed only 10% area of the drylands as degraded (Safriel and Adeel, 2005). This was followed by the programmes Land Degradation Assessment in Drylands (LADA) and the Global Assessment of Land Degradation and Improvement (GLADA). The former developed a methodology for measuring degradation in terms of the capacity of the land to provide goods and services over a period of time for its beneficiaries. Seven different goods and services were identified for measurement and trend analysis: biomass production, yearly biomass increments, soil health, water quality and quantity, biodiversity, economic value of land use and social services of the land and its use. Using existing maps and databases and incorporating satellite sensor-derived products like Net Primary Productivity (NPP, i.e., rate of CO_2 fixation by vegetation without losses through vegetation respiration, as calculated from satellite-derived normalized difference vegetation index [NDVI]), it constructed a series of indicator maps at a global level showing spatio-temporal distribution of NPP, rainfall, aridity index, Fournier index, land cover factor, soil erodibility and land management index (IIASA, 2009).

GLADA initially considered land degradation as a long-term loss of ecosystem function and measured it in terms of NPP trends from the GIMMS NDVI at 8 km resolution, and Rainfall Use Efficiency (RUE, calculated as ratio of NPP to rainfall) such that the RUE of degraded areas is lower than that of the non-degraded areas. Subsequently, GLADA used NDVI as a proxy for NPP to produce the global trends in greenness from 1981 to 2006. The study found that (a) globally about half of the areas experienced no significant change in NDVI, (b) globally significantly negative change in NDVI due to human activities took place in about 4% area, (c) negative change due to climate change covered 5% area, (d) significant positive change in NDVI due to human activities took place in about 10% area, and (e) climate-related positive change in NDVI occurred in about 15% area (Bai et al., 2008, 2010). The NPP mapping is now taken up by another project called Global LAnd Degradation Information System (GLADIS), which is implemented in WebGIS, and is tasked to provide status and trend analysis of the capacity of the ecosystem to provide services through graphs and maps (Nachtergaele et al., 2010; Nkonya et al., 2011). The major global maps being produced by GLADIS are on: ecosystem service status index (ESSI), land degradation index (LDI), biophysical degradation index (BDI) and land degradation impact index (LDII). It then explores the link between population pressure, poverty and land degradation. The linkages suggest that globally the degraded areas are inhabited by 42% of the very poor, 32% of the moderately poor and 15% of the non-poor sections of the population (Nachtergaele et al., 2010). Like the previous efforts, GLADIS also provides a broad estimation of degradation scenarios, despite the use of new technologies, and its results are difficult to translate into any action or policy strategy. The overemphasis on vegetation index as a proxy for land degradation is a major debatable point in both GLADA and GLADIS. However, GLADIS has shown a new method of linking the biophysical and socioeconomic factors, combining aspects of biomass, soil health, water resources, biodiversity, economic production, as well as sociocultural milieu through indices that may draw the attention of researchers.

From a detailed global review of economic assessment of desertification, a recent study found that the main focus of valuation thus far was agricultural productivity, especially the direct cost of soil erosion, and that the indirect costs like human suffering caused by degradation, loss of biodiversity, and so on, received less attention, which needs to be remedied through consideration both of on-site and off-site costs (Nkonya et al., 2011). The study also suggested the consideration of all the values of use, non-use and option values in economic valuation, avoidance of double counting, and inclusion of the cost of action and inaction against degradation in a framework that recognizes the following issues: (1) desertification is highly site-specific and so needs cost assessment at the microlevel; (2) site-specific analysis must be upscaled to the national and international levels to estimate the global costs; (3) the choice of discount rates, reflecting the time preference and the time horizon for global assessment need to be considered; (4) rural poverty is a cause and a result of land degradation, and so the poverty effects need to be included in the cost estimate; and (5) uncertainties of climate change issues need to be factored properly. Some country-level case studies (India, Kenya, Niger, Peru, Uzbekistan), undertaken to test the above framework, found that the cost of action was lower than the cost of inaction, and that the public investments in addressing land degradation in developing countries were very little (Nkonya et al., 2011). Although criticized for overestimating, the study by Dregne and Chou (1992) is perhaps still the only available estimate of the cost of land degradation at the global level (see, e.g. Table A.5 of Nkonya et al. [2011], pp. 149–150).

7.8.3 Recent Desertification Mapping in India

India is one of the major participants in the Asian Regional Thematic Programme Network on Desertification Monitoring and Assessment (TPN-1) activity of UNCCD, which proposed in 2003 an elaborate structure of the procedures to be followed for mapping and monitoring of desertification at 1:50,000 to 1:250,000 scales for national level mapping, and at 1:2.5 million to 1:5 million scales for regional level mapping. For this purpose, information was collected at different tiers for agro-climatic regions, land use systems, degradation processes and severity of degradation (Anonymous, 2003). Based on the TPN-1 guidelines, India developed its first national atlas on land degradation in 2007, using data from the Advanced Wide Field Sensor (AWiFS) on board the Indian Remote Sensing Satellite, IRS-P6-Resourcesat (SAC, 2007). The data were acquired during the years 2003, 2004 and 2005. A detailed methodology was developed from the earlier-tested method of the Central Arid Zone Research Institute (CAZRI), Jodhpur, which had suggested field indicators for identifying the degree of degradation under wind erosion, water erosion, waterlogging, salinization and vegetation degradation in the Thar Desert area (Singh et al., 1992). Additional indicators for the other areas of the country were also tested and developed (Tables 7.1–7.5 [Anonymous, 2003]). Altogether 12 land use/land

TABLE 7.1
Criteria for Nationwide Mapping of the Severity of Water Erosion in India

Status	Desertification Classes		
	Slight	**Moderate**	**Severe**
	Non-Arable Land		
Type of erosion	Sheet erosion and/or single rills (depth = 0.5 m and width = 0.4–0.9 m)	Rill erosion, and/or formation of gullies (depth = 0.6–3.0 m and width = 1.0–3.5 m)	Network of gullies/ ravines (in gullies: depth 3 m = 10 m and width = 3.5–20.0 m; in ravines: depth = >10 m, width = 20–40 m)
Density of channels, linear km per sq. km	<0.5	0.5–1.5	1.5–3.0
Removal of top soil horizon, %	<25	25–50	>50
	Arable Land		
Removal of top soil horizon	<25	25–50	>50
Loss of yield of main crop, %	<25	25–50	>50

TABLE 7.2

Criteria for Nationwide Mapping of the Severity of Wind Erosion in India

Status	Desertification Classes		
	Slight	Moderate	Severe
	Non-Arable Land		
Sand sheet in cm	<30 cm per hummock up to 100 cm over plains	<50–150 cm/stable dune and sandy hummock (east of 300 mm isohyet); <90–300 cm/ reactivated sand/plant roots up to 40–100 cm (west of 300 mm isohyet)	<1–4 m dunes/100–300 interdunal sand/ Barchans 2–4 cm (mostly west of 300 mm isohyet); 2–5 m active/drifting dunes: very severe
Percent area covered with sand dunes	<30	30–70	>70
Percent area covered with sod-forming plants	50–30	30–10	<10
	Arable Land		
Removal of top horizon, %	<25	25–50	>50
Blowouts, percentage of area	<5	5–10	>10
Loss of yield of main crops, %	<25	25–50	>50

TABLE 7.3

Criteria for Nationwide Mapping of the Severity of Waterlogging in India

Status	Desertification Classes		
	Slight	Moderate	Severe
Waterlogging	Seasonal (affecting one crop); 4–6 months	Affecting two crops; >6 months of submergence	Inland marshes

cover categories, 13 degradation processes and 2 severity classes (high and low) were identified for mapping. It was revealed that by 2004–2005 India had 105.5 m ha area (32.07% of the country's area, excluding the islands) under different land degradation categories. Notable among these, water erosion covered 33.56 m ha, followed by vegetation degradation (31.66 m ha) and wind erosion (17.56 m ha). The total area affected by desertification within arid region was 34.89 m ha, while that in the semi-arid region was 31.99 m ha (Table 7.6 [SAC, 2007]). An example of the mapping style

TABLE 7.4
Criteria for Nationwide Mapping of the Severity of Salinity/Alkalinity in India

Status	Desertification Classes		
	Slight	Moderate	Severe
Soil salinity/alkalinity	4–8 dS/m; <15	8–30 dS/m; 15–40	>30 dS/m; >40
Soil salinization, solid residue, %	0.20–0.40	0.40–0.60	>0.60
Salinity of groundwater, g/L	3–6	6–10	10–30
Salinity of irrigation water, g/L	0.5–1.0	1.0–1.5	>1.5
Seasonal salt accumulation, t/ha	16–30	30–45	45–90
Loss of yield of main crop, %	<15	15–40	>40

TABLE 7.5
Criteria for Nationwide Mapping of the Severity of Vegetation Degradation in India

Status Criteria	Desertification Classes		
	Slight	Moderate	Severe
Plant community	Climax or slightly changed	Long lasting secondary	Ephemeral secondary
Percentage of climax species	> 75	75–25	< 25
Decrease of total plant cover, %	< 25	25–75	> 75
Loss of forage, %	< 25	25–75	> 75
Loss of current increment of wood, %	< 25	25–75	> 75

TABLE 7.6
Current Status of Desertification in the Drylands of India (Area in m ha)

Type of Degradation	Arid	Semi-Arid	Dry Sub-Humid	Total
Water erosion	3.67	17.67	4.87	26.21
Wind erosion	16.16	1.60	0.01	17.77
Vegetation degradation	1.97	8.43	7.23	17.63
Salinity/alkalinity	3.03	0.71	0.23	3.97
Waterlogging	0.00	0.07	0.59	0.66
Barren/rocky	0.00	1.16	0.00	1.16
Manufactured (mining, etc.)	0.00	0.08	0.03	0.11
Mass movement	3.66	0.50	0.30	4.46
Frost shattering	6.39	1.77	1.31	9.47
Frost heaving	0.01	0.00	0.00	0.01
Total	34.89	31.99	14.57	81.45

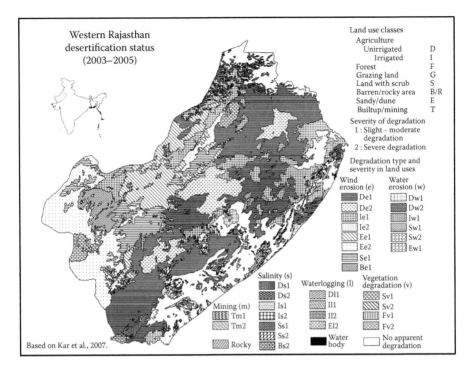

Western Rajasthan
desertification status
(2003–2005)

Land use classes
Agriculture
    Unirrigated   D
    Irrigated   I
Forest   F
Grazing land   G
Land with scrub   S
Barren/rocky area   B/R
Sandy/dune   E
Builtup/mining   T

Severity of degradation
1 : Slight - moderate
    degradation
2 : Severe degradation

Degradation type and
severity in land uses

Wind erosion (e): De1, De2, Ie1, Ie2, Ee1, Ee2, Se1, Be1

Water erosion (w): Dw1, Dw2, Iw1, Sw1, Sw2, Ew1

Salinity (s): Ds1, Ds2, Is1, Is2, Ss1, Ss2, Bs2

Mining (m): Tm1, Tm2

Rocky

Waterlogging (l): Dl1, Il1, Il2, El2

Vegetation degradation (v): Sv1, Sv2, Fv1, Fv2

Water body

No apparent degradation

Based on Kar et al., 2007.

FIGURE 7.4   Desertification status in western Rajasthan, India (2003–2005).

in the arid western part of Rajasthan state is provided in Figure 7.4, which shows the degradation types and their severity in major land uses (Kar et al., 2007).

The above effort was preceded by several assessments of land degradation in various parts of the country. The first mapping of desertification was carried out in 1977 by CAZRI at 1:2 M scale for the UN Conference on Desertification (UNCOD) at Nairobi. Subsequently several attempts were made at the country level to assess the extent of land degradation, using different methodologies (e.g. GLASSOD and its variants, perceived degree of degradation from terrain features, wasteland mapping through remote sensing, revenue records on land types, etc.). The estimates therefore varied from 53 m ha to 188 m ha. A harmonization of these different efforts suggested the likelihood of 120.7 m ha being affected by different forms of degradation by the end of the twentieth century, out of which 104.2 m ha was in arable lands (Anonymous, 2010).

7.8.4   RECENT MAPPING IN CHINA

The mapping under TPN-1 by China revealed that during 2000–2005 about 27.46% of the country's total area (263.62 m ha) was affected by different processes of desertification (Ho, 2000). Desertification due to wind erosion covered 19.16% area (183.94 m ha), while that due to water erosion covered 2.70% area (25.93 m ha), salinization 1.81% area (17.38 m ha), and freezing-thawing 3.79% area (36.37 m ha).

The respective figures for 1990–1999 were: 19.51% (187.31 m ha), 2.76% (26.48 m ha), 1.80% (17.29 m ha) and 3.79% (36.40 m ha). Provided an identical set of criteria, methodology and image data of similar resolutions were used for the two assessments, this translates into an overall reduction in desertification area. It was deduced from the study that the trend of desertification was reversed from an average annual expansion rate of 10,400 km^2 in the late twentieth century to an average annual contraction rate of 7585 km^2 during 1999–2004 (CCICCD, 2006). Earlier studies, based principally on the interpretation of aerial photographs and field information to identify areas that might have been affected by human action alone, suggested that by the last quarter of the 1990s about 86.16 m ha area of the country was affected by desertification, of which water erosion contributed 45.7%, wind erosion 44.1% and salinization, and so on, 10.2% (Zhu, 1998). The average annual rate of desertification was measured as 0.156 m ha between the late 1950s and the mid-1970s, which increased to 0.210 m ha between the mid-1970s and late 1980s, and to 0.246 m ha between the late 1980s and mid-1990s (Zhu, 1994; Wang and Wu, 1998). A detailed study in the Ordos grassland of Inner Mongolia found a very strong relationship of the spread of desertification with population growth, cultivation in sandy terrain and firewood collection (Zhu and Liu, 1995). Despite some recent works questioning those earlier assessments (Lu and Ju, 2001), the above historical studies provide important narratives on the progress of understanding land degradation processes during the second half of the twentieth century. A summary of the debate is provided by Yang et al. (2005). Desertification maps are now being prepared by many other participating countries for UNCCD and other reporting.

7.9   FROM DESERTIFICATION TO LAND DEGRADATION NEUTRALITY AND ZERO NET LAND DEGRADATION

The results of the assessment and monitoring of desertification at regional to global levels, as discussed above, do not show any significant improvement in the situation. Large areas continue to be under moderate to severe degradation, inhabited by the poorer section of the society (Nachtergaele et al., 2010). A GIS-based evaluation (Gibbs and Salmon, 2015) of some of the major global estimates of areas under desertification, carried out through different means, showed wide variability not only in area statistics of the different mapping efforts, but also in the spatial occurrences. The UNCCD (2011), in its submission to the UN Conference on Sustainable Development at Rio de Janeiro, Brazil (also called Rio+20), held from 20 to 22 June, 2012, suggested that drought and desertification in the drylands was transforming 12 million hectares of land into 'new man-made deserts' every year, an area that could potentially produce 20 million tons of grain every year. UNCCD (2011) also suggested that 78% of the degraded land occurred in non-dryland areas, and, therefore, a time had come to commit for building a land degradation neutral (LDN) world, and to set sustainable development targets toward zero net land degradation (ZNLD).

While elaborating on the need for LDN, UNCCD (2014) argued through graphics that land degradation and climate change can potentially form a feedback loop, especially when land degradation, triggered by poor land management practices with or without higher food production from the land, gets insufficiently addressed due to the reduced mitigation and adaptive capacity of the stakeholders, which can lead in turn to the increased emissions of greenhouse gases, triggering accentuation of the climate change processes. This can result in biodiversity loss due to the loss of habitats and changes in species abundance, and may result in a reduction in biodiversity and ecosystem services, as well as reduced carbon sinks, which in turn accentuate land degradation and climate change. As the cycle continues there is little scope for moving out of the feedback loop, unless the problems of land degradation, climate change and biodiversity losses are tackled in a synergistic manner.

Concerned at the serious challenges to sustainable development, narrated in the UNCCD (2011) document, and recognizing that good land management can contribute to many economic and social good, including food security, poverty eradication, and so on, the Rio+20 outcome document 'The Future We Want', as endorsed and adopted by the General Assembly of the United Nations in its resolution of 27 July 2012 (UN, 2012) agreed to the need of achieving a LDN-world in the context of sustainable development. UNCCD (2015) has defined LDN as a state whereby the amount and quality of land resources necessary to support ecosystem functions and services and enhance food security remains stable or increases within specified temporal and spatial scales and ecosystems. ZNLD, according to UNCCD (2012), is the achievement of land degradation neutrality, whereby land degradation is either avoided or offset by land restoration. Safriel (2017) reports that after proposing the offsetting principle, UNCCD found it somewhat vague and controversial as the principle could be construed as a possible license to degrade land somewhere and offset or compensate it somewhere else. By 2015, therefore, UNCCD distanced itself from the offsetting principle in favour of a focus on maintaining and improving land productivity through sustainable management and restoration of the land resources. The target date for ZNLD was set by UNCCD (2012) as 2030.

Safriel (2017) has traced in some details the reasons for a recent shift in concept in the UNCCD from 'desertification' in the drylands to LDN with a pan-world perspective. According to Safriel, the alleged failure of the earlier UNCCD efforts to effectively control degradation and its consequences on climate change and biodiversity loss is a major cause for the shift in thinking. Safriel (2017) argues that the LDN concept appeals to a wider section of scientists and policy makers more than the 'desertification' concept because the former proposes to work as an offsetting principle to land degradation, especially through proposing neutralization of the sum of degradation area through restoration of an equivalent sum of degraded area, and measuring the net balance through quantification of production and productivity of the land, as well as quantification of social benefits like food security and poverty eradication. The UN system found it to be a better alternative to address the land degradation problem globally, rather than to combat desertification in the drylands with little benefits. As Safriel (2017) informs, evaluation of the pilot projects and strategies to implement LDN at the UNCCD level, however, revealed lacunae in identification of indicators, target setting and their relevance to actual LDN goals, which could be exploited for misuse of the LDN concept, and hence LDN strategies need cautious steps.

Elaborating on the ZNLD concept, Stavi and Lal (2015) inform that if the loss of a fertile land is offset by restoration of an already degraded land, and if the annual rate of reclamation equals the degradation, then a ZNLD target is achieved. They suggest that land and soil restoration work should include not only croplands, range-lands and woodlands, but also 'natural and semi-natural lands' that do not generate direct economic revenues. Stavi and Lal (2015) feel that ZNLD would decrease the environmental footprints of agriculture, but will continue to support food security.

One is not sure if the concept and the implementation targets of ZNLD, as mentioned above, are too simplistic and utopian. In a hypothetical scenario, if the loss of land quality at one fertile cropland site is sought to be compensated by restoration work at say 20 degraded range pasture lands, the potential economic benefit from the restored pasture lands may equal the potential economic loss from the fertile cropland, but in the process the owner of the cropland may be driven to poverty and desperation, while the owner of the naturally degraded pasture lands, say the government, may like to put the restored land under commercial use, which may trigger larger social good, but the cropland owner may feel deserted. The ZNLD target for the area, though, will be achieved and the extension workers promoting the concept will seek newer areas for adoption. In a hypothetical nondemocratic set up, this ZNLD concept may be twisted to grab a good quality agricultural land for putting under some nonagricultural use, and if the production from the land fails and the land quality deteriorates due to conversion, the compensating restoration work may be carried out somewhere else under the administrative control of the state.

The concept of 'zero' in land degradation/desertification lexicon is not restricted to ZNLD alone, nor is ZNLD credited for the first use of the term 'zero'. From a summary of the proceedings of the first United Nations Conference on Desertification at Nairobi in 1977 (Biswas, 1978), we find that one of the conference objectives was described as 'to arrest and even reverse the process of desertification with the aim of achieving zero desert growth within the next 10 to 15 years' (p. 6). Again, in the same document of Biswas (1978), the global cost of corrective measures to prevent net losses of land through desertification was estimated at $400 million annually, 'the minimum amount of funds required to achieve and maintain "zero desert growth"' (p. 31). Possibly not being very sure about achieving a zero net balance in degradation, the UN General Assembly resolution of 27 July 2012 left out ZNLD from its document (UN, 2012).

According to Seitz et al. (2017), there is no 'one size fits all' solution for LDN. The land restoration efforts, according to them, need to understand the relationship between climate, land management and nonlinear ecosystem dynamics, and the efforts must be based on adequate knowledge and skill in a particular ecosystem through long-term field experiments in it, as well as be in a position to predict windows of opportunities and risk in land management there through a non-rigid managerial approach. Caspari et al. (2015), while evaluating the existing methods of land degradation assessment for their applicability in LDN projects, suggested that the assessment efforts should also be able to integrate the results from numerous sustainable land management (SLM) technologies for a proper understanding of the actual land transformations taking place. They also argued for the need to have a robust conceptual framework of LDN, which should acknowledge adequately the complexity and diversity of the ecosystem, and for a participatory approach where the scientists should play an intermediate role of knowledge-brokers.

7.10   CONCLUSION AND PERSPECTIVE

The concept and focus of desertification has changed much over the last few decades. A major shift has taken place in addressing the issues of land degradation alone to that of sustainable livelihood through land conservation and people's participation. Consequently, the methods for approaching the problems and their assessment have undergone changes. A more integrated approach that involves both the biophysical and the socioeconomic variables should now help in better understanding of the facets of desertification and through it better management strategies to control it. New developments in the technologies of remote sensing and GIS have now made it easier to assess and monitor desertification, and have opened new directions in understanding of its processes and impacts. However, a largely acceptable framework and methodologies for assessment and monitoring is still eluding the scientific communities, and new methods are being proposed in quick succession at the UNCCD level. As a result, the areas identified under different categories of degradation and their causal factors are shifting. This is not only leading to a problem for monitoring in the absence of a defined base status, but is also adding to the confusion of the policy planners as to where the problem lies, what the nature of the problem is to be addressed and what the investment areas would be. Hopefully, with more emphasis on an integrated human-environment approach to understanding and assessment of degradation, a focus on sustainable land management, livelihood and well-being of the stakeholders, and on people's participation in addressing the issues, desertification research will get the kind of global support it has wanted for a long time.

REFERENCES

Adeel, Z. and Safriel, U. (2008). Achieving sustainability by introducing alternative livelihoods. *Sustainability Science*, 3, 125–133.

Al-Bakri, J. T., Brown, L., Gedalof, Z., Berg, A., Nickling, W., Khresat, S., Salahat, M. and Saoub, H. (2016). Modelling desertification risk in the north-west of Jordan using geospatial and remote sensing techniques. *Geomatics, Natural Hazards and Risk*, 7, 531–549.

Alibekov, L. and Alibekov, D. (2008). Causes and socio-economic consequences of desertification in Central Asia. In R. Behnke (Ed.), *The Socio-Economic Causes and Consequences of Desertification in Central Asia*, Springer, Berlin, 33–41.

Anonymous. (1979). *Facing the Future, Mastering the Probable and Managing the Unpredictable*. Inter-Futures Study Team, Organisation for Economic Cooperation and Development and International Atomic Energy Agency, Paris.

Anonymous. (2001). *India: National Action Programme to Combat Desertification in the Context of United Nations Convention to Combat Desertification (UNCCD). Volume I: Status of Desertification*, Ministry of Environment & Forests, Government of India, New Delhi.

Anonymous. (2003). *Proposed benchmark and indicator system for desertification monitoring and assessment in Asian region*. In Appendix II, TPN1 Meeting on Benchmark and Indicators Development and Mapping for Desertification Monitoring and Assessment, jointly with The International Symposium on Space Technology Applications for Sustainable Dryland Development and Desertification Monitoring, Beijing.

Anonymous. (2010). *Degraded and Wastelands of India: Status and Spatial Distribution*, Indian Council of Agricultural Research, and National Academy of Agricultural Sciences, New Delhi.

Anonymous. (2011). *Elucidation of the 4th National Report Submitted to UNCCD Secretariat, 2010,* Ministry of Environment & Forests, Government of India, New Delhi.

Anonymous. (2012). *Great Green Wall for the Sahara and the Sahel Initiative: National Strategic Action Plan.* Ministry of Environment, Federal Republic of Nigeria.

Armstrong, R. (2001). *Grapes of Wrath in Inner Mongolia.* US Embassy, Beijing. http:/www /usembassy-china.org.cn/English/sandt/MongoliaDust-web.htm (accessed on 8 May, 2001).

Aubreville, A. (1949). *Climats, Forêts et Désertification de l'Afrique Tropicale,* Societe d' Éditions Géographiques, Maritimes et Coloniales, Paris.

Baartman, J. E. M., Godert, W. J., van Lynden, W. J., Reed, M. S., Ritsema, C. J. and Hessel, R. (2007). *Desertification and Land Degradation: Origins, Processes and Solutions – A Literature Review,* Report 4, Scientific Reports Series, ISRIC, the Netherlands, 2007.

Babaev, A. G., Kharin, N. G. and Orlovsky, N. S. (1993). *Assessment and Mapping of Desertification Processes: A Methodological Guide,* Desert Research Institute, Academy of Sciences of Turkmenistan, Ashkabad.

Bai, Z. G., de Jong, R. and van Lynden, G. W. J. (2010). An Update of GLADA – Global Assessment of Land Degradation and Improvement, *Report 2010/8.* ISRIC, Wageningen, 2010.

Bai, Z., Dent, D. L., Olsson, L. and Schaeoman, M. E. (2008). Global Assessment of Land Degradation and Improvement. 1. Identification by Remote Sensing. *Report 2008/01.* ISRIC, Wageningen, 2008.

Bauer, S. and Stringer, L. C. (2008). *Science and Policy in the Global Governance of Desertification: An Analysis of Institutional Interplay under the United Nations Convention to Combat Desertification.* Global Governance Working Paper No. 35, The Global Governance Project, Amsterdam.

Bayarjargal, Y., Karnieli, A., Bayasgalan, M., Khudulmur, S., Gandush, C. and Tucker, C. J. (2006). A comparative study of NOAA-AVHRR derived drought indices using change vector analysis. *Remote Sensing of Environment,* 105, 9–22.

Bharara, L. P. (1993). Livestock migration and pastoralism in Rajasthan Desert. In Sen, A. K. and Kar, A. (Eds.), *Desertification and its Control in the Thar, Sahara and Sahel Regions.* Scientific Publishers, Jodhpur, 219–247.

Biswas, M. R. (1978). *United Nations Conference on Desertification in Retrospect.* IIASA Professional Paper. IIASA, Luxemburg.

Blaschke, T., Hay, G. J., Kelly, M., Lang, S., Hofmann, P., Addink, E., Feitosa, R. Q., van der Meer, F., van der Werff, H., van Coillie, F. and Tiede, D. (2014). Geographic object-based image analysis – Towards a new paradigm. *ISPRS Journal of Photogrammetry and Remote Sensing,* 87, 180–191.

Boer, M. M. and Puigdefabregas, J. (2005) Assessment of dryland condition using spatial anomalies of vegetation index values. *International Journal of Remote Sensing,* 26, 4045–4065.

Bollasina, M., Ming, Y. and Ramaswamy, V. (2011). Anthropogenic aerosols and the weakening of the South Asian Summer Monsoon. *Science,* 334, 502–505.

Bollasina, M. and Nigam, S. (2011). Modeling of regional hydroclimate change over the Indian subcontinent: Impact of the expanding Thar Desert. *Journal of Climate,* 24, 3089–3106.

Bolt, K., Ruta, G. and Sarraf, M. (2005). *Estimating the Cost of Environmental Degradation,* The World Bank, Washington.

Bolwig, S., Cold-Ravnkilde, S. M., Rasmussen, K., Breinholt, T. and Mortimore, M. (2009). *Achieving Sustainable Natural Resource Management in the Sahel after the Era of Desertification: Markets, Property Rights, Decentralisation and Climate Change,* Report 2009:07. Danish Institute for International Studies, Copenhagen.

Brown, L. R. (1970). *Seeds of Change: The Green Revolution and Development in the 1970s.* Praeger, New York.

Butler, H. J., Shao, Y., Leys, J. F. and McTainsh, G. H. (2007). *Modelling Wind Erosion at National and Regional Scale using the CEMSYS Model: National Monitoring and Evaluation Framework*, National Land & Water Resources Audit, Canberra.

Callot, Y., Marticorena, B. and Bergametti, G. (2000). Geomorphological approach for modelling the surface features of arid environments in a model of dust emissions: Application to the Sahara Desert. *Geodinamica Acta*, 13, 245–270.

Caspari, T., van Lynden, G. and Bai, Z. (2015). *Land Degradation Neutrality: An Evaluation of Methods.* Report No. (UBA-FB) 002163/E of Federal Environment Agency, Germany. Umweltbundesamt, Dessau-Roslau.

CCICCD. (2006). *China National Report on the Implementation of the United Nation's Convention to Combat Desertification*, China National Committee for the Implementation of the UNCCD, Beijing.

Chambers, R. (1984). Beyond Green Revolution: A selective essay. In T. P. Bayliss-Smith, T. P. and Wanmali, S. (Eds.), *Understanding Green Revolutions: Agrarian Change and Development Planning in South Asia*, Cambridge University Press, Cambridge, 362–379.

Chand, R. and Raju, S. S. (2008). *Instability in Indian Agriculture during Different Phases of Technology and Policy.* Discussion Paper: NPP 01/2008, National Centre for Agricultural Economics and Policy Research, New Delhi.

Charney, J., Stone, P. H. and Quirk, W. J. (1975). Drought in the Sahara: A biophysical feedback mechanism. *Science*, 187, 434–435.

Claessens, L., Schoorl, J. M., Verburg, P. H., Geraedts, L. and Veldkamp, A. (2009). Modelling interactions and feedback mechanisms between land use change and landscape processes. *Agriculture, Ecosystems and Environment*, 129, 151–170.

Coppin, P., Jonckheere, I., Nackaerts, K., Muys, B. and Lambin, E. (2004). Digital change detection methods in ecosystem monitoring: A review. *International Journal of Remote Sensing*, 25, 1565–1596.

CSE. (2007). *Report on the Pollution in Bandi River by Textile Industries*, Centre for Science and Environment, New Delhi.

CSFD. (2011). *The African Great Green Wall Project: What Advice Can Scientists Provide?* CSFD – French Scientific Committee on Desertification, Agropolis International, Montpellier.

Dahl, B. E. and McKell, C. M. (1986). Use and abuse of China's deserts and rangelands. *Rangelands*, 8, 267–271.

Danish, K. W. (1995). International environmental law and the 'bottom-up' approach: A review of the desertification convention. *Indiana Journal of Global Legal Studies*, 3(1), 133–176.

Davis, D. K. (2005). Indigenous knowledge and the desertification debate: Problematising expert knowledge in North Africa. *Geoforum*, 36, 509–524.

Dey, S., Tripathi, S. N., Singh, R. P. and Holben, B. N. (2004). Influence of dust storms on the aerosol optical properties over the Indo-Gangetic basin. *Journal of Geophysical Research* 2004, 109 (D20211), 1–13, doi:10.1029/2004JD004924.

Dhir, R. P. (2003). Thar Desert in retrospect and prospect. *Proceedings, Indian National Science Academy*, 69A, 167–184.

Diep, L., Archer, D. and Gueye, C. (2016). Decentralisation in West Africa: The implications for urban climate change governance: The cases of Saint-Louis (Senegal) and Bobo-Dioulasso (Burkina Faso). *IIED Working Paper*, IIED, London.

Dipu, S., Prabha, T. V., Pandithurai, G., Dudhia, J., Pfister, G. P., Rajesh, K. and Goswami, B. N. (2013). Impact of elevated aerosol layer on the cloud macrophysical properties prior to monsoon onset. *Atmospheric Environment*, 70, 454–467.

Downing, T. E. and Ludeke, M. (2002). International desertification: Social geographies of vulnerability and adaptation. In Reynolds, J. F. and Stafford Smith, D. M., (Eds.) *Global Desertification: Do Humans Cause Deserts?* Dahlem University Press, Berlin, 233–252.

Dregne, H., Kassas, M. and Rozanov, B. (1991). A new assessment of the world status of desertification. *Desertification Control Bulletin*, 20, 6–18.

Dregne, H. E. (1983) *Desertification of Arid Lands.* Harwood Academic Publishers, Chur.

Dregne, H. E. (1991). Global status of desertification. *Annals or Arid Zone*, 30, 179–185.

Dregne, H. E. (2000). Desertification: Problems and challenges. *Annals or Arid Zone* 39, 363–371.

Dregne, H. E. and Chou, N.-T. (1992). Global desertification: Dimensions and costs. In Dregne, H. E. (Ed.) *Degradation and Restoration of Arid Lands*, Texas Tech. University, Lubbock, 249–281.

DSD. (2009). *Integrated Methods for Monitoring and Assessing Desertification/Land Degradation Processes and Drivers (Land Quality).* Draft White Paper of DSD Working Group 1, version 2, 19 August 2009. Drylandscience for Development Consortium (DSD).

Eklundh, L. and Olsson, L. (2003). Vegetation index trends for the African Sahel 1982–1999. *Geophysical Research Letters*, 30, 131–134.

Enne, G. and Zucca, C. (2000). *Desertification Indicators for the European Mediterranean Region: State of the Art and Possible Methodological Approaches*, ANPA – National Environmental Protection Integrated Strategies, Promotion, Communication Department: Rome, and NRD – Nucleo di Ricerca sulla Desertficazione, Universita degli Studi di Sassari, Sassari.

FAO. (1984). *Provisional Methodology for Assessment and Mapping of Desertification*, Food and Agriculture Organisation of the United Nations, Rome.

Fensholt, R. and Rasmussen, K. (2011). Analysis of trends in the Sahellian 'rain-use-efficiency' using GIMMS NDVI, RFE and GPCP rainfall data. *Remote Sensing of Environment*, 115, 438–451.

Fleskens, L., Kirkby, M. J. and Irvine, B. J. (2016). The PESERA-DESMICE modeling framework for spatial assessment of the physical impact and economic viability of land degradation mitigation technologies. *Frontiers in Environmental Science*, 4, 31, doi:10.3389/fenvs.2016.00031.

Fryrear, D. W., Saleh, A., Bilbro, J. D., Schomberg, H. M., Stout, J. E. and Zobeck, T. M. (1998). Revised Wind Erosion Equation (RWEQ). *Technical Bulletin 1, Wind Erosion and Water Conservation Research Unit*, USDA-ARS, Southern Plains Area Cropping Systems Research Laboratory.

Gautam, R., Hsu, N. C., Lau, K.-M. and Kafatos, M. (2009). Aerosol and rainfall variability over the Indian monsoon region: Distributions, trends and coupling. *Annales Geophysicae*, 27, 3691–3703.

Geist, H. J. and Lambin, E. F. (2004). Dynamic causal patterns of desertification. *BioScience*, 54, 817–829.

Ghose, B., Singh, S. and Kar, A. (1977). Desertification around the Thar: A geomorphological interpretation. *Annals of Arid Zone*, 16, 290–301.

Gibbes, C., Adhikari, S., Rostant, L., Southworth, J. and Qiu, Y. (2010). Application of object based classification and high resolution satellite imagery for Savanna ecosystem analysis. *Remote Sensing*, 2, 2748–2772.

Gibbs, H. K. and Salmon, J. M. 2015. Mapping the world's degraded lands. *Applied Geography*, 57, 12–21.

Glantz, M. H. and Orlovsky, N. S. (1983). Desertification: A review of the concept. *Desertification Control Bulletin*, 9, 12–22.

Goudie, A. S. (2014). Desert dust and human health disorders. *Environmental International*, 63, 101–113.

Grunblatt, J., Ottichilo, W. K. and Sinange, R. K. (1992). A GIS approach to desertification assessment and mapping. *Journal of Arid Environments*, 23, 81–102.

Hazell, P. B. R. (2009). *The Asian Green Revolution*. IFPRI Discussion Paper 00911. International Food Policy Research Institute, Washington, DC.

Hellden, U. (2008). A coupled human-environment model for desertification simulation and impact studies. *Global and Planetary Change*, 64, 158–168.

Hellden, U. and Tottrup, C. (2008). Regional desertification: A global synthesis. *Global and Planetary Change*, 64, 169–176.

Herrmann, S. M. and Hutchinson, C. F. (2005). The changing contexts of the desertification debate. *Journal of Arid Environments*, 63, 538–555.

Ho, P. (2000). China's rangelands under stress: A comparative study of pasture commons in the Ningxia Hui autonomous region. *Development and Change*, 31, 385–412.

Hoffmann, C., Funk, R., Li, Y. and Sommer, M. (2008). Effect of grazing on wind driven carbon and nitrogen ratios in the grasslands of Inner Mongolia. *Catena*, 75, 182–190.

Ibanez, J., Valderrama, J. M. and Puigdefabregas, J. (2008). Assessing desertification risk using system stability condition analysis. *Ecological Modelling*, 213, 180–190.

IIASA. (2009). *Land Degradation Assessment in Drylands Project (LADA) Final Report: Compilation of Selected Global Indicators of Land Degradation*, PR. No. 39701, International Institute of Applied System Analysis.

Jodha, N. S. (1985). Population growth and the decline of common property resources in Rajasthan, India. *Population and Development Review*, 11, 247–264.

Kaneko, D., Yang, P., Chang, N. B. and Kumakura, T. (2010). Developing a desertification assessment system using a photosynthesis model with assimilated multi satellite data. *International Archives of the Photogrammetry, Remote Sensing and Spatial Information Science*, 38, 547–552.

Kar, A. (1986). Physical environment, human influences and desertification in Pushkar-Budha Pushkar lake region, India. *The Environmentalist*, 6, 227–232.

Kar, A. (1993a). Neotectonic influences on morphological variations along the coastline of Kachchh, India. *Geomorphology*, 1993, 8, 199–219.

Kar, A. (1993b). Aeolian processes and bedforms in the Thar Desert. *Journal of Arid Environments*, 25, 83–96.

Kar, A. (1996). Morphology and evolution of sand dunes in the Thar Desert as key to sand control measures. *Indian Journal of Geomorphology*, 1, 177–206.

Kar, A. (2011a). Geoinformatics in spatial and temporal analysis of wind erosion in Thar Desert. In, Anbazhagan, S., Subramanian, S. K. and Yang, X., (Eds.) *Geoinformatics in Applied Geomorphology*, CRC Press, Boca Raton, FL, 39–62.

Kar, A. (2011b). Caring for the Thar. *Geography and You*, 11(March–April), 64–69.

Kar, A. (2012). GCM-derived future climate of arid western India and implications for land degradation. *Annals of Arid Zone*, 51, 147–169.

Kar, A. (2014). Agricultural land use in arid western Rajasthan: Resource exploitation and emerging issues. *Agropedology*, 24, 179–196.

Kar, A., Ajai and Dwivedi, R. S. (2016). Desertification. In Dwivedi, R. S. and Roy, P. S., (Eds.), *Geospatial Technology: For Integrated Natural Resources Management*, Yes Dee Publishing, Chennai, 295–320.

Kar, A., Moharana, P. C., Raina, P., Kumar, M., Soni, M. L., Santra, P., Ajai, Arya, A. S. and Dhinwa, P. S. (2009). Desertification and its control measures. In Kar, A., Garg, B. K., Kathju S. and Singh, M. P. (Eds.) *Trends in Arid Zone Research in India*, Central Arid Zone Research Institute, Jodhpur, 1–47.

Kar, A., Moharana, P. C. and Singh, S. K. (2007). Desertification in arid western India. In Vittal, K. P. R., Srivastava, R. L., Joshi, N. L., Kar, A., Tewari, V. P., Kathju, S., (Eds.), *Dryland Ecosystems: Indian Perspective*, Central Arid Zone Research Institute, and Arid Forest Research Institute, Jodhpur, 1–22.

Kar, A. and Takeuchi, K. (2003). Towards an early warning system for desertification. In, *Early Warning Systems*. UNCCD Ad Hoc Panel, Committee on Science and Technology. UN Convention to Combat Desertification, Bonn, 37–72.

Kar, A. and Takeuchi, K. (2004). Yellow dust: An overview of research and felt needs. *Journal of Arid Environments*, 59, 167–187.

Kharin, N. G., Nechaeva, N. T., Nikolaev, V. N., Babaeva, T., Dobrin, L. G., Babaev, A., Orlovsky, N. S., Redzhepbaev, K., Kirsta, B. T., Nurgeldyev, O. N., Batyrov, A. and Svintsov, I. P. (1985). *Methodological Principles of Desertification Processes Assessment and Mapping (Arid Lands of Turkmenistan Taken as Example)*, Desert Research Institute, Turkmen SSR Academy of Sciences, Askhabad.

Kolkovsky, S., Hulata, G., Simon, Y., Segev, R. and Koren, A. (2003). Integration of agri-aquaculture systems – The Israeli experience. In Gooley, G. J. and Gavine, F. M. (Eds.), *Integrated Agri-Aquaculture Systems, A Resource Handbook for Australian Industry Development*, Rural Industrial Research and Development Corporation, RIRDC Publication, Kingston, Australia, 14–23.

Kust, G. S. and Andreeva, O. V. (1998). Application of desertification assessment methodology for soil degradation mapping in the Kalmyk Republic of the Russian Federation. *Desertification Control Bulletin*, 32, 2–12.

Lam, D. K., Remmel, T. K. and Drezner, T. D. (2011). Tracking desertification in California using remote sensing: A sand dune encroachment approach. *Remote Sensing*, 3, 1–13.

Lambin, E. F., Turner, B. L. II, Geist, H. J., Agbola, S. B., Angelsen, A., Bruce, J. W., Coomes, O. et al. (2001). The causes of land use and land cover change: Moving beyond the myths. *Global Environmental Change*, 11, 261–269.

Lavauden, L. (1927). Les forêts du Sahara. *Revue des Eaux et Forêts* 6, 265–277, 7, 329–341.

Lu, Q. and Ju, H. (2001). Root causes, processes and consequence analysis of sandstorms in northern China in 2000. In Yang, Y., Squires, V. R. and Lu, Q. (Eds.). *Global Alarm: Dust and Sandstorms from the World's Drylands*, UNCCD, Bonn, 241–254.

Mainguet, M. (1994). *Desertification: Natural Background and Human Mismanagement*. Springer-Verlag, Berlin.

Marticorena, B. and Bergametti, G. (1995). Modeling the atmospheric dust cycle: 1. Design of a soil-derived dust emission scheme. *Journal of Geophysical Research*, 100(D8), 16415–16430.

Marticorena, B., Bergametti, G., Aumont, B., Callot, Y., N'Doume, C. and Legrand, M. (1997) Modeling the atmospheric dust cycle: 2. Simulation of Saharan dust sources. *Journal of Geophysical Research*, 102(D4), 4387–4404.

Ministry of Rural Development. (2014). The Mahatma Gandhi National Rural Employment Guarantee Act 2005 (website). Department of Rural Development, Ministry of Rural Development, GOI. http://www.nrega.nic.in/netnrega/home.aspx. (Accessed 18 April, 2017).

Molinari, P. (2014). A geographic information system (GIS) with integrated models: A new approach for assessing the vulnerability and risk of desertification in Sardinia (Italy). *Global Bioethics*, 25, 27–41.

Morgan, R. P. C. (2005). *Soil Erosion and Conservation*, 3rd ed., Blackwell.

Morgan, R. P. C. and Duzant, H. (2008). Modified MMF (Morgan-Morgan-Finney) model for evaluating effects of crops and vegetation cover on soil erosion. *Earth Surface Processes and Landforms*, 32, 90–106.

Mortimore, M. and Turner, B. (2005). Does the Sahelian small holders' management of woodland, farm trees, and rangeland support the hypothesis of human-induced desertification? *Journal of Arid Environments*, 63, 567–595.

Mortimore, M. J. (1998). *Roots in the African Dust*. Cambridge University Press, Cambridge.

Mouat, D., Lancaster, J., Wade, T., Wickham, J., Fox, C., Kepner, W. and Ball, T. (1997). Desertification evaluated using an integrated environmental assessment model. *Environmental Monitoring and Assessment*, 48, 139–156.

Munson, S. M., Belnap, J. and Okin, G. S. (2011). Responses of wind erosion to climate-induced vegetation changes on the Colorado Plateau. *Proceedings, National Institute of Science*, 108, 3854–3859.

Murthy, C. S., Seshasai, M. V. R., Kumari, B. and Roy, P. S. (2006). Agricultural drought assessment at disaggregated level using AWiFS/WiFS data of Indian remote sensing satellites. *Geocarta International*, 22, 127–140.

Nachtergaele, F., Petri, M., Biancalani, R., Van Lynden, G. and Van Velthuizen, H. (2010). Global Land Degradation Information System (GLADIS), Beta Version: An Information Database for Land Degradation Assessment at Global Level, Land Degradation Assessment in Drylands. *Technical Report No. 17*, FAO, Rome.

Narain, P. and Kar, A. (Eds). (2005). *Drought in Western Rajasthan: Impact, Coping mechanism and Management Strategies*, Central Arid Zone Research Institute, Jodhpur.

NASA. (2012). *Mission Overview*. https://www.nasa.gov/mission_pages/Grace/overview/index.html (accessed 18 April, 2017).

Niyogi, D., Chang, H.-I., Chen, F., Gu, L., Kumar, A., Menon, S. and Pielke, R. A. (2007). Potential impacts of aerosol-land-atmosphere interactions on the Indian monsoonal rainfall characteristics. *Natural Hazards*, 42, 345–359.

Nkonya, E., Gerber, N., Baumgartner, P., Von Braun, J., De Pinto, A., Graw, V., Kato, E., Kloos, J. and Walter, T. (2011). *The Economics of Desertification, Land Degradation, and Drought: Towards an Integrated Global Assessment*, ZEF-Discussion Papers on Development Policy No. 150.

Odingo, R. S. (1992). Implementation of the Plan of Action to Combat Desertification (PACD) 1978–1991. *Desertification Control Bulletin*, 21, 6–14.

Okin, G. S., Gillette, D. A. and Herrick, J. E. (2006). Multi-scale controls on and consequences of aeolian processes in landscape change in arid and semi-arid environments. *Journal of Arid Environments*, 65, 253–275.

Oldeman, L. R. and Van Lynden, G. W. J. (1996). Revisiting the GLASOD Methodology, *Working Paper 96/03*, International Soil Reference and Information Centre, Wageningen.

Orr, B. J. (2011). Scientific review of the UNCCD provisionally accepted set of impact indicators to measure the implementation of strategic objectives 1, 2 and 3. White Paper – Version 1, UNCCD, Bonn.

Patra, P. K., Behera, S. K., Herman, J. R., Maksyutov, S., Akimoto, H. and Yamagata, T. (2005). The Indian summer monsoon rainfall: Interplay of coupled dynamics, radiation and cloud microphysics. *Atmospheric Chemistry and Physics Discussions*, 5, 2879–2895.

PESERA. (2007). *PESERA User's Manual*. Revised for the ENVASSO Project by R. J. A. Jones in collaboration with B. Irvine and M. Kirkby. Joint Research Centre, European Union, Ispra.

Pickup, G. (1988). Modelling arid zone soil erosion at the regional scale. In Warner R. F., (Ed.), *Fluvial Geomorphology of Australia*, Academic Press Australia, Sydney, 105–127.

Pickup, G. (1996). Estimating the effects of land degradation and rainfall variation on productivity in rangelands: An approach using remote sensing and models of grazing and herbage dynamics. *Journal of Applied Ecology*, 33, 819–832.

Pingali, P. L. (2012). Green Revolution: Impacts, limits, and the path ahead. *PNAS*, 109, 12302–12308.

Poesen, J. W., Vandaele, K. and Van Wesemael, B. (1996). Contribution of gully erosion to sediment production on cultivated lands and rangelands. *IAHS Publication*, no. 236, 251–266.

Prince, S. D. (2002). Spatial and temporal scales for detection of desertification. In Reynolds, J. F. and Stafford Smith, D. M. (Eds.), *Global Desertification: Do Humans Cause Deserts?* Dahlem University Press, Berlin, 23–40.

Puigdefabregas, J. (1995). Desertification: Stress beyond resilience, exploring a unifying process structure. *Ambio*, 24, 311–313.

Puigdefabregas, J. (1998). Ecological impacts of global change on drylands and their implications for desertification. *Land Degradation and Development*, 9, 393–406.

Rasmy, M., Gad, A., Abdelsalam, H. and Siwailam, M. (2010). A dynamic simulation model of desertification in Egypt. *Egyptian Journal of Remote Sensing and Space Sciences*, 13, 101–111.

Renard, K. G., Foster, G. R., Weesies, G. A., McCool, D. K. and Yoder, D. C. (1997). Predicting Soil Erosion by Water: A Guide to Conservation Planning with the Revised Universal Soil Loss Equation, *Agricultural Handbook 703*, USDA.

Reynolds, J. F., Grainger, A., Stafford-Smith, D. M., Bastin, G., Garcia-Barrios, L., Fernandez, R. J., Janssen, M. A. et al. (2011). Scientific concepts for an integrated analysis of desertification. *Land Degradation and Development*, 22, 166–181.

Reynolds, J. F. and Stafford Smith, D. M. (Eds.). (2002). Global Desertification: Do Humans Cause Desertification? *Dahlem Workshop Report 88*, Dahlem University Press, Berlin.

Reynolds, J. F., Stafford Smith, D. M., Lambin, E. F., Turner, B. L. II, Mortimore, M., Batterbury, S. P. J., Downing, T. E. et al. (2007). Global desertification: Building a science for dryland development. *Science*, 316, 847–851.

Rhee, J., Im, J. and Carbone, G. J. (2010). Monitoring agricultural drought for arid and humid regions using multi-sensor remote sensing data. *Remote Sensing of Environment*, 114, 2875–2887.

Ribot, J. C. (2002). *Democratic Decentralization of Natural Resources*. World Resources Institute, Washington, DC.

Rodell, M., Velicogna, I. and Famiglietti, J. S. (2009). Satellite-based estimates of groundwater depletion in India. *Nature*, 460, 999–1002.

Roder, A. and Hill, J. (Eds.) (2005). *Proceedings of the 1st International Conference on Remote Sensing and Geoinformation Processing in the Assessment and Monitoring of Land Degradation and Desertification, Trier, Germany, 2005*, Remote Sensing Department, University of Trier, Germany. http://ubt.opus.hbz-nrw.de/volltexte/2006/362.

Roder, A. and Hill, J. (Eds.) (2009). *Recent Advances in Remote Sensing and Geoinformation Processing for Land Degradation Assessment*, CRC Press, Boca Raton, FL.

SAC. (2007). *Desertification and Land Degradation Atlas of India*, Space Applications Centre, Ahmedabad.

Safriel, U. (2006). Dryland development, desertification and security in the Mediterranean. In Kepner, W. G., Rubio, J. L., Mouat, D. A. and Pedrazzini, F. (Eds.), *Desertification in the Mediterranean Region: A Security Issue*, NATO Security through Science Series, vol. 3, Springer, 227–250.

Safriel, U. (2017). Land Degradation Neutrality (LDN) in drylands and beyond – Where has it come from and where does it go. *Silva Fennica*, 51(1B), article ID 1650, https://doi.org/10.14214/sf.1650.

Safriel, U. and Adeel, Z. (Coordinating Lead Authors) (2005). Dryland systems. In Hassan, R., Scholes, R. and Ash, N. (Eds.). *Ecosystems and Human Well-being: Current State and Trends*, Volume 1, Millennium Ecosystem Assessment, Island Press, Washington, 623–662.

Safriel, U. and Adeel, Z. (2008). Development paths of drylands: Thresholds and sustainability. *Sustainability Science*, 3, 117–123.

Salvati, L., Forino, G. and Zitti, M. (2011). Socio-economic factors and land degradation: Mediterranean perspectives. *Annals of Arid Zone*, 50, 279–294.

Santini, M., Caccamo, G., Laurenti, A., Noce, S. and Valentini, R. (2010). A multi-component GIS framework for desertification risk assessment by an integrated index. *Applied Geography*, 30, 394–415.

Schlesinger, W. H., Reynolds, J. F., Cunnighan, G. L., Huenneke, L. F., Jarrell, W. M., Virginia, R. A. and Whitford, W. G. (1990). Biological feedbacks in global desertification. *Science*, 247, 1043–1048.

Seaquist, J. W., Hickler, T., Eklundh, L., Ardo, J. and Heumann, B. W. (2009). Disentangling the effects of climate and people on Sahel vegetation dynamics. *Biogeosciences*, 6, 469–477.

Seitz, D., Fleskens, L. and Stringer, L. (2017). Learning from non-linear ecosystem dynamics is vital for achieving land degradation neutrality. *Land Degradation and Development*, doi:10.1002/ldr.2732.

Shruthi, R. B. V., Kerle, N. and Jetten, V. (2011). Object-based gully feature extraction using high spatial resolution imagery. *Geomorphology*, 134, 260–268.

Singh, R. P., Roy, S. and Kogan, F. (2003). Vegetation and temperature condition indices from NOAA AVHRR data for drought monitoring over India. *International Journal of Remote Sensing*, 24, 4393–4402.

Singh, S., Kar, A., Joshi, D. C., Kumar, S. and Sharma, K. D. (1994). Desertification problem in western Rajasthan. *Annals or Arid Zone*, 33, 191–202.

Singh, S., Kar, A., Joshi, D. C., Ram, B., Kumar, S., Vats, P. C., Singh, N., Raina, P., Kolarkar, A. S. and Dhir, R. P. (1992). Desertification mapping in western Rajasthan. *Annals or Arid Zone*, 31, 237–246.

Sommer, S., Zucca, C., Grainger, A., Cherlet, M., Zougmore, R., Sokona, Y., Hill, J., Della Peruta, R., Roehrig, J. and Wang, G. (2011). Application of indicator systems for monitoring and assessment of desertification from national to global scales. *Land Degradation and Development*, 22, 184–197.

Stavi, I. and Lal, R. (2015). Achieving zero net land degradation: Challenges and opportunities. *Journal of Arid Environments*, 112, 44–51.

Stein, M. A. (1921). *Serindia: Detailed Report of Explorations in Central Asia and Westernmost China*. vol. 1, Oxford University Press, London.

Stocking, M. and Murnaghan, N. (2000). *Land Degradation: Guidelines for Field Assessment*, Overseas Development Group, University of East Anglia, UK.

Stringer, L. C., Reed, M. S., Fleskens, L., Thomas, R. J., Le, Q. B. and Lala-Pritchard, T. (2017). A new dryland development paradigm grounded in empirical analysis of dryland systems science. *Land Degradation and Development* (accepted article). Doi:10.1002/ldr.2716.

Sundriyal, Y. P., Shukla, A. D., Rana, N., Jayangodaperumal, R., Srivastava, P., Chamyal, L. S., Sati, S. P. and Juyal, N. (2013). Terrain response to the extreme rainfall event of June 2013: Evidence from the Alaknanda and Mandakini River valleys, Garhwal Himalaya, India. *Episodes*, 38(3), 1–10.

Swift, J. (1996). Desertification: Narratives, winners and losers. In Leach, M. and Mearns, R. (Eds.) *The Lie of the Land: Challenging Received Wisdom in African Environmental Change and Policy*, James Currey, Oxford, 73–90.

Symeonakis, E. and Drake, N. (2004). Monitoring desertification and land degradation over sub-Saharan Africa. *International Journal of Remote Sensing*, 25, 573–592.

Tegen, I., Werner, M., Harrison, S. P. and Kohfeld, K. E. (2004). Relative importance of climate and land use in determining present and future global soil dust emission. *Geophysical Research Letters*, 31(L05105): 1–4, doi:10.1029/2003GL019216.

Thomas, D. S. G. (1997). Science and the desertification debate. *Journal of Arid Environments*, 37, 599–608.

Thomas, D. S. G., Knight, M. and Wiggs, G. F. S. (2005). Remobilization of southern African desert dune systems by twenty-first century global warming. *Nature*, 435, 1218–1221.

Thomas, D. S. G. and Middleton, N. (1994). *Desertification: Exploding the Myth*, John Wiley, London.

Tiffen, M., Mortimore, M. and Gichuki, F. (1994). *More People Less Erosion: Environmental Recovery in Kenya*, John Wiley & Sons, Chichester.

Tombolini, I., Colantoni, A., Renzi, G., Sateriano, A., Sabbi, A., Morrow, N. and Salvati, L. (2016). Lost in convergence, found in vulnerability: A spatially-dynamic model for desertification risk assessment in Mediterranean agro-forest districts. *Science of the Total Environment*, 569–570, 973–981.

Trimble, S. W. and Crosson, P. (2000). US soil erosion rates – Myth and reality. *Science*, 289, 248–250.

Tucker, C. J., Dregne, H. E. and Newcombe, W. W. (1991). Expansion and contraction of the Sahara Desert from 1980–1990. *Science*, 253, 299–301.

Umutoni, C., Ayantunde, A., Turner, M. and Sawadogo, G. (2016). Community participation in decentralized management of natural resources in the southern region of Mali. *Environment and Natural Resources Research*, 6, http://dx.doi.org/10.5539/enrr.v6n2p1.

UN. (1992). Managing fragile ecosystems: Combating desertification and drought. In Chapter 12, Agenda 21, United Nations Conference on Environment and Development, Rio de Janeiro, A/CONF.151/4 (Part II), United Nations, New York. 46–66.

UN. (2012). *Resolution adopted by the General Assembly on 27 July 2012 – The future we want*. Report A/RES/66/288: United Nations.

UNCCD. (1995). *United Nations Convention to Combat Desertification*, Interim Secretariat for the Convention to Combat Desertification, Geneva.

UNCCD. (2007). Decision 3/COP.8: The 10-year strategic plan and framework to enhance the implementation of the Convention. (available online at http://www2.unccd.int/sites/default/files/relevant-links/2017-01/Decision%203COP8%20adoption%20of%20The%20Strategy%20%281%29_0.pdf. Accessed 18 April, 2017).

UNCCD. (2011). *Towards a Land Degradation Neutral World. The submission of the UNCCD Secretariat to the preparatory process for the Rio+20 Conference*. Revised version 18 November 2011. UNCCD Secretariat, Bonn,.

UNCCD. (2012). *Zero Net Land Degradation. A Sustainable Development Goal for Rio+20*. UNCCD Secretariat Policy Brief. UNCCD Secretariat, Bonn.

UNCCD. (2014). *Land Degradation Neutrality: Resilience at Local, National and Regional Levels*. UNCCD Secretariat, Bonn.

UNCCD. (2015). *Report of the Intergovernmental Working Group on the follow-up of the outcomes of the United Nations Conference on Sustainable Development (Rio+20)*. Advance Draft-01 June 2015. Available at: http://www.unccd.int/Lists/Site DocumentLibrary/Rio+20/IWG%20on%20rio%2020/ADVANCE%20DRAFT%20 IWG%20Report_01_June_2015.pdf.

UNCOD. (1977). *Round-up, Plan of Action and Resolutions*, United Nations Conference on Desertification, Nairobi.

UNEP. (1997). *World Atlas of Desertification*, 2nd ed. United Nations Environment Programme, Nairobi, and Edward Arnold, London.

Van Rompaey, A. J. J., Vieillefont, V., Jones, R. J. A., Montanarella, L., Verstraeten, G., Bazzoffi, P., Dostal, T. et al. (2003). Validation of Soil Erosion Estimates at European Scale, *European Soil Bureau Research Report 13*, Office for Official Publications of the European Communities, Luxembourg.

Veron, S. R., Paruelo, J. M. and Oesterheld, M. (2006) Assessing desertification. *Journal of Arid Environments*, 66, 751–763.

Vrieling, A. (2006). Satellite remote sensing for water erosion assessment: A review. *Catena*, 65, 2–18.

Wang, T. and Wu, W. (1998). Combating desertification in China, archive.unu.edu/env/Land/iran-1/06-WangTao%20Paper.doc.

Wang, X., Chen, F. and Dong, Z. (2006). The relative role of climatic and human factors in desertification in semi-arid China. *Global Environmental Change*, 16, 48–57.

Warren, A. (2002). Land degradation is contextual. *Land Degradation and Development*, 13, 449–459.

Warren, A. (2010). Sustainability in aeolian systems. *Aeolian Research*, 1, 95–99.

Warren, A. and Khogali, M. (1992). *Assessment of Desertification and Drought in the Sudano-Sahelian Region (1985–1991)*. United Nations Sudano-Sahelian Office, New York.

Warren, A. and Olsson, L. (2003) Desertification: Loss of credibility despite the evidence. *Annals or Arid Zone*, 42, 271–288.

WCED (Ed). (1987). *Our Common Future*. Oxford University Press, Oxford, 1987.

Wessels, K. J., Prince, S. D., Malherbe, J., Small, J., Frost, P. E. and VanZyl, D. (2007). Can human-induced land degradation be distinguished from the effects of rainfall variability? A case study in South Africa. *Journal of Arid Environments*, 68, 271–297.

Williams, D. M. (1997). Grazing the body: Violations of land and limb in Inner Mongolia. *American Ethnologist*, 24, 763–785.

Winslow, M. D., Vogt, J. V., Thomas, R. J., Sommer, S., Martius, C. and Akhtar-Schuter, M. (2011). Editorial: Science for improving the monitoring and assessment of dryland degradation. *Land Degradation and Development*, 22, 145–149.

Wolman, M. G. and Gerson, R. (1978). Relative scales of time and effectiveness of climate in watershed geomorphology. *Earth Surface Processes*, 3, 189–208.

Yang, X., Zhang, K., Jia, B. and Ci, L. (2005). Desertification assessment in China: An overview. *Journal of Arid Environments*, 63, 517–531.

Zhu, Z. (1994). The status and prospect of desertification in China. *Journal of Geographical Science*, 650–659.

Zhu, Z. (1998). Concept, cause and control of desertification in China. *Quaternary Sciences*, 5, 145–155.

Zhu, Z. and Liu, S. (1983). *Combating Desertification in Arid and Semi-Arid Zones in China*. Institute of Desert Research, Lanzhou.

Zhu, Z. and Liu, Y. (1995). Sand dune stabilization in China. In Sen, A. K. and Kar, A. (Eds.), *Land Degradation and Desertification in Asia and the Pacific Region*, Scientific Publishers, Jodhpur, 273–293.

Zobeck, T. M., Parker, N. C., Haskell, S. and Guoding, K. (2000). Scaling up from field to region for wind erosion prediction using a field-scale wind erosion model and GIS. *Agriculture, Ecosystems and Environment*, 82, 247–259.

Zonn, I. (1995). Desertification in Russia: Problems and solutions. *Environmental Monitoring and Assessment*, 37, 347–363.

8 Drought Forecasting
Artificial Intelligence Methods

Jan Adamowski and Anteneh Belayneh

CONTENTS

8.1 INTRODUCTION

Throughout history, most human settlements, however small or prosperous, have had to contend with drought. Droughts are a natural feature of climate and can occur in all climatic zones. Drought can broadly be defined as a negative departure from the normal precipitation over a period of time in a given area. However, different definitions of drought exist. Meteorological drought refers more specifically to a precipitation deficit, and is the focus of our predictions; agricultural drought refers to a deficit in soil moisture for plant growth; and hydrological drought refers to a deficiency in bulk water availability (Forzieri et al., 2014). Drought has mostly been linked to changes in the precipitation regime but other climatic factors, such as high temperatures and high winds can increase the severity of an event (Forzieri et al., 2014).

Droughts often become highly visible when they are associated with famine, an acute shortage of food. Prolonged drought is but one of many possible causes of famine; others include overproduction, poverty, the ravages of war, and destruction of crops and grazing by fire, diseases, locusts and other pests (Whitmore, 2000). Yet, for the most part droughts can occur without resulting in famine. Indeed, famines have frequently taken place in the absence of drought conditions (Glanz, 1994). A series of famines in India between 1860 and 1877 were widely blamed on economic and administrative policies rather than just by drought conditions (Srivastava, 1968). Often drought, which has been described as a 'creeping' phenomenon, combines with other societal and environmental conditions to produce famine-like conditions. Drought has also been blamed for environmental degradation and desertification, prompting mass migration of established communities and internal unrest. While drought may play an important factor in each of the aforementioned processes, it often becomes one of many intervening factors (Whitmore, 2000). There is little surprise when drought-related crop failures occur in developing countries, yet developed countries such as the United States and Canada have not been able to prevent their agricultural systems from also being affected by drought. Thus, no country can claim to be immune from the impacts of drought.

8.2 DROUGHT AS A NATURAL HAZARD

A natural hazard is defined as a threat of a naturally occurring event that will have a negative effect on people or the environment (Mishra and Singh, 2010). Drought is a type of natural hazard that is further aggravated by human societal pressures associated with a growing global water demand (Mishra and Singh, 2010). Occurrences of drought are dependent on atmospheric conditions as well as the hydrologic processes which feed moisture to the atmosphere. Once dry hydrologic conditions are established, positive feedback mechanisms of droughts set in, whereby moisture depletion from upper soil layers lowers evapotranspiration rates, which in turn lower the relative humidity of the atmosphere. A lower relative humidity corresponds to a lower chance of precipitation. Only disturbances that carry enough moisture from outside the dry region will be able to produce sufficient rainfall to end drought conditions (Mishra and Singh, 2010).

Drought ranks first among all natural hazards when measured in terms of the number of people affected (Byun and Wilhite, 1999). They differ from other natural hazards in several ways. First, the onset and the end of a drought are difficult to determine. The impacts of a drought increase slowly and often accumulate over a considerable period, and may linger years after the end of the drought. Second, defining drought conditions is difficult and may lead to confusion due to the lack of a universal definition. Third, unlike other natural hazards, the impacts of drought are non-structural and spread over large geographical areas, resulting in difficulty in the quantification of impact and for the provision of relief (Mishra and Singh, 2010). Human activities can directly trigger a drought, unlike most other natural hazards, by exacerbating drought conditions through overfarming, excessive irrigation, deforestation and overuse of available water resources.

Droughts produce a complex web of impacts that span many sectors of a society, including the regional and global economy, reaching well beyond the area experiencing drought. Droughts are a widespread phenomenon since half the Earth's terrestrial surfaces are susceptible to them (Kogan, 1997). Further, almost all major agricultural lands are located in areas susceptible to drought (Mishra and Singh, 2010).

The impacts of drought are complex due to the difficulty that arises in identifying the inception and the end of a given drought (Belayneh et al., 2014). Approximately 22% of the economic damage caused by natural disasters, globally, and 33% of the damage in terms of the number of persons affected can be attributed to drought (Keshavarz et al., 2013). A 2012 drought that hit the Midwest region of the United States resulted in agricultural losses that were over $20 billion (Kam et al., 2014). In Europe, the average annual economic damage because of droughts has reached $8.14 billion (Forzieri et al., 2014). The ramifications of drought are all the starker in sub-Saharan Africa where rain-fed agriculture is the backbone for most countries in the region. Apart from the economic impacts caused by droughts in sub-Saharan Africa, recent droughts in 2008–2009 and 2010–2011 in the Horn of Africa were responsible for famine and a great humanitarian crisis (Mwangi et al., 2014). Table 8.1 lists some of the major droughts throughout history and their impacts on the regions in which they occurred.

Given the potential impacts of droughts, it is essential that they are forecasted accurately with sufficient lead time to help mitigate some of the consequences. Drought forecasts can be done using either physical/conceptual models or data-driven models. The latter have become increasingly popular in hydrologic forecasting due to minimum information requirements, rapid development times, and have been found to be accurate in various hydrological applications (Adamowski, 2008). Most data-driven drought forecasts focus on meteorological drought, which is usually a precursor to agricultural and hydrological drought (Kam et al., 2014). They use precipitation, either as the sole input or in combination with other meteorological elements. A combination of meteorological variables could include precipitation and temperature or precipitation and soil moisture.

Forecasters have several drought indices to choose from depending on their area of thematic or application interest, including the Palmer Drought Index, the Crop Moisture Index and the Standardized Precipitation Index. The latter index is

TABLE 8.1
Major Droughts and Their Impacts in the Last 150 Years

Location	Time Frame	Impact
United States	1988–1989	$60 billion in economic damages
United States	1930–1936	80% of United States experienced drought
Horn of Africa	2011–2012	12 million people affected
China	1876–1879	Approximately 1 million deaths
Africa	1981–1984	Up to 20,000 died of starvation
Australia	2002–2007	Federal government paid $4.5 billion in drought assistance
Canada	2001–2002	$5.8 billion in economic damage

Note: All dollar figures are in U.S. dollars.

recommended by the World Meteorological Organization for meteorological drought monitoring because it requires only precipitation as an input. This makes it easier to implement, making it possible to describe drought on multiple timescales that are analogous to agricultural and hydrological drought conditions (WMO, 2009). These data-driven indices are described in more detail below, and their strengths and their limitations compared and assessed.

8.2.1 THE PALMER DROUGHT SEVERITY INDEX (PDSI)

The Palmer Drought Severity Index (PDSI) was developed by Wayne Palmer in 1965 and was the first comprehensive effort to assess the total moisture status of a region (Mishra and Singh, 2010). The index is based on the water balance, using a concept of supply and demand over a two-layer soil model. The basis of the PDSI is the difference between the amount of precipitation required to retain a normal water balance level and the actual precipitation (Bordi and Sutera, 2007). Several coefficients are calculated which define local hydrological norms related to temperature and precipitation. The calculation of the coefficients above depends heavily on the soil water capacity of the underlying layer (Bordi and Sutera, 2007). Since the inception of the PDSI, modified versions such as the Palmer Hydrologic Drought Index (PHDI) have evolved. The PHDI is commonly used for water supply monitoring.

There are several limitations of the PDSI, which include an inherent timescale making PDSI more suitable for agricultural impacts and less so for hydrologic droughts. The PDSI assumes that all precipitation comes in the form of rainfall, making values obtained during winter months and at high elevations often questionable. The PDSI also assumes that runoff only occurs after all soil layers have been saturated, leading to an underestimation of runoff (Mishra and Singh, 2010). In addition, the PDSI can be slow to respond to developing and diminishing drought patterns (Mishra and Singh, 2010).

8.2.2 CROP MOISTURE INDEX

The crop moisture index (CMI) was developed by Wayne Palmer in 1968 to evaluate short-term moisture conditions across major crop producing regions. The CMI identifies potential agricultural droughts. It is not intended to assess long-term droughts. Computation of the CMI involves the use of weekly values of temperature and precipitation to determine a simple moisture budget. The CMI responds to rapidly changing conditions, which may ultimately provide misleading information about long-term conditions.

Another limitation of the CMI is its sensitivity to potential evapotranspiration. A rise in the CMI may occur with an increase in potential evapotranspiration, thereby indicating wetter moisture conditions; however, there is no natural case where an increase in potential evapotranspiration would produce wetter moisture conditions. The second limitation of the CMI is that it is not a long-term drought-monitoring tool, due to its rapid response to changing short-term conditions. CMI is best suited for measuring agricultural drought during warm seasons (Heim, 2002).

8.2.3 STANDARDIZED PRECIPITATION INDEX (SPI)

The standardized precipitation index (SPI) was developed by McKee et al. (1993). It offers many advantages. The index is based on precipitation alone, making its evaluation relatively simple (Cacciamani et al., 2007). Secondly, the index makes it possible to describe drought on timescales that typically encapsulate the four types of drought described above. As mentioned earlier, indices developed over the short term are useful in monitoring meteorological drought and agricultural drought. Likewise, indices developed over long-term timescales are useful for monitoring hydrologic and socioeconomic drought (Cacciamani et al., 2007). A third advantage of the SPI is its standardization which makes it particularly suited to compare drought conditions among different time periods and regions with different climates (Cacciamani et al., 2007).

The SPI index is based on an equi-probability transformation of aggregated monthly precipitation into a standard normal variable (Cancelliere et al., 2007). The computation of the index requires fitting a probability distribution to aggregated monthly precipitation series (3, 6, 12, 24, 48 months). The probability density function is then transformed into a normal standardized index whose values classify the category of drought characterizing each place and timescale (Cacciamani et al., 2007). The SPI can only be computed when sufficiently long (at least 30 years) and preferably continuous time-series of monthly precipitation data are available (Cacciamani et al., 2007).

The SPI may be used for monitoring both dry and wet conditions (Morid et al., 2006). Positive SPI values indicate greater than median precipitation and negative values indicate less than median precipitation. The 'drought' categories of the SPI range are arbitrarily split into 'near normal', 'moderately dry', 'severely dry' and 'extremely dry' (Morid et al., 2006). Each of these categories is defined by a range of SPI values. For example, the 'near normal' category is defined by a range of SPI values from 0.5 to −0.5, while the 'severely dry' category is defined by a range of SPI values from −1.5 to −2. A more negative SPI value is indicative of a more severe drought condition.

One disadvantage of the SPI index is that it is not always easy to find a probability distribution that models the raw precipitation data. Another disadvantage is that it is not always possible to access reliable time-series data to produce a robust estimate of the distribution parameters. Moreover, the application of the index in arid regions on timescales of less than 3 months can result in misleading SPI values (Cacciamani et al., 2007). To overcome the possible lack of a probability distribution that models the raw precipitation data, several probability distributions can be used simultaneously. However, the use of different probability distributions affects the SPI values as the index is based on the fitting of a distribution to precipitation time series. The gamma distribution (McKee et al., 1993; Edwards and McKee, 1997; Mishra and Singh, 2010), Pearson Type III distribution (Guttman, 1999), lognormal, extreme value and exponential distributions have been widely applied to simulations of precipitation distributions.

8.2.4 COMPARISON OF DROUGHT INDICES

Of the aforementioned data-driven drought indices, the PDSI (Kim and Valdes, 2003; Morid et al., 2006; Cutore et al., 2009; Hwang and Carbone, 2009; Karamouz et al., 2009) and the SPI (McKee et al., 1993; Tsakiris and Vangelis, 2004; Mishra and Desai, 2005, 2006; Bordi and Sutera, 2007; Cacciamani et al., 2007; Cancielliere et al., 2007; Mishra et al., 2007; Bacanli et al., 2008) have found widespread application in the field of drought forecasting. The main strength of these two drought indices lies in their standardization. For the purposes of comparing drought conditions of different areas often having different hydrological balances, the most important characteristic of a drought index is its standardization (Bordi and Sutera, 2007). Standardization of a drought index ensures independence from geographical position as the index in question is calculated with respect to the average precipitation in the original place (Cacciamani et al., 2007).

Many studies have sought to compare SPI and PDSI for forecasting droughts (Hayes, 1996; Heim, 2002). One of the key differences between the two indices is that the special characteristics of the PDSI vary from site to site while those of the SPI do not. Another difference is the complex structure with an exceptionally long memory that characterizes the PDSI while the SPI is an easily interpreted, simple moving average process. This characteristic makes the SPI useful as the primary drought index because of its simplicity as well as its probabilistic and spatial invariance, making it more ideal for risk and decision-making analysis. The SPI is also more representative of short-term precipitation than the PDSI and is thus a better indicator for soil moisture variation and soil wetness (Mishra and Singh, 2010). Moreover, the SPI provides a better spatial standardization than does the PDSI with respect to extreme drought events (Lloyd-Hughes and Saunders, 2002). Lastly, the SPI has been found to be more effective than the PDSI in detecting the onset of a drought event (Hayes, 1996).

8.3 DATA-DRIVEN METHODS FOR DROUGHT FORECASTING

The first data-driven drought forecasts to be developed used stochastic models such as Autoregressive Integrative Moving Average (ARIMA) models (Mishra and Desai, 2005). These models were also prevalent in other hydrologic forecasting applications such as streamflow forecasting. ARIMA models were used to forecast the SPI drought index by Mishra and Desai (2005) and Mishra et al. (2007). The ARIMA models could forecast the SPI drought index on multiple timescales but the precision of the models was limited due to their inability to forecast the nonlinear components of the drought time series. ARIMA and other stochastic models are essentially linear, and are limited in their ability to forecast nonlinear data. Given this limitation, from the mid-2000s onward forecasters began to turn to artificial neural networks (ANNs) for hydrological forecasting. In addition to forecasting the SPI using ARIMA models, Mishra and Desai (2006) compared the results of these models with those of ANNs in the Kansabati River Basin of India. Their results indicated that ANN models were more effective at forecasting the SPI due to their ability to model nonlinear components of a time series.

8.3.1 ARTIFICIAL NEURAL NETWORKS

ANNs are nonlinear, data-driven models, based on artificial intelligence (machine-learning) computation techniques, which can provide powerful solutions to many complex modelling problems. They have many features which make them attractive for use in forecasting. One such feature is their rapid development. ANNs are easy to develop as they do not require very detailed knowledge about the physical characteristics of the study area. Another feature is their rapid execution time. In addition, they have parsimonious data requirements compared to other traditional models. Such parsimony is ideal for developing countries where hydrologic data may be sparse or incomplete. ANNs can mimic a large class of nonlinear functions, and therefore they are ideal candidates to develop empirical (regression type) models. ANNs have been widely used to model time series in various fields of applications such as dynamic systems, nonlinear signal processing, pattern recognition, identification and classification (BuHamra et al., 2003).

ANN models used in most forecasting studies have had a feed-forward three-layered architecture, typically consisting of three layers: the input layer, the output layer and the hidden layer. Hidden neurons with appropriate nonlinear functions are used to process the information received by the input nodes. To build a model for forecasting, the neural network is processed through three stages: the training stage where the network is trained to predict future data, based on past and present data; the testing stage where the network is tested to stop training or to keep training; and the validation stage where the network ceases training and is used to forecast future data and to calculate different measures of error. A typical three-layered ANN model is described below (Kim and Valdes, 2003):

$$y'_k(t) = f_0 \left[\sum_{j=1}^{m} w_{kj} \cdot f_n \left(\sum_{i=1}^{N} w_{ji} x_i + w_{j0} \right) + w_{k0} \right] \qquad (8.1)$$

where N is the number of samples, m is the number of hidden neurons, $x_i(t) =$ the ith input variable at time step t; $w_{ji} =$ weight that connects the ith neuron in the input layer and the jth neuron in the hidden layer; $w_{j0} =$ bias for the jth hidden neuron; $f_n =$ activation function of the hidden neuron; $w_{kj} =$ weight that connects the jth neuron in the hidden layer and k^{th} neuron in the output layer; $w_{k0} =$ bias for the kth output neuron; $f_0 =$ activation function for the output neuron; and $y_k(t)$ is the forecasted k^{th} output at time step t (Kim and Valdes, 2003).

ANNs have been extensively used for drought forecasting purposes (Mishra and Desai, 2006; Morid and Smakhtin, 2007; Bacanli et al., 2008; Barros and Bowden, 2008; Cutore et al., 2009; Karamouz et al., 2009; Marj and Meijerink, 2011; Mishra and Nagarajan, 2012). Apart from ANNs, several other artificial intelligence methods have emerged as effective in the field of drought forecasting and hydrologic forecasting in general. These methods include support vector machines (SVM), fuzzy systems and generic algorithms (Nourani et al., 2014). These artificial intelligence methods have proven to be very flexible and able to handle a large amount of data, allowing forecasters to effectively model the nonlinear aspects of hydrologic phenomena.

Another artificial intelligence technique that has been used in hydrologic forecasting is the adaptive neuro-fuzzy inference system (ANFIS). ANFIS is a system that integrates ANN models with fuzzy logic and has been used in drought forecasting by Chou and Chen (2007) and Bacanli et al. (2008). ANFIS models are useful in situations where the underlying physical relationships are not fully understood (Bacanli et al., 2008). The fuzzy inference system contains three components: a rule base containing fuzzy if-then rules; a database which defines the membership function; and an inference system which combines the fuzzy rules and produces the system results. The main difficulty with the fuzzy inference system is that no consistent way exists of defining the membership function or the fuzzy rules. However, by combining the system with ANN models under a single framework, it is possible to use the learning capability of ANN models for automatic fuzzy rule generation and parameter optimization.

8.3.2 SUPPORT VECTOR REGRESSION

Another artificial intelligence method that has been used for drought forecasting is support vector regression (SVR). SVR was developed by Vapnik (1995) and is based on the structural risk-minimization principle. This principle theoretically minimizes the expected error of a learning machine and therein reduces the problem of overfitting. The structural risk minimization principle has been shown to be superior to the empirical risk minimization principle which is used by many conventional neural networks (Cao and Tay, 2001). SVR has been used in several hydrological forecasting studies. Khan and Coulibaly (2006) found that an SVR model performed better than ANNs in 3–12 month predictions of lake water levels. Kisi and Cimen (2009) used SVRs to estimate daily evaporation and discovered that SVR models offer an effective method for forecasting daily evapotranspiration. A study by Belayneh and Adamowski (2012) used SVR models to forecast the SPI in the Awash River Basin and found out that the SVRs are effective at forecasting the SPI drought index on multiple timescales. In contrast to ANNs, which seek to minimize training error, SVMs attempt to minimize the generalization error (Cao and Tay, 2001). The purpose of an SVR model is to estimate a functional dependency $f(\vec{x})$ between a set of sampled points $x = \{\vec{x}_1, \vec{x}_2, ..., \vec{x}_n\}$ taken from R^n and target values $Y = \{y_1, y_2, ..., y_n\}$ with $y_i \in R$ (the input and target vectors [*x and y variables*] refer to the monthly records of a given time series). A more detailed description of the development of an SVR model can be found in Cimen (2008).

8.3.3 COMPARISON OF ANN AND SVR MODELS

In general, the performances of SVR and ANN models are comparable. Theoretically, SVR models should perform better than ANN models because they adhere to the structural risk minimization principle instead of the empirical risk minimization principle and should therefore not be as susceptible to local minima or maxima. However, some studies have shown that the results from SVR and ANN models are in fact comparable. For example, Shin et al. (2005) and Chevalier et al. (2011)

found that the application of ANN models in time series forecasting was comparable to those of SVR models, especially as the size of the training set was increased. Chevalier et al. (2011) also found that SVR models were superior in the training phase, while ANN models were superior in the evaluation phase. Witten et al. (2011) also found that ANN models are comparable to SVR models because they can learn to ignore irrelevant attributes. These authors also agree that there is no universally superior learning method. In the context of drought forecasting, Belayneh et al. (2014) found that forecasts of the SPI drought index using both ANN and SVR models were comparable in terms of the evaluation criteria used.

Apart from being nonlinear, drought data, especially precipitation for meteorological drought forecasts, is also nonstationary. In other words, the probability distribution of the data changes over time. The ability of artificial intelligence techniques such as ANNs and SVMs to deal with nonstationary data is limited and, in order to adequately cope with this problem, researchers have increasingly turned to hybrid or coupled models. These coupled models combine data processing schemes with artificial intelligence techniques. One of the most popular and influential data-preprocessing tools in hydrologic forecasting is the use of wavelet transforms. These have been used in conjunction with artificial intelligence models for the purposes of evaluating rainfall-runoff models (Lane, 2007), to forecast river flow (Adamowski, 2008; Adamowski and Sun, 2010), to forecast groundwater levels (Adamowski and Chan, 2011), to forecast urban water demand (Adamowski and Prasher, 2012), and for the purposes of drought forecasting (Kim and Valdes, 2003; Özger et al., 2011; Belayneh and Adamowski, 2012; Mishra and Singh, 2012). They are described in detail in the next section.

8.3.4 Wavelet Transforms

The wavelet transform is a mathematical tool that provides a time-frequency representation of a signal in the time domain (Partal and Kisi, 2007). Wavelet transforms have become useful tools for analyzing local variation within a given time series; coupled models have been proposed for forecasting time series based on a wavelet transform preprocessing (Adamowski, 2008). Wavelet transforms provide useful decompositions of an original time series, allowing these decompositions to capture useful information at various resolution levels and improve the forecast ability of artificial intelligence models (Adamowski, 2008). Coupled wavelet and artificial intelligence models have been used extensively in hydrologic forecasting. A review of the applications of these models can be found in Nourani et al. (2014).

Wavelet analysis begins by selecting a mother wavelet (ψ). The continuous wavelet transform (CWT) is defined as (Nason and Von Sachs, 1999):

$$W(\tau,s) = \frac{1}{\sqrt{|s|}} \int_{-\infty}^{\infty} x(t)\psi^* \left(\frac{t-\tau}{s} \right) dt \qquad (8.2)$$

where s is the scale parameter; τ is the translation and * corresponds to the complex conjugate (Nourani et al., 2014). The CWT produces a continuum of all scales as the output with each scale corresponding to the width of the wavelet; hence, a larger scale indicates that more of a time series is used in the calculation of the coefficient than in smaller scales. The CWT is useful for processing different images and signals; however, it is not often used for forecasting because it takes time to compute. Instead, in forecasting applications, the discrete wavelet transform (DWT) is more frequently used. The DWT requires less computation time and is simpler to implement. DWT scales and positions are usually based on powers of two (dyadic scales and positions). This is achieved by modifying the wavelet representation to (Cannas et al., 2005):

$$\psi_{j,m}(m) = \frac{1}{\sqrt{\left|s_0^j\right|}} \sum_k \psi\left(\frac{k - m\tau_0 s_0^j}{s_0^j}\right) x(k) \tag{8.3}$$

where j and m are integers that control the scale and translation, respectively, $s_0 > 1$ is a fixed dilation step and τ_0 is a translation factor that depends on the dilation step. Discretizing the wavelet results in the time-space scale being sampled at discrete levels. The DWT has high-pass and low-pass filters. The original time series passes through both these filters and detailed coefficients and approximation series are obtained.

Kim and Valdes (2003) were the first to use wavelet analysis for the purposes of drought forecasting. In their study they coupled dyadic wavelet transforms and ANN models to forecast the PDSI in the Conchos River Basin of Mexico (Kim and Valdes, 2003). The coupling of ANNs and wavelets improved the forecast accuracy of the Palmer index for forecast lead times by up to 6 months in the Conchos River Basin. In addition, a study by Belayneh and Adamowski (2012) compared the ability of ARIMA, ANN, SVR and WANN (coupled wavelet ANN) models to forecast long-term (6 and 12 months lead time) drought in the Awash River Basin of Ethiopia. The study forecast the SPI and found that the hybrid WANN model had the best forecast accuracy. In addition to ANN models, wavelets have been coupled with various other artificial intelligence models to forecast drought. Belayneh et al. (2014) included a wavelet-SVR (WSVR) conjunction model in addition to the WANN model to forecast the SPI in the Awash River Basin. The study forecast the SPI 12 and SPI 24, as indicators of long-term drought conditions. The coupled WANN and WSVR models had better results compared to models that did not use any wavelet preprocessing.

Shirmohammadi et al. (2013) coupled wavelets with ANFIS techniques (WANFIS) for meteorological drought forecasting. The study evaluated the ability of WANFIS models to predict drought and compared them to WANN models and found that WANFIS models were more accurate. Özger et al. (2011) also compared the ability of WANN models to forecast drought with a wavelet fuzzy logic conjunction model. They evaluated the ability of the wavelet fuzzy logic model to forecast drought over long lead times in Texas. The study concluded that the wavelet fuzzy logic model was more accurate than the WANN model (Özger et al., 2011). Mishra and Singh (2012) investigated the relationship between meteorological variables and hydrological drought properties using the PHDI. The relationship between meteorological variables and the PHDI was investigated using a wavelet-Bayesian regression model, which improves the modelling capability of a traditional Bayesian regression model (Mishra and Singh, 2012). Deng et al. (2011) attempted to predict seasonal drought by forecasting changes in soil moisture. Wavelet-based denoising was applied to preprocess the original chaotic soil water signal; the results of the prediction showed improvement of the model in comparison to ANN and ANFIS models.

8.4 UNCERTAINTY ANALYSIS

Within hydrological forecasting studies, there has been a recent trend toward using ensemble models. An ensemble model uses two or more artificial intelligence models to achieve partial solutions to a given problem and combines these partial solutions to obtain a more complete prediction (Helmy et al., 2013). An ensemble model may reduce the risk of the selection of a poorly performing machine learning model because a set of models having similar training performances may have different generalization performances for the testing datasets (Helmy et al., 2013). An ensemble model may also provide an efficient approach for certain applications in which the amount of data is too large for building a single model (Helmy et al., 2013). It is also proven to be effective in the absence of adequate training data by building different models using resampling techniques (Erdal and Karakurt, 2013; Helmy et al., 2013). The ensemble model became popular due to its higher generalization capability compared to individual machine learning techniques (Erdal and Karakurt, 2013).

One ensemble technique that has been used in hydrologic forecasting is the bootstrap technique. The bootstrap is a resampling technique with replacement that is used for statistical interpretation (Tiwari and Chatterjee, 2010). The bootstrap is used to estimate statistical characteristics such as bias, variance, distribution functions and confidence intervals, and thus provides an excellent application for combination with machine learning techniques. The bootstrap technique has been coupled with ANN models for several forecasting applications over the last decade. Shu and Burn (2004) used bootstrap ANN (BANN) ensembles to study flood frequency and Tiwari and Chatterjee (2010) used BANN ensembles to forecast floods. Tiwari and Chatterjee (2010) also coupled BANN models with wavelet transforms to forecast uncertainty of floods. Li et al. (2010) used bootstrap-support vector regression (BSVR) models for the purposes of streamflow prediction. Tiwari et al. (2012) coupled wavelet analysis with BANN models to forecast daily river discharge, and noted the superiority of these models in forecasting river flow compared to traditional ANN models. Tiwari and Adamowski (2013) coupled BANN models with wavelet transforms for the purposes of forecasting urban water demand. These studies demonstrated that the hybrid WBANN model reduces the uncertainty associated with the forecasts, and that the performance of WBANN forecasted bands was more accurate and reliable than BANN confidence bands (Tiwari and Adamowski, 2013).

Another ensemble technique that has become increasingly popular in hydrologic forecasting is the boosting technique. Boosting is an ensemble method that attempts to improve the performance of a given learning algorithm (Schapire, 1990; Freund and Schapire, 1996). The purpose of boosting is to produce a sequence of models so that each subsequent model concentrates more on the training cases that are not well predicted by the previous model. The main difference between the bootstrap and boosting algorithms is that the distribution of the training set changes adaptively in the latter based on the performance of the previously created network, while the bootstrap algorithm changes the distribution of the training set stochastically. Zaier et al. (2010) compared the effectiveness of BANN models with boosting-ANN (BS-ANN) ensembles in the estimation of ice thickness on lakes. Li et al. (2010) used BSVR models for the purposes of streamflow prediction. Shu and Burn (2004) compared the ability of BANN and BS-ANN models in estimating the index flood and the 10-year flood quantile. Finally, Erdal and Karakurt (2013) built boosting and bootstrap ensembles using a benchmark SVR model for the purposes of streamflow forecasting.

In addition, both the bootstrap and boosting techniques have been used for drought forecasting. Belayneh et al. (2013a,b) coupled both techniques with artificial intelligence models to forecast the SPI drought index. These ensemble models also included wavelet preprocessing. The results of both studies found that coupling artificial intelligence models with the bootstrap and boosting techniques improves their forecast accuracy and reliability. Furthermore, preprocessing the SPI time series with wavelet transforms improves the forecast accuracy of the coupled ensemble models.

This trend of ensemble drought forecasts has continued in recent years. Dehghani et al. (2014) coupled ANN models with a Monte Carlo simulation to forecast the standardized hydrological drought index (SHDI) in the Karoon River in southwestern Iran (2014). In another recent study, Forzieri et al. (2014) coupled a hydrological model with an ensemble of bias-corrected climate simulations and a water use scenario.

8.5   OTHER COUPLED MODELS

Apart from the coupling of artificial intelligence models with statistical techniques such as the bootstrap or boosting technique, researchers have increasingly begun coupling data-driven models to large-scale climatic indices. The aim of these coupled models is to use the information conveyed by large-scale climatic indices (such as the North Atlantic Oscillation) to improve the forecasting ability of drought indices. Climatic indices can characterize the state of the oceans and atmosphere, and coupling them with drought indices can provide useful information for forecasting precipitation and drought conditions under changing climatic conditions. Another advantage of coupling artificial intelligence models with climatic indices lies in the ability of artificial intelligence models to simulate nonlinear interactions between the various variables with climatic indices.

One example of such a model was that developed by Cutore et al. (2009), who forecasted the Palmer index in Sicily using neural networks, the North Atlantic Oscillation (NAO) and the European Blocking (EB) climatic indices. The results of their study indicated that the Palmer index was correlated with the NAO and EB during the autumn and winter months. Including the two indices resulted in improvements in the forecast accuracy of the Palmer index. In Australia, Barros and Bowden (2008) forecasted the SPI in the Murray–Darling Basin by integrating multivariate linear regression with sea surface temperature anomalies over the Indian and Pacific oceans. Their forecasts involved long lead times and their coupled forecasts of the SPI could explain up to 60% of the drought variance within the Murray–Darling basin (Barros and Bowden, 2008). Another, more recent study, in the Maharloo Basin of Iran, attempted to couple ANN models with atmospheric circulation factors, such as the NAO, the Pacific/North American (PNA) and El Niño to forecast drought (Sigaroodi et al., 2014). These studies indicate a growing need to establish a relationship between climatic indices and precipitation and drought indices for drought forecasts.

8.6   FUTURE WORK

While data-driven forecasts of drought have evolved over the past two decades, there are still areas that need to be expanded upon. While the potential links between drought indices and climatic indices have been an area of recent interest, preprocessing the drought indices with wavelet transforms is an avenue for further research. Wavelet-based models can also be coupled to physically based models such as the soil and water integrated model (SWIM). Another area of potential improvement is in the selection of appropriate wavelet transforms for drought forecasts. Selection of wavelet transforms and decomposition levels are linked to the characteristics of a given data series. A more rigorous investigation of wavelet properties could lead to the selection of specific wavelets depending on the type of forecast being undertaken.

Coupling artificial intelligence models with statistical techniques such as the bootstrap and boosting algorithms should also be more extensively researched. These ensemble techniques have proven to be effective in producing more reliable forecasts in other hydrological forecasting studies. Their application to drought forecasts is not as extensive and should be investigated. In addition, these ensemble models should also incorporate wavelet transforms. Future studies should also attempt to quantify time shift error since it is associated with common forecasting problems within regression models. Data-driven hydrologic forecasts often produce phase-shift errors where a time lag exists between the output results and the observed data. Procedures to correct this time shift error have been studied for data-driven models (Zimmer, 2011; Forzieri et al., 2014). Time-shift correction techniques need to be incorporated more extensively in data-driven drought forecasts to ensure the outputs of these models are more reliable and have greater applicability.

8.7   CONCLUSION

Data-driven drought forecasts have steadily evolved over the past two decades. Initially these forecasts relied on models that were linear in nature, such as ARIMA models. Over the course of the last two decades, researchers have increasingly turned to artificial intelligence models to overcome the limitations of linear models. In addition, researchers have increasingly coupled these artificial intelligence models with wavelet analysis to address the issues of nonstationarity that are present in drought time series. The more recent trend is the use of ensemble models to make forecasts more robust and reduce the uncertainty present in using a single predictive model. This avenue of research needs to be further investigated with respect to drought as it has been shown to be effective in forecasting other hydrologic applications, especially when coupled with wavelet analysis. Finally, data-driven models have been used in conjunction with climatic indices. The incorporation of climatic indices improves the predictive capacity of drought indices and illustrates that researchers have begun to combine data-driven models with climatic models to include atmospheric and oceanic interactions in the forecasting of droughts.

REFERENCES

Abrahart, R., Heppenstall, A., See, L. M. (2007). Timing error correction procedure applied to neural network rainfall—Runoff modelling. *Hydrological Sciences Journal. 52*(3), 414–431. doi: 10.1623/hysj.52.3.414.

Adamowski, J. (2008). Development of a short-term river flood forecasting method for snowmelt driven floods based on wavelet and cross-wavelet analysis. *Journal of Hydrology. 353*(3–4), 247–266. doi: 10.1016/j.jhydrol.2008.02.013.

Adamowski, J., Chan, H.F. (2011). A wavelet neural network conjunction model for groundwater level forecasting. *Journal of Hydrology. 407*, 28–40.

Adamowski, J., Prasher, S., (2012). Comparison of machine learning methods for runoff forecasting in mountainous watersheds with limited data. *J. Water Land Dev. 17*, 89–97. doi: 10.2478/v10025-012-0012-1.

Adamowski, J., Sun, K. (2010). Development of a coupled wavelet transform and neural network method for flow forecasting of non-perennial rivers in semi-arid watersheds. *Journal of Hydrology. 390*, 85–91.

Bacanli, U.G., Firat, M., Dikbas, F. (2008). Adaptive Neuro-Fuzzy Inference System for drought forecasting. *Stochastic Environmental Research and Risk Assessment. 23*(8), 1143–1154.

Barros, A., Bowden, G. (2008). Toward long-lead operational forecasts of drought: An experimental study in the Murray-Darling River Basin. *Journal of Hydrology. 357*(3–4), 349–367. doi: 10.1016/j.jhydrol.2008.05.026.

Belayneh, A., Adamowski, J. (2012). Standard precipitation index drought forecasting using neural networks, wavelet neural networks and support vector regression. *The Journal of Applied Computational Intelligence and Soft Computing. 2012* (2012), Article ID 794061, http://dx.doi.org/10.1155/2012/794061.

Belayneh, A., Adamowski, J., Khalil, B. (2013). A boosting ensemble approach for drought forecasting. *NABEC 2013 Northeast Agricultural & Biological Engineering Conference*. Altoona, PA, June 16–19 (online at 2013http://nabec.asabe.org/index _files/program_books/NABECProgram2013.pdf).

Belayneh, A., Adamowski, J., Khalil, B. (2013). Forecasting drought via bootstrap and machine learning methods. *CSCE 3rd Specialty Conference on Disaster Prevention and Mitigation*, Montreal, QC, Canada, May 29–June 1, 2013.

Belayneh, A., Adamowski, J., Khalil, B., Ozga-Zielinski, B. (2014). Long-term SPI drought forecasting in the Awash River Basin in Ethiopia using wavelet neural network and wavelet support vector regression models. *Journal of Hydrology. 508*, 418–429. doi: 10.1016/j.jhydrol.2013.10.052.

Bordi, I., Sutera, A. (2007). Drought monitoring and forecasting at large scale, Chapter 1 In G. Rossi et al. (Eds.) *Methods and Tools for Drought Analysis and Management*. Pp. 3–27, Springer.

BuHamra, S., Smaoui, N., Mahmoud, G. (2003). The Box-Jenkins analysis and neural networks: Prediction and time series modelling. *Applied Mathematical Modelling*. 27, 805–815.

Byun, H.R., Wilhite, D.A. (1999). Objective quantification of drought severity and duration. *Journal of Climatology*. 12, 2747–2756.

Cacciamani, C., Morgillo, A., Marchesi, S., Pavan, V. (2007). Monitoring and forecasting drought on a regional scale: Emilia-Romagna region. *Water Science and Technology Library*. 62(1), 29–48.

Cancelliere, A., Di Mauro, G., Bonaccorso, B., Rossi, G. (2007). Stochastic forecasting of drought indices. Chapter 5, in G. Rossi et al. (Eds.) *Methods and Tools for Drought Analysis and Management*. Pp. 83–100, Springer, Dordrecht.

Cancelliere, A., Di Mauro, G., Bonaccorso, B., Rossi, G. (2007). Drought forecasting using the Standardized Precipitation Index. *Water Resources Management*. 21(5), 801–819.

Cannas, B., Fanni, A., Sias, G., Tronci, S., Zedda, M.K. (2005). River flow forecasting using neural networks and wavelet analysis. *Geophysical Research Abstracts*. 7, 08651. SRef-ID: 1607-7962/gra/EGU05-A-08651.

Cao, L., Tay, F. (2001). Financial forecasting using support vector machines. *Neural Computing and Applications*. 10, 184–192.

Chevalier, R., Hoogenboom, G., McClendon, R., Paz, J. (2011). Support vector regression with reduced training sets for air temperature prediction: A comparison with artificial neural networks. *Neural Computing and Applications*. 20, 151–159.

Chou, F., Chen, B. (2007). Development of drought early warning index: Using neuro-fuzzy computing technique. In: *8th International Symposium on Advanced Intelligence Systems 2007*, Korea. Paper No:A1469.

Cimen, M. (2008). Estimation of daily suspended sediments using support vector machines. *Hydrol. Sci. J. 53*(3), 656–666.

Cutore, P., Di Mauro, G., Cancelliere, A. (2009). Forecasting Palmer Index using neural networks and climatic indexes. *Journal of Hydrologic Engineering*. 14(6). doi: http://dx.doi.org/10.1061/(ASCE)HE.1943-5584.0000028.

Dehghani, M., Saghafian, B., Saleh, F.N., Farokhnia, A., Noori, R. (2014). Uncertainty analysis of streamflow drought forecast using artificial neural networks and Monte-Carlo simulation. *International Journal of Climatology*. 34(4), 1169–1180. doi: 10.1002/joc.3754.

Deng, J., Chen, X., Du, Z., Zhang, Y. (2011). Soil water simulation and predication using stochastic models based on LS–SVM for red soil region of China. *Water Resour. Manage*. 25, 2823–2836.

Edwards, D.C., McKee, T.B. (1997). Characteristics of 20th century drought in the United States at multiple scales. *Atmospheric Science Paper*. 634.

Erdal, H.I., Karakurt, O. (2013). Advancing monthly streamflow prediction accuracy of CART models using ensemble learning paradigms. *Journal of Hydrology*. 477, 119–128.

Forzieri, G., Feyen, L., Rojas, R., Flörke, M., Wimmer, F., Bianchi, A. (2014). Ensemble projections of future streamflow droughts in Europe. *Hydrology and Earth System Sciences*. 18(1), 85–108. doi: 10.5194/hess-18-85-2014.

Freund, Y., Schapire, R.E. (1996). Experiments with a new boosting algorithm. In: *Proceedings of the Thirteenth International Conference on Machine Learning*. Morgan Kaufmann, Burlington, MA, 148–156.

Glantz, M.H. (1994). *Drought Follows the Plow*. University Press, Cambridge, Great Britain.

Guttman, N.B. (1999). Accepting the standardized precipitation index: A calculation algorithm. *Journal of American Water Resource Association*. 35(2): 311–322.

Hayes, M. (1996). *Drought Indexes*. National Drought Mitigation Center, University of Nebraska–Lincoln, p. 7 (available from University of Nebraska–Lincoln, 239LW Chase Hall, Lincoln, NE 68583).

Heim, R.R. Jr. (2002). A review of twentieth-century drought indices used in the United States. *Bull. Amer. Meteor. Soc. 83*, 1149–1165.

Helmy, T., Rahman, S.M., Hossain, M.I., Abdelraheem, A. (2013). Non-linear heterogenous ensemble model for permeability prediction of oil reservoirs. *Arab Journal of Science and Engineering. 38*, 1379–1395.

Hwang, Y., Carbone, G.J. (2009). Ensemble forecasts of drought indices using a conditional residual resampling technique. *Journal of Applied Meteorology and Climatology. 48*, 1289–1301.

Kam, J., Sheffield, J., Yuan, X., Wood, E.F. (2014). Did a skillful prediction of sea surface temperatures help or hinder forecasting of the 2012 Midwestern U.S. drought? *Environmental Research Letters. 9*(3), 034005. doi: 10.1088/1748-9326/9/3/034005.

Karamouz, M., Rasouli, K., Nazil, S. (2009). Development of a hybrid index for drought prediction: Case study. *Journal of Hydrologic Engineering. 14*, 617–627.

Keshavarz, M., Karami, E., Vanclay, F. (2013). The social experience of drought in rural Iran. *Journal of Land Use Policy. 30*, 120–129.

Khan, M.S., Coulibaly, P. (2006). Application of support vector machine in lake water level prediction. *Journal of Hydrologic Engineering. 11*(3), 199–205.

Kim, T., Valdes, J. (2003). Nonlinear model for drought forecasting based on a conjunction of wavelet transforms and neural networks. *Journal of Hydrologic Engineering. 8*, 319–328.

Kisi, O., Cimen, M., (2009). Evapotranspiration modelling using support vector machines. *Hydrological Science Journal. 54*(5), 918–928.

Kogan, F. (1997). Global drought watch from space. *Bull. Am. Meteorol. Soc. 78*, 621–636.

Lane, S. (2007). Assessment of rainfall–runoff models based upon wavelet analysis. *Hydrological Processes. 21*, 586–607.

Li, P., Kwon, H., Sun, L., Lall, U., Kao, J.J. (2010). A modified support vector machine based prediction model on streamflow at the Shimen Reservoir, Taiwan. *International Journal of Climatology. 30*(8), 1256–1268.

Lloyd-Hughes, B., Saunders, M.A. (2002). A drought climatology for Europe. *International Journal of Climatology. 16*, 1197–1226.

Marj, A., Meijerink, A. (2011). Agricultural drought forecasting using satellite images, climate indices and artificial neural network. *International Journal of Remote Sensing. 32*(24), 9707–9719.

McKee, T., Doesken, N., Kleist, J. (1993). The relationship of drought frequency and duration to time scales, *Paper Presented at 8th Conference on Applied Climatology*. American Meteorological Society, Anaheim, CA.

Mishra, A., Desai, V. (2005). Drought forecasting using stochastic models. *Stochastic Environmental Research and Risk Assessment. 19*(5), 326–339.

Mishra, A., Desai, V. (2006). Drought forecasting using feed-forward recursive neural network. *Ecological Modelling. 198*(1–2), 127–138.

Mishra, A., Desai, V., Singh, V. (2007). Drought forecasting using a hybrid stochastic and neural network model. *Journal of Hydrologic Engineering. 12*(6), 626–638.

Mishra, A., Singh, V. (2010). A review of drought concepts. *Journal of Hydrology. 391*(1–2), 202–216.

Mishra, A., Singh, V. (2012). Simulating hydrological drought properties at different spatial units in the United States based on wavelet–Bayesian regression approach. *Earth Interactions. 16*(17), 1–23. doi: 10.1175/2012EI000453.1.

Mishra, S., Nagarajan, R. (2012). Forecasting drought in Tel River Basin using feed-forward recursive neural network. *2012 International Conference on Environmental, Biomedical and Biotechnology IPCBEE*. vol. 41.

Morid, S., Smakhtin, V., Moghaddasi, M. (2006). Comparison of seven meteorological indices for drought monitoring in Iran. *International Journal of Climatology. 26*(7), 971–985.

Morid, S., Smakhtin, V., Bagherzadeh, K. (2007). Drought forecasting using artificial neural networks and time series of drought indices. *International Journal of Climatology. 27*(15), 2103–2111.

Mwangi, E., Wetterhall, F., Dutra, E., Di Giuseppe, F., Pappenberger, F. (2014). Forecasting droughts in East Africa. *Hydrology and Earth System Sciences. 18*(2), 611–620. doi: 10.5194/hess-18-611-2014.

Nason, G.P., Von Sachs, R. (1999). Wavelets in time-series analysis. *Philosophical Transactions of the Royal Society A: Mathematical, Physical and Engineering Sciences. 357*(1760), 2511–2526.

Nourani, V., Hosseini Baghanam, A., Adamowski, J., Kisi, O. (2014). Applications of hybrid wavelet–artificial intelligence models in hydrology: A review. *Journal of Hydrology. 514*, 358–377. doi: 10.1016/j.jhydrol.2014.03.057.

Özger, M., Mishra, A., Singh, V. (2011). Long lead time drought forecasting using a wavelet and fuzzy logic combination model: A case study in Texas. *Journal of Hydrometeorology. 13*(1), 284–297. doi: 10.1175/JHM-D-10-05007.1.

Partal, T., Kisi, O. (2007). Wavelet and neuro-fuzzy conjunction model for precipitation forecasting. *Journal of Hydrology. 342*(1–2), 199–212.

Schapire, R. (1990). The strength of weak learnability. *Mach. Learn. 5*, 197–227.

Shin, K., Lee, T., Kim, H. (2005). An application of support vector machines in bankruptcy prediction model. *Expert Systems with Applications. 28*, 127–135.

Shirmohammadi, B., Moradi, H., Moosavi, V., Semiromi, M., Zeinali, A. (2013). Forecasting of meteorological drought using wavelet–ANFIS hybrid model for different time steps (Case Study: Southeastern part of east Azerbaijan Province, Iran). *Nat. Hazards. 69*, 389–402.

Shu, C., Burn, D. (2004). Artificial neural network ensembles and their application in pooled flood frequency analysis. *Water Resources Res. 40* (W09301). http://dx.doi .org/10.1029/2003WR002816.

Sigaroodi, S., Chen, Q., Ebrahimi, S., Nazari, A., Choobin, B. (2014). Long-term precipitation forecast for drought relief using atmospheric circulation factors: A study on the Maharloo Basin in Iran. *Hydrology and Earth System Sciences. 18*(5), 1995–2006. doi: 10.5194/hess-18-1995-2014.

Srivastava, H. (1968). *The History of Indian Famines from 1858–1918.* Sri Ram Mehra and Co., Agra.

Tiwari, M., Adamowski, J. (2013). Urban water demand forecasting and uncertainty assessment using ensemble wavelet-bootstrap-neural network models. *Water Resources Research. 49*, 6486–6507, doi:10.1002/wrcr.20517.

Tiwari, M., Chatterjee, C. (2010). Development of an accurate and reliable hourly flood forecasting model using wavelet-bootstrap-ANN (WBANN) hybrid approach. *Journal of Hydrology. 394*, 458–470.

Tiwari, M., Song, K, Chatterjee, C., Gupta, M. (2012). Improving reliability of river flow forecasting using neural networks, wavelets and self-organising maps. *J. Hydroinform. 15*(2), 486–502.

Tsakiris, G., Vangelis, H. (2004). Towards a drought watch system based on spatial SPI. *Water Resources Management. 18*(1), 1–12.

Vapnik, V. (1995). *The Nature of Statistical Learning Theory*, Springer Verlag, New York.

Whitmore, J.S. (2000). *Drought Management on Farmland.* Kluwer Academic Publishers, Dordrecht, the Netherlands.

Witten, I.H., Frank, E., Hall, M.A. (2011). *Data Mining: Practical Machine Learning Tools and Techniques.* Morgan Kaufman, Burlington, MA.

WMO (World Meteorological Organisation). Press release December 2009, WMONo. 872, 2009.

Zaier, I., Shu, C., Ouarda, T., Seidou, O., Chebana, F. (2010). Estimation of ice thickness on lakes using artificial neural network ensembles, *Journal of Hydrology. 383*(3–4), 330–340.

Zimmer, M., Wernli, H. (2011). Verification of quantitative precipitation forecasts on short time-scales: A fuzzy approach to handle timing errors with SAL. *Meteorologische Zeitschrift. 20*(2), 95–105. doi: 10.1127/0941-2948/2011/0224.

9 Estimating CO$_2$ Emissions from Subsurface Coal Fires

S. Taku Ide, Claudia Kuenzer and Franklin M. Orr, Jr.

CONTENTS

9.1 INTRODUCTION—COAL FIRES: CAUSE, OCCURRENCE, IMPLICATIONS

Subsurface coal fires occur in many countries. In this chapter, the term 'subsurface coal fire' refers to an underground burning or smouldering coal seam. Coal seams can ignite naturally at their outcrop through spontaneous combustion, lightning or through anthropogenic influence. Spontaneous combustion, also sometimes termed self-ignition, arises due to the exothermal process of carbon oxidation. When carbon and oxygen react to form carbon dioxide, heat is released. If ventilation is limited and the heat cannot be released from the inner surface of the coal, combustion can start at temperatures as low as 80°C (Lohrer et al., 2005).

The risk for self-ignition is increased by the following factors: a large coal volume from which heat cannot escape easily; high content of volatiles in the coal; large internal surface area (in fractures, for example); access to sufficient oxygen; and high surrounding air temperatures (vanDijk et al., 2011). However, coal fires can also be triggered by lightning strikes, forest and peat fires, and careless human behaviour in mining areas. The geophysical characteristics of coal fires are described in detail in Sternberg (2004), Gielisch (2007), Zhang and Kuenzer (2007), Zhang et al. (2007), and Ide et al. (2011). Coal fire-induced geomorphology is presented in Kuenzer and Stracher (2011).

Coal fire locations in China (Zhang et al. 2004; Kuenzer, 2005; Yang et al. 2008) (see Figure 9.1), India (Prakash and Gupta, 1999; Kuenzer et al., 2008d; Kuenzer et al., 2012), the United States (Stracher et al. 2007) and South Africa (Stracher and Taylor, 2004) have been extensively researched during the past decades, but fires also occur in other places, including in Russia, Australia, Venezuela and Indonesia. Coal fires are therefore of international interest and a problem of global magnitude.

FIGURE 9.1   Environmental impacts of coal fires. (a and b) Strong gas emissions through surface cracks and sinkholes above a subsurface coal fire in Wuda, China. (c) Coal fire-induced bedrock collapse. The seam, which was covered by sandstone, has burnt, and the sandstone surface has collapsed. (d) A team of geologists is drilling boreholes into a bedrock layer and subsurface coal fire to investigate the fire's burning depth and inside fire temperatures. (Photographs by C. Kuenzer.)

Coal fires not only lead to the destruction of the valuable economic resource, coal. They can also lead to unpredictable land subsidence due to the volume loss underground and to the release of greenhouse gases (CO_2, CH_4) and toxic gases (e.g. SO_2, NO_x, CO, N_2O) (Hower et al., 2009). These gases trigger vegetation deterioration (Kuenzer and Voigt, 2003), are a threat to the health of local inhabitants and contribute to the global annual greenhouse gas (GHG) budget (Kuenzer et al., 2007a; vanDijk et al., 2011).

The fires have been investigated by different geo-engineering and geo-scientific disciplines, employing a variety of techniques including in situ mapping (Kuenzer et al., 2008b; Ide et al., 2010); borehole core analyses and geologic modelling (Wessling et al., 2008; Ide, 2011); remote sensing-based investigations (Li et al., 2005; Hecker et al., 2007; Kuenzer et al., 2008a,b,c; Düzgün et al., 2011); geomagnetic, geoelectric, and micro-seismic surveying (Schauman et al., 2008); and gas composition analyses (Litschke, 2005; Hower et al., 2009, 2013; Ide, 2011; Engle et al., 2012).

Many questions have been answered in the past few years. It is known that coal fires lead to strong anomalies in airborne and spaceborne thermal earth observation data (Tetzlaff, 2004; Zhang, 2004). Remote sensing allows the monitoring of coal fire dynamics, and the mapping of secondary features such as cracks, sinkholes and trenches, as well as pyrometamorphic rocks. Kuenzer et al. (2007b) demonstrated that it is possible to detect formerly unknown fires, exclusively from remotely sensed data, and to later validate the findings in the field. The location of former and current fire centres underground can be derived from geomagnetic data, as strongly heated bedrock changes its magnetization when reaching the Curie point (Gielisch, 2007; Ide et al., 2011). Field surveys and modelling have shown that fires can expand and 'move' over distances ranging from several tenths of a metre to hundreds of metres underground within the course of only 1 year. They can have different stages of burning intensity, and can remain dormant for some time before flaring up again (Kuenzer et al., 2008b). Coal fire induced cracks in the bedrock surface usually precede the fire-front (Wolf and Bruining, 2007; Ide et al., 2010). Even exotic fields of inquiry, such as coal fire-related mineral formation, are well studied (Stracher et al., 2014).

Several hundred coal fire-related publications exist, but few investigators have published reliable, quantitative figures on the GHG emissions from this hazard. The validity of the often-quoted (especially in the news media) estimate, that coal fires in China alone contribute at least 2–3% to all annual human-induced CO_2 emissions (van Genderen and Guan, 1997), would appear to be questionable, and not based on scientific evidence nor was the original estimate published in any peer-reviewed journal. Instead, it was most likely based on exaggerated coal-loss numbers from China, which were calculated not from the amount burned, but from the coal 'lost' due to the fires – for each ton of burned coal, approximately 10 tons of nearby coal became inaccessible. Even though Cassells (1997) concluded that these numbers were not realistic, they have been cited widely. Kuenzer et al. (2007b) demonstrated that the true value number must be at least a magnitude lower – probably ranging between 0.1 and 0.3%, but even this estimate comes with significant uncertainty. It is currently impossible to measure or indirectly derive accurate numbers for a country as vast as China or for the world. However, scientifically backed approximations are already an advance in this frontier science research.

Important coal fire-related questions currently on the research agenda include:

- What quantities of GHG are emitted by coal fires at global, national or even local scales?
- What is the contribution of coal fires to the annual human-induced CO_2 emissions budget?
- What are the best approaches to quantify coal fires (amount of coal burnt, amount of GHG release)?
- Which mechanisms could be utilized to increase the interest of private companies to extinguish coal fires?

This chapter examines pathways toward answering some, but not all, of these urgent questions.

9.2 QUANTIFYING EMISSIONS OF GREENHOUSE GASES FROM COAL FIRES

9.2.1 EXTINGUISHING COAL FIRES: THE FUEL-CO_2-ENERGY NEXUS

Coal fires can be extinguished via three different ways: first, by depleting the fire of oxygen, for example, by covering the burning material with overburden material (loess, sand, etc.); second, by hindering fire spread through excavation and the removal of the combustible fuel (coal); and third, by lowering the temperature in the system, typically by quenching the fire with water or foams.

In the past decade, the assessment of methods to extinguish coal fires was high on the agenda not only of academic researchers, but also of those employed in larger energy companies and engineering clusters. One reason was the goal of the Kyoto Protocol (Grubb et al., 1997) to improve the overall flexibility and economic efficiency of making emissions cuts. It introduced three mechanisms for emissions reduction, which are presented in detail in Rosenqvist et al. (2003). One of those mechanisms was the limited trading system 'clean development mechanism' (CDM), a mechanism whereby non-Annex I parties (nonindustrialized countries) can create 'certified emission reductions' (CERs) through developing projects that reduce the net emissions of GHGs. Annex I parties (governments as well as private entities) can assist in financing these projects and purchase the resulting credits to achieve compliance with their own reduction commitments. Thereby, a CER is a unit of GHG reduction that has been generated and certified under the provisions of the Kyoto Protocol for the CDM. Credits adhering to the guidelines of CDM have been generated and have been bankable since 2000. For the CDM, it was especially important to determine the level of emissions that would have occurred without the investment. Only then could proper credit be given for the difference between the (lower) actual emissions and the baseline level that would have occurred otherwise (Rosenqvist et al., 2003). The baseline definition of a coal fire's GHG emissions requires that at least one of the three major parameters for coal fire quantification be known (with the assumption that they will be stable over time): the amount of coal burning per year (tons of coal), the amount of energy released per year (megawatts) and the amount of GHGs emitted per year (tons of CO_2) (see Figure 9.2).

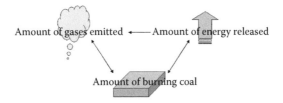

Amount of gases emitted ⟵ Amount of energy released

Amount of burning coal

FIGURE 9.2   Relationship between the three major quantitative coal fire parameters: Amount of burning coal (fuel), amount of gases emitted (as CO$_2$ equivalents) and amount of energy released. (From Kuenzer C., Zhang J., Jing L., Huadong G., Dech S. (2013) Thermal infrared remote sensing of surface and underground coal fires. In: Kuenzer C., Dech S. (Eds.) *Thermal Infrared Remote Sensing. Remote Sensing and Digital Image Processing*, vol 17. Springer, Dordrecht.)

From a very simplified perspective, the solution seems easy. Either one measures GHG emissions directly, or one retrieves the amount indirectly via the other quantities. If the amount of coal burning in a coal fire is known, it is possible to calculate how much gas is emitted. Another option is if the amount of energy released by a fire is known, one knows the amount of subsurface coal that is burning. However, for these indirect methods of gas emissions estimation, two transfer functions (relationships) would have to be established:

- A transfer function between the amount of burning coal and the amount of emitted GHGs.
- A transfer function between coal fire energy release and the amount of burning coal (again the first transfer function is needed; thus, this pathway contains more uncertainty).

Under laboratory conditions, it is possible to measure gas emissions quantitatively from a small volume of burning coal. However, this is not possible under realistic in situ conditions in a coal fire area. The technical difficulties of measuring the three quantities in situ are presented in Table 9.1.

Uncertainties related to the underground occurrence of coal fires make them especially difficult to assess in a quantitative way. Variations in the physical and chemical characteristics of coal, heterogeneities in coal layering, differences in overburden bedrock characteristics, varying crack and vent pathway densities above the burning seam, related variations in the degree of combustion, uncertainties in measuring techniques, and the problem of spatial, and especially, temporal, transfer of results are key parameters that complicate GHG emissions quantification.

Local estimates are thus usually based on very detailed (and very costly) in situ measurements and the development of complex 3-D models from direct gas measurements or the exact estimation of the amount of underground burnt coal. Regional approaches then must rely on groupings of individual fires and the extrapolation of the results. Thus, the country-wide approach of GHG emissions estimation necessarily yields an approximate result that could be significantly lower or higher than actual emissions for that country's coal fire-related emissions; the exact amount most likely cannot be determined.

TABLE 9.1

Difficulties for Direct In Situ Measurements of the Three Parameters Relevant for Coal Fire Quantification: Amount of Burning Coal, Amount of Energy Release, Amount of Emitted Gases

	Technical Difficulties of Direct Measurements
Amount of burning coal	Direct measurements of burning coal: • Not possible because the coal burns underground. >> Only approximation via a 3D model possible.
Amount of energy release	For in situ temperature and calorimetric measurements on a dense grid, the following limitations need to be considered: • To define energy release of a coal fire zone on the surface, the underground coal fire outline on the surface needs to be known (available in situ mapping result) • Energy release can undergo strong fluctuations • Similar burning volumes of coal might lead to different energy release depending on overlying bedrock characteristics and crack systems • It is not possible to map a whole coal fire area (regional scale) of several km^2 in this way (too costly and time consuming) >> Remote sensing as an option
Amount of emitted gases	In situ gas measurements on a dense grid: • Gas emissions undergo strong fluctuations • Gas is emitted through cracks and vents but also through the normal overlaying 'soil'. • It is not possible to measure emitted gas over a whole coal fire area (regional scale) of several km^2 in this way (too costly and time consuming) >> Reasonable for one fire and then extrapolating

9.2.2 CASE STUDY 9.1: EMISSIONS QUANTITIES FOR COAL FIRE AREAS AND PROVINCES IN CHINA

Studies by Kuenzer et al. (2007b) and vanDijk et al. (2011) estimated GHG emissions from coal fires that are burning in three Chinese provinces. This chapter does not repeat the details of calculations here – the interested reader is referred to the two mentioned sources. Based on these publications, it is estimated that coal fire related emissions for the year 2009 in each of the provinces were

- 7.44 Mt of CO_2 equivalent in Ningxia Province
- 2.35 Mt of CO_2 equivalent in Inner Mongolia Province
- 39.00 Mt of CO_2 equivalent in Xinjiang Province

The calculations were based on extremely detailed data obtained from provincial mining authorities, including information on the individual coal fields, the number of existing fires, size of the fires based on above-ground thermal mapping, coal characteristics (rank, carbon content, volatiles), as well as local experts' estimates of lost coal per fire. This type of data has been collected by hundreds of mine engineers in the provinces, working in state-funded mines. In addition to local and regional emissions estimate approaches, coal fire emissions quantification should always integrate this type of local knowledge if data are available and accessible.

Extrapolating these provincial numbers then to the whole country, based on knowledge of coal deposits, information on other fires, as well as national Chinese mining statistics, led vanDijk et al. (2011) to estimate that maximum GHG emissions (including a strong CH_4 component with a 21-fold global warming potential, [GWP] (Litschke, 2005)) amounted to 58 Mt of CO_2 equivalents in 2009. With respect to the global annual fossil fuel related CO_2 emissions budget of 28 Gt (IEA, 2010), the fires in China therefore contributed a maximum of 0.22%.

One promising approach for indirect coal fire quantification via the fuel-CO_2-energy nexus is based on space-borne Earth observation. Techniques available allow detection of coal fire-related thermal anomaly clusters at the Earth's surface, and calculation of the energy release of these clusters. Indirectly, it might then be possible to calculate back to the amount of coal burned (and thus the GHG amount released) (Tezlaff, 2004). This approach can be combined with in situ data (e.g. in situ mapped coal fire outlines, information on coal quality) to enhance overall accuracy. However, even without in situ data, it could work by automated thermal anomaly extraction, as presented in Kuenzer et al. (2007a), clustering, and energy-release derivation exclusively based on 60 m resolved thermal data. The approach could never be followed further in depth, as in 2003 Landsat-7 ETM+ had a scan mirror failure and good thermal data were no longer available. However, a new Landsat-8, also including a thermal band, was launched in February 2013. It is expected that remote sensing-based coal fire quantification efforts will be possible using analytics available in the new satellite (Figure 9.3).

9.2.3   CASE STUDY 9.2: THE SAN JUAN BASIN, SOUTHWEST COLORADO

In the remainder of this chapter, we consider ways to estimate local CO_2 emissions from specific coal fires, using a coal fire in the San Juan Basin in southwest Colorado as an example. We refer to that fire as the North Coalbed Fire (NCF), to distinguish it from other currently active fires in the region. The NCF was discovered in 1998 on the Southern Ute Indian Reservation. To set the stage for the emissions estimates, we describe the geological setting and summarize a variety of field observations at the NCF. We then offer a conceptual picture of the NCF that is consistent with field observations, and we conclude with estimates of CO_2 emissions made using three independent methods based on that conceptual picture.

9.2.3.1   Geologic Setting and Field Observations: The San Juan Basin and the North Coalbed Fire

Figure 9.4 shows the location of the NCF on the northwest edge of the San Juan Basin, which is located near the Four Corners, and spans Colorado and New Mexico. The San Juan Basin is a coal-bed methane (CBM)-bearing area that is shaped like a pie plate, with a flat central basin with monocline outcrops at the edges that dip toward the centre of the basin. Most of the commercial production of CBM comes

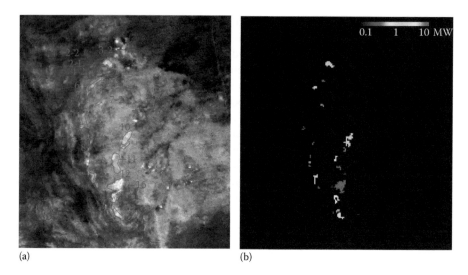

(a)   (b)

FIGURE 9.3   Coal fire-related energy release. (a) Thermal Landsat-7 ETM+ summer nighttime image acquired in 2002. (b) Energy release in MW for the individual coal fires as mapped in situ (red fire outlines overlain on thermal satellite data). Left: Polygons, defining an area, need to be available so that the algorithm can calculate a fire's energy release. The subset shows 20 × 20 km and represents the Wuda coal mining syncline. (From Tetzlaff, A. (2004). Coal fire quantification using Aster, ETM and Bird instrument data. PhD Thesis, Geosciences, Maximilians-University, Munich, 155 pp. Van Dijk, P., Zhang, J., Jun, W., Kuenzer, C. and Wolf, K.H. (2011). Assessment of the contribution of in situ combustion of coal to greenhouse gas emission; based on a comparison of Chinese mining information to previous remote sensing estimates. *International Journal of Coal Geology*, Special Issue RS/GIS, DOI: 10.1016/j .coal.2011.01.009; pp. 108–119.)

from the Central Basin. While most of the CBM is captured by the production wells there, field measurements have demonstrated that some of the CBM migrates to the outcrop and presumably escapes to the atmosphere.

The NCF is a small fire, with an areal extent of approximately 600 m × 200 m. The fire is approximately 20 m deep. Its approximate location in the San Juan Basin is bounded by the red dotted rectangle, and it is also labelled the 'North Coalbed Fire' in Figure 9.4a. The cross-section shown in Figure 9.4b indicates the relationship of the burning coal seam (the lower coal, black solid line) and the topography. Both the local topography and the lower coal seam dip and flatten toward the southeast in the direction of the Central Basin. The local topography slopes between 5° and 9° to the southeast, and the 8 m (approximately 25 ft) thick lower coal seam dips 6° to 15° in the same direction (Condon, 1988). Near the outcrop, approximately 15 m of sandstones and shales overlie the lower coal seam where the fire is currently active. There is a thin (0.5 m) layer of topsoil above the sandstone at the surface. The competent and extensive Pictured Cliffs sandstone unit underlies the lower coal seam.

Figure 9.5 shows the location of zones of burned coal; areas of unburned coal; and the area of elevated temperature, indicating current or recent combustion. This image was obtained by measuring the effects of the fire on both the magnetic

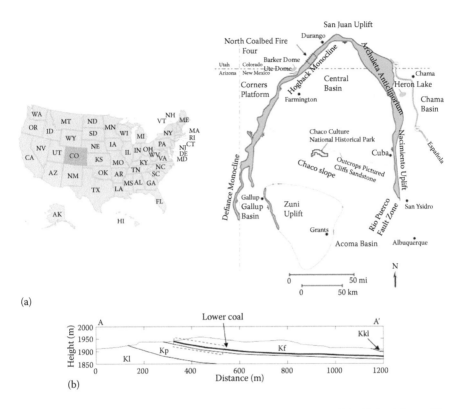

(a)

(b)

FIGURE 9.4   (a) Top left: The location of Colorado in red inside of the United States. Top right: San Juan Basin and its characteristic geologic features. The NCF location is highlighted by the dotted box (red) in the northwestern corner of the basin along the Hogback Monocline. The shaded area around the perimeter of the San Juan Basin (green) denotes outcrops of Pictured Cliffs sandstone. (Adapted from Lorenz, J. and Cooper, S. (2003). Tectonic setting and characteristics of natural fractures in Mesaverde and Dakota reservoirs of the San Juan Basin. *New Mexico Geology*, 25, 3–14.) (b) Bottom: Cross-section A-A' showing surface topography and representative subsurface stratigraphy in the vicinity of the NCF. The coalbed fire is located near the coal outcrop inside of the red dotted box. Kl = Lewis Shale, Kp = Pictured Cliffs sandstone, Kf = Fruitland Formation, Kkl = Kirtland Shale. (From Ide, T.S., Crook, N., and Orr, F.M., Jr. (2011). Magnetometer measurements to characterize a subsurface coal fire. *International Journal of Coal Geology*, 87, 190–196 and Ide, T.S., Pollard, D., and Orr, F.M., Jr. (2010). Fissure formation and subsurface subsidence in a coalbed fire. *International Journal of Rock Mechanics & Mining Sciences*, 47, 81–93.)

susceptibility and orientations of magnetite minerals contained within sandstone and shale layers above the coal. Areas not affected by the fire show small positive magnetic susceptibility and alignment, most likely due to the presence of magnetite particles laid down by sedimentary processes that created the sandstones and shales. When rocks are heated above the Curie temperature (585°C) of magnetite, any previous magnetic alignment is lost. When the heated zone cools subsequently, however, the magnetic orientations are restored, but are now aligned with the Earth's current magnetic field. These changes are large enough to allow detection of the current

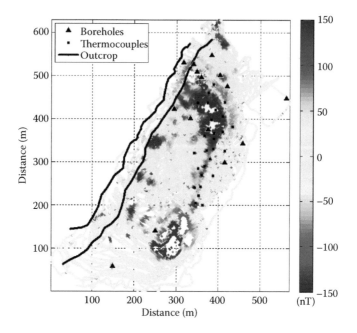

FIGURE 9.5   A map of magnetic anomaly data after diurnal fluctuations, ambient magnetic field, spikes, and asymmetries have been removed. Blue = previously burned and cooled; red = currently active fire zones; and green/yellow = unaltered zones. (From Ide, T.S. and Orr F.M. Jr. (2011). Comparison of methods to estimate the rate of CO_2 emissions and coal consumption from a coal fire near Durango, CO. *International Journal of Coal Geology*, 86, 95–107. Ide, T.S., Crook, N., and Orr, F.M., Jr. (2011). Magnetometer measurements to characterize a subsurface coal fire. *International Journal of Coal Geology*, 87, 190–196.)

location of the combustion zone. Several investigators have used magnetometer surveys to characterize underground coal fires (Hooper, 1987; Bandelow and Gielisch, 2004; Sternberg, 2004; Gielisch, 2007; Schaumann et al., 2008; Ide and Orr, 2011) although in some of these investigations, the resolution of the survey is not reported.

Magnetic orientation and susceptibility data were collected using a pack-mounted caesium vapour magnetometer. Those data were processed to reduce the data to pole, and to remove the ambient magnetic field, diurnal magnetic fluctuations, spikes due to the presence of metal objects and dropped data. The results are shown in Figure 9.5: blue regions show where the overburden was heated and subsequently cooled, the red regions where some portion of the overburden is above the Curie temperature of magnetite at present, and the light green regions where the overburden has not been affected by the coal fire. Ide (2011) and Ide et al. (2011) discuss details of the magnetic anomaly surveys. The locations of the burned and hot zones shown in Figure 9.5 are completely consistent with observations of cuttings from boreholes drilled to install thermocouples, measured compositions of gases present in the fissures, and areas of snowmelt after winter storms (Ide, 2011; Ide et al., 2011).

The most obvious visual indicator of the presence of a coal fire is a series of surface fissures (Figure 9.6). While an effort was made to map every fissure at the NCF,

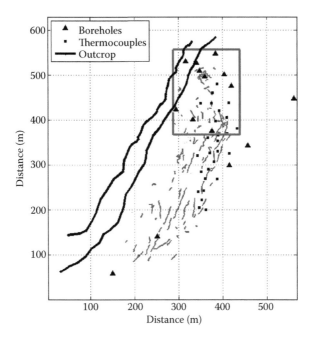

FIGURE 9.6   A distribution map of the surface fissures recorded at the NCF. Blue fissures are those at ambient temperatures, while the red fissures emit gases at elevated temperatures. The red box in the figure shows the location of the Crestal Extension Fire. (From Ide, T.S. and Orr, F.M. Jr. (2011). Comparison of methods to estimate the rate of CO_2 emissions and coal consumption from a coal fire near Durango, CO. *International Journal of Coal Geology*, 86, 95–107.)

there are most certainly some that were not mapped, either because we failed to see them or because they were overlain by topsoil. Some are at ambient temperature of typically –5°C to 30°C depending on the season and others emit hot combustion product gases often exceeding 1000°C. These are indicated using blue and red lines, respectively, in Figure 9.6. The hottest zone at the NCF is labelled the Crestal Extension Fire, the region bounded by the red box in Figure 9.6. Fissures mapped in Figure 9.6 were those that were observed in 2007. Since then, more fractures have developed to the northeast of the uppermost red fissures in Figure 9.6, which is consistent with the idea that fissures open ahead of existing combustion zones. Field measurements to estimate the rate of CO_2 emissions were performed at the Crestal Extension Fire. The fissures mapped in Figure 9.6 form because of combustion of the coal. A schematic cross-section of a coal fire is shown in Figure 9.7. Coal fires like this one most likely start at the outcrop, where the coal is exposed to air, and a source of ignition such as a lightning strike can ignite a fire that then propagates underground. As coal is burned, the structural support for the overburden is converted to gas and ash, and as the burned zone grows, that area can no longer support the weight of the overburden above. As a result, the roof of the burned zone sags and eventually collapses. Analysis of the stresses that result indicates that tensile stresses concentrate near the hinge point of the collapse (Ide et al., 2010; Ide, 2011). These stresses cause fissures to open at the location of existing fractures in the overburden.

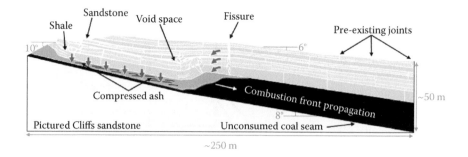

FIGURE 9.7   A schematic diagram of the NCF, showing how the fire has burned into the formation from the outcrop. After some distance of the coal seam has burned, it can no longer support the weight of the overburden and subsidence occurs. Pre-existing joints near the subsidence can open to form fissures when they are subjected to opening mode stresses. (From Ide, T.S., Pollard, D., and Orr, F.M., Jr. (2010). Fissure formation and subsurface subsidence in a coalbed fire. *International Journal of Rock Mechanics & Mining Sciences*, 47, 81–93.)

Those fissures that open ahead of the combustion front provide an exit path for hot combustion product gases, while fissures that formed earlier provide inlets for air to sustain combustion. A continuing process of this sort creates the lines of more or less parallel fissures observed at the NCF (cf. Figure 9.6).

Figure 9.8 shows the sequence of events observed as a new fissure breaks through to the surface at the NCF. The first evidence of a new fissure is a line of moist ground (Figure 9.8a). Sometime thereafter, precipitation of elemental sulphur is observed (Figure 9.8b). Over time, as the fissure width grows, tarry materials are deposited (Figure 9.8c), and the temperature of the gases emitted increases, presumably as the rock at the side of the fissure is heated. Figure 9.8d,e shows

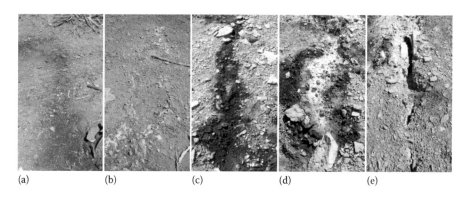

(a)   (b)   (c)   (d)   (e)

FIGURE 9.8   A sequence showing how the fissures form over the NCF. From left to right: (a) Formation of a line of moisture at the surface, (b) sulphur precipitation around the fissure, (c) evolution of tar, (d) ammonium chloride precipitation, and finally (e) emission of hot combustion gases from the fissure. (From Ide, T.S. (2011). Anatomy of subsurface coal fires: A case study of a coal fire on the Southern Ute Indian Reservation. Unpublished PhD dissertation, Stanford University.)

that as the temperature increases further, crystals of ammonium chloride precipitate close to the edges of the fissure.

Detailed analysis of fissures at the NCF shows that fissure widths and configurations depend on the orientation and the magnitude of the opening mode stresses exerted on the pre-existing joints (Ide et al., 2010). Some fissures can be large enough at the surface for a person to crawl into them. Observations made inside of the fissures suggested that many of the surface fissures were connected in the subsurface, and boreholes drilled near the fissures confirmed that the fissures were connected to an extensive network of underground layer of rubble, ash, and unburned coal and char. Video images captured by an underground borehole camera showed that the rubble zones above the coal seam provided high permeability conduits through which air can reach the combustion zone and produced combustion gases can migrate to fissures ahead of the combustion zone.

Figure 9.9 shows another view of fissure formation at a fossil coal fire exposed by erosion in a streambed near the NCF. That outcrop corroborates how a fissure can open above active combustion zones. The unburned coal is at the bottom of the photograph. The right portion of the coal seam has burned, with a layer of ash in place of the coal, and thermally altered rocks above the burned zone. The person at the right of the photograph (circled) standing directly next to the ash and the thermally altered rocks is approximately 1.5 m tall. The resulting collapse of the overburden created a large fissure that extends to the surface from the coal seam. The sketch at the right side of Figure 9.9 shows the location of the large fissure and several smaller ones on the face of the outcrop. Numerical simulations of the stress settings showed that the magnitude of the collapse and the aperture of the fissure measured at the outcrop are consistent with the idea that fissures formed when a tensile stress was exerted on a pre-existing fracture (Ide et al., 2010).

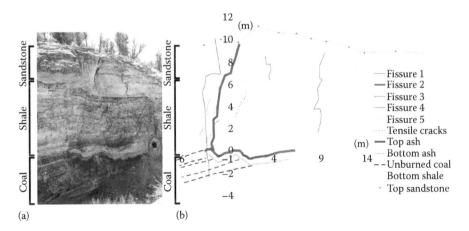

FIGURE 9.9   (a) Outcrop near the NCF that shows a fissure that opened up as a result of a subsurface collapse, and (b) major features that are mapped using the LaserRange Finder. Fissure 2 is the largest opening mode feature at the outcrop. (From Ide, T.S., Pollard, D., and Orr, F.M., Jr. (2010). Fissure formation and subsurface subsidence in a coalbed fire. *International Journal of Rock Mechanics & Mining Sciences*, 47, 81–93.)

9.2.3.2 Conceptual Picture of a Coal Fire

Figure 9.10 shows schematically a sequence of combustion steps consistent with the propagation of the combustion zone and the observations of fissure formation described above. In Zone 1, cool air is drawn into the subsurface through an ambient-temperature fissure. Hot, low-density gases escaping from exit fissures, which act as chimneys, provide the driving force to draw air into the combustion zone through the cool fissures. In Zone 1, the temperatures have begun to cool as heat is transferred to cool air entering the subsurface rubble zone above the coal seam. This zone may also contain some char, coal that has been heated sufficiently that volatiles have been driven off and pyrolysis of some parts of the coal has occurred. Because the temperature is relatively low in Zone 1, the rate of char oxidation is low.

In Zone 2 of Figure 9.10, the flowing air is warmed by the residual heat trapped in the subsurface as it flows toward the combustion zone through the fractured region that lies immediately atop the coal seam. Here heated air contacts more char and ash, the remnants of the coal after it has undergone devolatilization in the presence of O_2 and/or pyrolysis if the heating occurred in an environment lacking O_2. We use the term 'thermal decomposition process' to refer to both devolatilization and pyrolysis of the coal. Volatiles refer to steam, gas phase tar, CO_2, CO, CH_4, H_2, and other hydrocarbon and aromatic components. As the temperature of the coal increases, steam, tar and sulphur are produced first. Other species evolve subsequently. A more detailed description of this process can be found in Ide (2011). In Zone 2, O_2-char combustion may occur. If O_2 is present with the volatiles, it is more likely that the O_2 will be consumed in reactions with the combustible volatiles because the homogeneous gas phase combustion reactions occur at rates that are orders of magnitude larger than the heterogeneous and exothermic char combustion reactions. But as the upstream portion of Zone 2 in Figure 9.10 illustrates, in regions where char and O_2 react, the O_2-char reaction provides heated combustion products that flow downstream to heat and devolatilize more coal. Two sources of combustible volatiles exist in the fractured zone. The first source is the coal, which produces combustible volatiles because of thermal decomposition. The second source is CH_4 in the native gases that flow from the Central San Juan Basin toward the Hogback Monocline. Production histories of wells that are drilled just down dip of the NCF suggest that the up-dip flow of native gas may be significant.

Figure 9.10 shows that there are three sources of the CO_2 that is measured at the surface, which are: native CO_2 flow that is flowing up-dip from the Central San Juan Basin CO_2 produced from oxidizing the native CH_4 flowing from the Central Basin, and CO_2 produced from the coal combustion reaction. The isotopic signature, $\delta^{13}C$ (a measure of the ratio of stable isotopes $^{13}C{:}^{12}C$, reported in parts per thousand (per mil, ‰)), of the CO_2 at the surface can be expressed as a function of the mole fraction of each of the three sources. Because the isotope signatures of the three sources are unique, the isotope measurements, together with mass balance equations, can be used to derive the contribution of each of the components, in Ide and Orr (2011). Results obtained showed that on average approximately 44.0% of the CO_2 came from burning native San Juan Basin CH_4, 42.8% of the CO_2 came from native San Juan Basin CO_2 and CO_2 from coal combustion accounted for 13.2%.

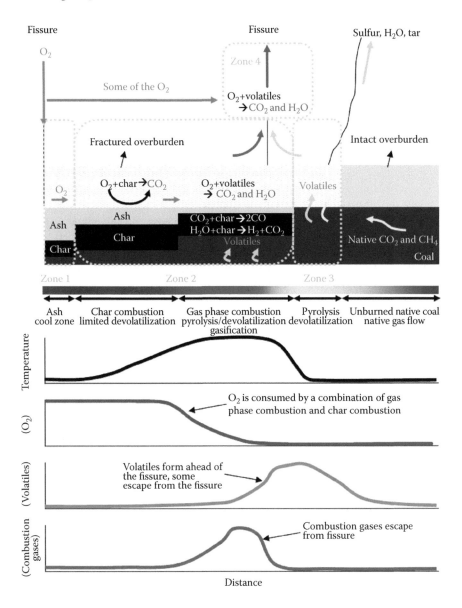

FIGURE 9.10 A conceptual picture of the workings of the NCF, divided into four zones. Cool air is drawn into the fractured zone in the subsurface immediately above the affected coal seam, where most of the homogeneous combustion reactions take place. Combustion gases that are produced in this fractured zone are vented up hot fissures that act as chimneys. Char gasification and combustion reactions can take place. Ahead of the fire, the coal may produce sulphur, tar and moisture. Profiles of temperatures, O$_2$ concentration, volatile species concentration and combustion gas concentration are shown below. (From Ide, T.S., Crook, N., and Orr, F.M., Jr. (2011). Magnetometer measurements to characterize a subsurface coal fire. *International Journal of Coal Geology*, 87, 190–196.)

The energy that is produced in Zone 2 can be transferred ahead of the combustion zone as well as deeper into the coal seam. The affected areas are defined as Zone 3 in Figure 9.10. When the coal seam temperature increases, gas-phase volatiles are produced. Unlike in Zone 2, where the combustible volatiles react with the O_2 in the fractured zone, gases that are formed in Zone 3 flow downstream without reacting because the O_2 has been consumed in the combustion reactions in Zone 2, leaving little or no O_2 to react with the gases that evolve from the coal in Zone 3. Steam and water production dominate in regions that are ahead of the active combustion zone. Some elemental sulphur will also evolve in reducing environments (Berkowitz, 1985). Steam and gaseous tar that is produced in these areas can migrate upwards through pre-existing fractures and condense at the surface as they cool. This picture is consistent with the observed moisture, tar and sulphur noted at incipient fissures at the NCF.

Zone 4 in Figure 9.10 is the region inside of the fissures. As gases migrate upward through the chimney, combustible volatile gases such as CO, H_2 and CH_4 that did not react in the subsurface due to the lack of oxygen can now react with additional air that is entrained through near-surface fractures. When temperatures inside of the fissures are higher than the autoignition temperatures of the combustible volatiles, they can burn without an ignition source, if the O_2 concentration limits are satisfied. Temperatures inside of the fissure, which can surpass 1000°C, exceed the autoignition temperatures of CO, H_2 and CH_4, which are 609°C, 500°C and 580°C, respectively. Blue flames, which likely result from CO and CH_4 combustion, were visible at the mouth of the fissure near the ground surface during field visits prior to sunrise. These additional combustion reactions keep the temperature of the gases in the fissure high, which makes the rocks glow. Moreover, these reactions enhance the temperature differential between the chimney temperature and the ambient air temperature. This, in turn, increases the buoyancy-driven flow that circulates air and combustion gases through the subsurface. Secondary combustion reactions in the fissures explain why CO, H_2 and CH_4 mole fractions are often nonexistent or at ppm levels at the top of the fissure where the gas samples are collected.

9.2.4 Estimating CO_2 Emissions

In this section, we describe three methods to estimate CO_2 emissions and coal consumption rates from the Crestal Extension Fire region (cf. Figure 9.6). In each approach, a relationship between coal consumption rates and CO_2 emissions rates is required. We use the ultimate analysis of the coal to relate CO_2 emissions rates to coal consumption: it contains 72.33% carbon by mass, with the remainder composed of ash, moisture, N_2, H_2, S and O_2. In the estimates that follow we assume that all the available carbon in the coal is transformed into CO_2.

In method 1, time-lapse, high-resolution surface deformation measurements were used estimate coal consumption rates. Data were acquired near the hottest fissures found over the NCF. The volume loss measured at the surface was assumed to be the coal volume lost in the subsurface due to the fire. Similar estimates based on synthetic aperture radar interferometry are described by Jiang et al. (2011). In method 2, measured velocities of exhaust gases from a fissure over an active region were combined with measured amounts of CO_2 seeping from the less fractured areas over the active fire. Method 3 is based on an analogy between coal fire fissures and natural convection chimneys. The dimensions and the thermocouple temperatures in the most active fissure over the site were used to set the chimney geometry and the thermal gradient. Ide and Orr (2011) and Ide (2011) give additional details of the estimates described here.

9.2.4.1 Method 1 – Subsidence Measurements

Surface deformations were measured at 171 locations over the NCF on two occasions spanning a 7-month interval. The first survey was carried out in April 2009, and the second in November 2009. The same locations (+/– 0.0125 m) were used in both surveys. The Nikon DTM 521 system used for the survey had a vertical resolution of 3.18×10^{-3} m (3.18 mm). A contour map shown in Figure 9.11 shows the differences in surface elevation between the two surveys. If we assume that the surface deforms in direct proportion to the amount of coal consumed in the subsurface, an estimate for the rate of coal combustion over the 7-month period can be obtained by calculating the surface deformation volume. This approach also assumes that

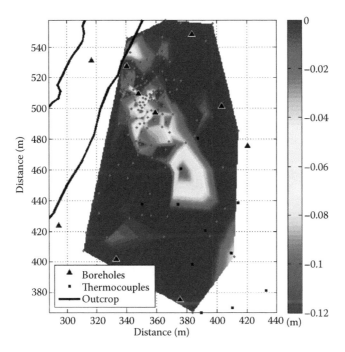

FIGURE 9.11 A contour map showing the magnitude of subsidence based on two topography measurements conducted 7 months apart. Subsidence was most active in previously burned areas and regions that are currently hot, based on the magnetometer surveys. The volume of subsidence can be used to infer the volume of coal that has been consumed in the subsurface. (From Ide, T.S. and Orr, F.M. Jr. (2011). Comparison of methods to estimate the rate of CO₂ emissions and coal consumption from a coal fire near Durango, CO. *International Journal of Coal Geology*, 86, 95–107.)

the deformation occurs on a continuous basis, and the surface deforms immediately when the coal is consumed. The overburden is also viewed as an elastic medium.

The surface deformation was calculated by overlaying a 0.5 m × 0.5 m grid on top of the contour map in Figure 9.11. An interpolated displacement value in the z-direction was assigned to each grid cell, and total deformation over the surveyed area was calculated by summing the deformation at each grid point. Surface volume change calculated using this method showed that over the 7-month period approximately 104 m³ of displacements occurred at the surface, or 178 m³ on a yearly basis. Since the density of the coal at the NCF is approximately 1400 kg/m³, this surface deformation translates to a coal consumption rate of 249 t/year for this portion of the fire. The ultimate analysis of the coal was used to convert the coal consumption rate to CO_2 emissions rate, which was approximately 660 t/yr.

9.2.4.2   Method 2 – Exhaust Gas Flow Velocity

A second estimate was obtained using a particle tracking approach illustrated in Figure 9.12. A camera sensitive to volatile organic compounds (VOC) was stationed adjacent to the fissure that was most actively venting combustion gases. A plume of the gases emitted was recorded, and particle positions were traced to determine their average velocity as they exited the fissure into the atmosphere. This average velocity was multiplied by the exhaust gas density, the CO_2 gas composition in the exiting gas, and the dimensions of the cross-sectional area of the fissure to obtain a CO_2

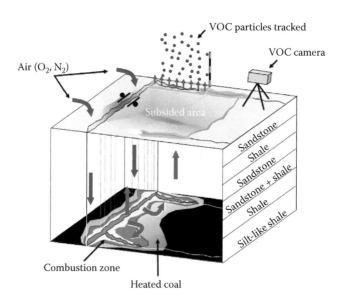

FIGURE 9.12   Natural convection draws air into the subsurface through one set of fissures and releases combustion gases from fissures above (and often ahead of) active underground coal fires. Volatile organic compounds (VOCs) contained in the combustion gas emissions can be used to measure the rate at which the coal fire is emitting CO_2. (From Ide, T.S. and Orr, F.M. Jr. (2011). Comparison of methods to estimate the rate of CO_2 emissions and coal consumption from a coal fire near Durango, CO. *International Journal of Coal Geology*, 86, 95–107.)

mass flow. While there are numerous fissures emitting gases at the NCF, only two of them were emitting gases at a rate that could be detected by the VOC camera. Of the two fissures that showed emissions activities that could be recorded, only the most active one was used to estimate the rate of CO_2 emissions. Thus, results obtained using this method must be an underestimate of true emissions from the NCF.

Based on the VOC video footage collected in 2009, the average velocity of the combustion gases was 1.66 m/s. The temperature of the gases emitted varied along the fissure from about 315 to 1080°C, depending on the location of the fissure. The corresponding exhaust gas density for this temperature range is approximately 0.281 to 0.610 kg/m³, with an average of 0.446 kg/m³. The cross-sectional area of the fissure was about 0.348 m² (11.6 m long, 0.03 m wide). The exhaust gases contained about 26% CO_2 by mass, and the resulting CO_2 emissions rate from the NCF was approximately 2112 t/year.

The large difference between this CO_2 emissions rate estimate and that obtained from the subsidence method arises because not all the CO_2 emissions came from coal consumption. In the first method, only the CO_2 emissions that result from coal consumption were estimated. The emissions rate from the fissure flow includes CO_2 from combustion reactions involving the native CH_4 flowing up dip from the Central Basin (cf. Figure 9.10) as well as the native San Juan Basin CO_2 that flows up dip with the CH_4.

9.2.4.3   Method 3 – Natural Convection Chimney Model

In the third method, we draw an analogy between the flow in the vertical fissures and a natural convection chimney. As Figure 9.13 illustrates, density driven flow allows air and combustion gases to circulate in and out of the subsurface. A chimney model shown in Figure 9.11 was used to represent this flow.

In this simple model, we assume that the coal fire has only one inlet and one exhaust vent, obviously a significant simplification of the flow at the NCF where

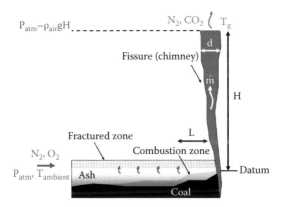

FIGURE 9.13   A diagram showing the natural convection chimney that was used to model the NCF. The geometry of the chimney, along with the temperature and pressure gradients define the rate at which air and combustion gases can flow from the subsurface to the surface. (From Ide, T.S. and Orr, F.M. Jr. (2011). Comparison of methods to estimate the rate of CO_2 emissions and coal consumption from a coal fire near Durango, CO. *International Journal of Coal Geology*, 86, 95–107.)

many inlets provide air and a modest number of fissures provide the exits. We also assume that gases rising through the chimney fissure are isothermal, so that a single, constant gas average density can represent the variable gas density within the combustion zone in the fissure, and that zero friction losses occur in the inlet channel that supplies air to the fire. Ide and Orr (2011) explore the validity of each of these assumptions in more detail.

Based on these assumptions, an expression for the exit gas velocity and the combustion gas mass flow rate can be formulated using a macroscopic mechanical energy balance (Rosner, 2000; Bird et al., 2001). The exit gas velocity from the chimney can be expressed as a function of fissure geometry, roughness coefficient of the fissure walls, pressure gradient induced by the thermal buoyancy force, and the combustion gas density. Temperature does not explicitly appear in the equation, but is implicitly accounted for through both the pressure gradient and gas density terms. The mass flow rate of the exit gases is simply the product of the gas velocity, the combustion gas density, and the cross-sectional area of the fissure (Ide and Orr, 2011).

A range of geometric and temperature values measured at the NCF was used in the energy balance equations to calculate a range of possible mass flow rates. There were 10,000 simulations performed to explore different combinations of the input variables. This resulted in a distribution of mass flow rates. With the 26% gas composition of CO_2 in the exhaust stream, this model gave a CO_2 emissions value of 1616 t/yr with a standard deviation of 350 t/yr (Ide, 2011; Ide and Orr, 2011). As in Method 2, this CO_2 emissions estimate includes coal combustion, native CH_4 combustion, and native CO_2 flowing up dip. Of the 1616 t of annual CO_2 emissions, 213 t were attributed to emissions from coal combustion.

9.2.5   FLUX ACCUMULATION CHAMBER MEASUREMENTS

Methods 2 and 3 described above do not account for the fact that CO_2 could be seeping at the NCF. In both methods, the CO_2 flux was assumed to occur through a single major fissure. To determine the seepages of CO_2 emissions from areas in between fissures, additional CO_2 mass flow rates were measured in 2008, using a flux accumulation chamber. The survey was conducted specifically over the hottest area at the NCF, the Crestal Extension Fire (cf. Figure 9.6). Fluxes were measured at nodes of a 50 ft × 50 ft grid (approximately 15 m × 15 m), that was interpolated over the surveyed area using Kriging. This resulted in an additional 3851 m^3/day of CO_2 emissions over the surveyed area. Using a density of 1.39 kg/m^3 for the CO_2 to account for both the ambient temperature and the altitude at which the survey was performed, the emissions were an additional 1954 t/yr of CO_2 emitted, an amount that is roughly equal to the emissions rate calculated in Methods 2 and 3 (2112 t/yr and 1616 t/yr, respectively).

9.2.6   COMPARISON OF RESULTS

Table 9.2 lists the contribution of CO_2 emissions that is produced by burning coal (13.2%), burning native CH_4 (44.0%), and from native CO_2 (42.8%), and the total

TABLE 9.2

Comparison of the Rates of CO$_2$ Emissions Attributed to Coal Combustion, Native San Juan Basin CH$_4$ Combustion, and Native San Juan Basin CO$_2$

| | CO$_2$ Emissions Rates (metric tons/yr) | | | |
Method	Coal Combustion	Native CH$_4$ Combustion	Native CO$_2$	Total
Subsidence	660	–	–	–
VOC only	279	929	904	2112
Chimney only	213	711	692	1616
Flux chamber only	258	860	836	1954
VOC + Flux chamber	537	1789	1740	4066
Chimney + Flux chamber	471	1571	1528	3570

Note: Results of coal combustion rates are similar between all three methods when flux chamber measurements are included. Total CO$_2$ emissions rates calculated from the VOC and the chimney methods are consistent.

CO$_2$ emissions. These results were calculated using the isotope-signature-based breakdown described earlier. Quantities of CH$_4$ burned or the native CO$_2$ contribution could not be obtained from the subsidence measurements, since the method was based on measuring changes in land volume rather than measuring the flow of CO$_2$. Table 9.2 shows that values obtained from the VOC method and the chimney methods are similar. The flux chamber measurements, which account for the CO$_2$ that is escaping into the atmosphere from the entire area over the active coal fire region, show emissions roughly equal to the rate of CO$_2$ that is escaping from a large fissure over the NCF. Thus, adding this measurement to both the VOC and chimney method better accounts for the total rate of CO$_2$ emissions at the Crestal Extension Fire. Once the CO$_2$ measured by the flux accumulator was accounted for, the CO$_2$ emissions rates from the coal were in line with the rate obtained by the subsidence method.

9.3   DISCUSSION

The simple estimates described here have obvious limitations. For example, the assumption that the loss of surface volume was exactly offset by the volume loss in the subsurface almost certainly leads to an underestimate of the amount of coal consumed in the subsurface. This assumption is plausible but not strictly satisfied for unmined coal seams, but it is not appropriate for situations in which the subsidence results from removal of significant quantities of coal by mining. At the NCF, borehole camera observations showed numerous fractures in a previously burned region. The existence of fractures on the order of 1 to 5 cm above the coal seam suggests that the overburden does not entirely compact void spaces left by the burned coal seam, at least not immediately. In fact, if subsidence were perfect then there would be no channels through which air could travel to get to the combustion zone under a very small pressure gradient. It is also possible that subsidence occurs sometime after combustion. Subsidence does not take place until the fire consumes a critical radius of coal (for a given overburden thickness). Hence, the observed surface subsidence may not reflect the present rate of coal consumption. Obtaining estimates of this sort also requires that high resolution surface deformation data be available, through precision GPS measurements such as those used at the small NCF, or perhaps, airborne synthetic aperture radar over larger areas.

An obvious limitation of the estimate based on the flow velocity and composition is that the measurements were performed for one fissure only, even though that fissure was the largest and most active vent. Additional measurements for other fissures would improve that estimate. At the least, the example given demonstrates the technique, but repeated observations would be useful to judge variability of the flow velocity. The flux chamber accumulation measurements made over the Crestal Fire Extension indicate that flow of CO_2 coming from areas other than fissures cannot be neglected. The pressure gradient induced by the thermal buoyancy and gravity is small enough that appreciable gas flow through low-permeability (on the order of 10 md) sandstones and shale is unlikely. Instead, it is suggested that transport through the overburden takes place along the network of fractures, in locations where fissures are not readily visible at the surface. It is possible that the 30 cm of soil that covers the site hides smaller fissures through which gases can escape. For large fires, obtaining sufficient areal coverage that recognizes the spatial variability of the fluxes will be an issue.

The estimation method using the natural convection chimney analogy is also limited by the fact that only one active fissure is considered. However, the distributions of exhaust gas velocity, CO_2 mass rate and coal consumption rate obtained from the suite of chimney simulations show that despite the range of possible parameter values, the resulting rate estimates are bounded within an order of magnitude. The exhaust gas velocity calculated for the dimensions and conditions measured at the coal fire agree reasonably well with the exhaust gas velocity measured by the VOC camera. Thus, the chimney estimates may prove useful for estimating fluxes based on measurements of fissure dimensions.

Finally, the isotope material balances used to determine what fraction of emitted CO_2 came from coal combustion showed that only 13.2% of the CO_2 in the exhaust came from coal, with the remainder roughly equally split between the native San Juan Basin CO_2 and the oxidized native San Juan Basin CH_4. This result suggests that methane combustion plays a significant role in this coal fire.

The rate of native gas (CH_4–CO_2) mixing required to produce the gas composition and the isotope signatures at the exhaust fissures is also consistent with observed production of native gases from wells near to and down dip of the NCF, at rates between 30 and 300 MCFD (thousand cubic feet per day) or 850 and 8500 m^3/day. Native gases that are not intercepted by these production wells migrate up dip toward the Hogback monocline where the NCF is currently located.

9.4 CONCLUSIONS

Subsurface coal fires propagate through a self-sustaining process in which coal burns, overburden rocks sag and collapse, and fissures form that provide outlets for combustion products gases and inlets for the air required to sustain combustion. The combustion process involves interplay of thermal decomposition of the coal to produce gases that react rapidly with oxygen, slower reactions of char and oxygen, and combustion of gases that result from partial oxidation in the fissures that transport combustion products to the surface. In some settings, at least, combustion of CH$_4$ flowing through the coal from greater depths can contribute to the combustion process and the resulting CO$_2$ emissions. The likelihood of this being the case can be determined by examining C^{13} isotope signatures of the native and combustion-product gases.

Estimates of CO$_2$ emissions from subsurface coal fires can be made on various scales. One promising method of estimating CO$_2$ emissions over a large area is using satellite imagery. Thermal anomaly data associated with underground fires may be used to derive the rate of coal consumed in the area. This approach is appropriate for making CO$_2$ emissions estimates from coal fires that extend over a large area, since detailed ground surveys are time-consuming and expensive. Fieldwork and in situ measurements are recommended to be undertaken at individual fires only when significant sources of CO$_2$ are detected using the remote sensing method. On the scale of an individual fire, more direct estimates can be made by techniques that include surface deformation measurements, measurements of the composition and flow velocity of exhaust gases emitted from fissures, model estimates of chimney flow in fissures, and flux chamber measurements over the surface area of the fire. These estimates are limited by the inevitable spatial variability of emissions, and hence the estimates almost certainly underestimate total emissions.

REFERENCES

Bandelow, F.K. and Gielisch, H.F. (2004). Modern exploration methods as key to fighting of uncontrolled coal fires in China. In: *2004 Denver Annual Meeting*, The Geological Society of America, Denver, CO.

Berkowitz, N. (1985). *The Chemistry of Coal*. Elsevier Science Publishing Company, New York, NY.

Cassels, C.J.S. (1997). Thermal modelling of underground coal fires in northern China. PhD thesis, International Institute for Aerospace Survey and Earth Sciences (ITC), The Netherlands, *ITC Dissertation No. 51*.

Condon, S.M. (1988). Joint patterns on the northwest side of the San Juan Basin (Southern Ute Indian Reservation), Southwest Colorado. In: *Geology and Coalbed Methane Resources of the Northern San Juan Basin, Colorado and New Mexico*. Rocky Mountain Association of Geologists.

Düzgün, S., Kuenzer, C., and Karacan, Ö. (2011). Applications of remote sensing and GIS of coal fires, mine subsidence, environmental impacts of coal mine closure and reclamation. *International Journal of Coal Geology, Special Issue RS/GIS*, DOI: 10.1016/j.coal.2011.02.001; pp. 1–3.

Engle, M.A., Radke, L.F., Heffern, E.L., O'Keefe, J.M., Hower, J.C., Smeltzer, C.D., Hower, J.M. et al. (2013). Gas emissions, minerals, and tars associated with three coal fires, Powder River Basin, USA. *Science of the Total Environment*, 420, 146–159.

Gielish, H. (2007). Detecting concealed coal fires. *Engineering Geology*, 18, 199–210.

Grubb, M., Vrolijk, C., and Brack, D. (1997). *The Kyoto Protocol: A Guide and Assessment*. Royal Institute of International Affairs Energy and Environmental Programme, London.

Hecker, C., Kuenzer, C., and Zhang, J. (2007). Remote sensing based coal fire detection with low resolution MODIS data. In: Stracher, G.B. (Ed.) *Geology of Coal Fires: Case Studies from Around the World*, Geological Society of America Reviews in Engineering Geology, v. XVIII, doi: 10.1130/2007.4118(15), pp. 229–239.

Hooper, R.L. (1987). Factors affecting the magnetic susceptibility of baked rocks above a burned coal seam. *International Journal of Coal Geology*, 9, 157–169.

Hower J.C., Henke, K., O'Keefe, J., Engle, M.A., Blake, D.R., and Stracher, G.B. (2009). The Tip Top Coalmine Fire, Kentucky: Preliminary investigation of the measurement of mercury and other hazardous gases from coal-fire gas vents: *International Journal of Coal Geology*, 80, 63–67.

Hower, J.C., O'Keefe, J.M., Henke, K.R., Wagner, N.J., Copley, G., Blake, D.R., Garrison, T. et al. (2013). Gaseous emissions and sublimates from the Truman Shepherd coal fire, Floyd County, Kentucky: A re-investigation following attempted mitigation of the fire. *International Journal of Coal Geology*, 116–117, 63–74.

Ide, T.S. (2011). Anatomy of subsurface coal fires: A case study of a coal fire on the Southern Ute Indian Reservation. Unpublished Ph.D. dissertation, Stanford University.

Ide, T.S., Crook, N., and Orr, F.M., Jr. (2011). Magnetometer measurements to characterize a subsurface coal fire. *International Journal of Coal Geology*, 87, 190–196.

Ide, T.S. and Orr, F.M., Jr. (2011). Comparison of methods to estimate the rate of CO_2 emissions and coal consumption from a coal fire near Durango, CO. *International Journal of Coal Geology*, 86, 95–107.

Ide, T.S., Pollard, D., and Orr, F.M., Jr. (2010). Fissure formation and subsurface subsidence in a coalbed fire. *International Journal of Rock Mechanics & Mining Sciences*, 47, 81–93.

Kuenzer, C. (2005). Demarcating coal fire risk areas based on spectral test sequences and partial unmixing using multi sensor remote sensing data. Ph.D. thesis, Technical University Vienna, Austria, 199 pp.

Kuenzer, C., Bachmann, M., Mueller, A., Lieckfeld, L. and Wagner, W. (2008a). Partial unmixing as a tool for single surface class detection and time series analysis. *International Journal of Remote Sensing*, DOI: 10.1080/01431160701469107, pp. 1–23.

Kuenzer, C., Hecker, C., Zhang, J., Wessling, S. and Wagner, W. (2008d). The potential of multi-diurnal MODIS thermal bands data for coal fire detection. *International Journal of Remote Sensing*, 29, 923–944, DOI: 10.1080/01431160701352147.

Kuenzer, C. and Stracher, G.B. (2011). Geomorphology of coal seam fires. *Geomorphology*, 138, doi:10.1016/j.geomorph.2011.09.004, 209–222.

Kuenzer, C. and Voigt, S. (2003). Vegetationsdichte als moeglicher Indikator fuer Kohlefloezbraende? Untersuchung mittels Fernerkundung und GIS. In: Strobl, J., Blaschke, T. and Griesebner, G. (Eds.), *Angewandte Geographische Informationsverarbeitung XV*. Beitraege zum 15. AGIT-Symposium Salzburg 2–4, July 2003, Heidelberg, Wichmann, pp. 256–261.

Kuenzer, C., Wessling, S., Zhang, J., Litschke, T., Schmidt, M., Schulz, J., Gielisch, H. and Wagner, W. (2007a). Concepts for green house gas emission estimating of underground coal seam fires, *Geophysical Research Abstracts*, 9, 11716, EGU 2007, 16–20 April 2007, Vienna.

Kuenzer, C., Zhang, J., Hirner, A., Bo, Y., Jia, Y. and Sun, Y. (2008b). Multitemporal in situ mapping of the Wuda coal fires from 2000 to 2005 – Assessing coal fire dynamics. *Spontaneous Coal Seam Fires: Mitigating a Global Disaster*. UNESCO Beijing, 2008. ERSEC Ecological Book Series, Vol. 4, pp. 132–148.

Kuenzer, C., Zhang, J., Li, J., Voigt, S., Mehl, H. and Wagner, W. (2007b). Detection of unknown coal fires: Synergy of coal fire risk area delineation and improved thermal anomaly extraction. *International Journal of Remote Sensing*, DOI: 10.1080/01431160701250432, 28, 4561–4585.

Kuenzer, C., Zhang, J., Tetzlaff, A., Voigt, S. and Wagner, W. (2008c). Automated demarcation, detection and quantification of coal fires in China using remote sensing data. *Spontaneous Coal Seam Fires: Mitigating a Global Disaster*. UNESCO Beijing, 2008. ERSEC Ecological Book Series, Vol. 4, pp. 362–380.

Li, J., Voigt, V., Kuenzer, C., Yang, B., Zhang, J., Zhang, Y., Kong, B. and Zhang, S. (2005). The progress in detecting coal fires on remote sensing – The first results of the joint Sino-German research project in innovative technologies for exploration, extinction and monitoring of coal fires in North China. *Proceedings of the Dragon Programme Mid-Term Result Conference*, 27.06.2005–01.07.2005, Santorini, Greece, pp. 3–12.

Litschke, T. (2005). Innovative Technologies for Exploration, Extinction and Monitoring of Coal Fires in North China. MSc thesis no. 1239595, University Duisburg-Essen, Institut für Geographie, Abteilung Geology.

Lohrer, C., Krause, U. and Steinbach, J. (2005a). Self-ignition of combustible bulk materials under various ambient conditions. *Process Safety and Environmental Protection*, 83/B2, 145–150. Special issue.

Lorenz, J. and Cooper, S. (2003). Tectonic setting and characteristics of natural fractures in Mesaverde and Dakota reservoirs of the San Juan Basin. *New Mexico Geology*, 25, 3–14.

Prakash, A. and Gupta, R.P. (1999). Surface fires in Jharia coalfield, India — Their distribution and estimation of area and temperature from TM data. *International Journal of Remote Sensing*, 20(10), 1935–1946.

Rosenquvist, A., Milne, A., Lucas, R., Imhoff, M. and Dobson, C. (2003). A review of remote sensing technology in support of the Kyoto Protocol. *Environmental Science and Policy*, 6, 441–455.

Schaumann G., Siemon, B. and Yu, C. (2008). Geophysical investigation of Wuda coal mining area, Inner Mongolia: Electromagnetics and magnetics for coal fire detection. In: *Spontaneous Coal Seam Fires: Mitigating a Global Disaster*. UNESCO Beijing, 2008ERSEC Ecological Book Series, Vol. 4, pp. 336–350.

Sternberg, R. (2004). Magnetic surveys over clinkers and coal seam fires in Western North Dakota. Geological Society of America Denver Annual Meeting. Session No. 15: Wild Coal Fires: Burning Questions with Global Consequences? *Geological Society of America Abstracts with Programs*, 36(5), 43.

Stracher, G.B., Kuenzer, C., Hecker, C., Zhang, J., Schroeder, P.A. and McCormack, J.K. (2014). Wuda and Ruqigou coalfield fires of Northern China. In: Stracher, G.B., Sokol, E.V., and Prakash, A. (Eds.), *Coal and Peat Fires: A Global Perspective, Volume 3: Case Studies – Coal Fires*, Elsevier Science.

Stracher, G.B., Lindsley-Griffin, N., Griffin, J.R., Renner, S., Schroeder, P., Viellenave, J., Masalehdani, M.N.-N. and Kuenzer, C. (2007). Revisiting the South Cañon Number 1 Coal Mine fire during a geologic excursion from Denver to Glenwood Springs, Colorado. In: Raynolds, R.G. (Ed.), *Roaming the Rocky Mountains and Environs: Geological Field Trips*, Geological Society of America Field Guide 10, pp. 101–110.

Stracher, G.B. and Taylor, T.P. (2004). Coal fires burning out of control around the world: Thermodynamic recipe for environmental catastrophe. *International Journal of Coal Geology*, 59/1–2, 7–17.

Tetzlaff, A. (2004). Coal fire quantification using Aster, ETM and Bird instrument data. PhD Thesis, Geosciences, Maximilians-University, Munich, 155 pp.

Van Dijk, P., Zhang, J., Jun, W., Kuenzer, C. and Wolf, K.H. (2011). Assessment of the contribution of in-situ combustion of coal to greenhouse gas emission; based on a comparison of Chinese mining information to previous remote sensing estimates. *International Journal of Coal Geology*, Special Issue RS/GIS, DOI: 10.1016/j.coal.2011.01.009; pp. 108–119.

Wessling, S., Kuenzer, C., Kessels, W. and Wuttke, M. (2008). Numerical modelling to analyze underground coal fire induced thermal surface anomalies. *International Journal of Coal Geology*, 74, 175–184, DOI 10.1016/j.coal.2007.12.005.

Wolf, K.-H.A.A. and Bruining, J. (2007). Modelling the interaction between underground coal fires and their roof rocks. *Fuel*, 86, 2761–2777.

Yang, B., Chen, Y., Li, J., Gong, A., Kuenzer, C. and Zhang, J. (2005). Simple normalization of multi-temporal thermal IR data and applied research on the monitoring of typical coal fires in Northern China. *Proceedings of the Geoscience and Remote Sensing Symposium*, IGARSS 25–29 July 2005, 8, 5725–5728.

Yang, B., Li, J., Chen, Y., Zhang, J. and Kuenzer, C. (2008). Automated detection and extraction of surface cracks from high resolution Quickbird imagery. *Spontaneous Coal Seam Fires: Mitigating a Global Disaster*. UNESCO Beijing, 2008, ERSEC Ecological Book Series, Vol. 4, pp. 381–389.

Zhang, J. (2004). Spatial and Statistical Analysis of Thermal Satellite Imagery for Extraction of Coal Fire Related Anomalies. PhD dissertation, TU-Vienna, 161 pp.

Zhang, J. and Kuenzer, C. (2007). Thermal surface characteristics of coal fires 1: Results of in situ measurements. *Journal of Applied Geophysics*, 63, 117–134, DOI:10.1016/j.jappgeo.2007.08.002.

Zhang, J., Kuenzer, C., Tetzlaff, A., Oettl, D., Zhukov, B. and Wagner, W. (2007). Thermal characteristics of coal fires 2: Results of measurements on simulated coal fires. *Journal of Applied Geophysics*, 63, 135–147, DOI:10.1016/j.jappgeo.2007.08.003.

Zhang, X., Zhang, J., Kuenzer, C., Voigt, S. and Wagner, W. (2004). Capability evaluation of 3-5µm and 8-12.5µm airborne thermal data for underground coalfire detection. *International Journal of Remote Sensing*, 25(12), 2245–2258.

10 Spatial Analysis of Flood Causes in the Hindu Kush Region

Atta-ur Rahman

CONTENTS

10.1 INTRODUCTION

The Hindu Kush is a region of high mountain systems, located to the west of Karakorum and the northwestern part of Pakistan. Sometimes colloquially referred to as the 'Roof of the World', it is separated from the Himalaya at the Pamir Knot, where the border of Pakistan, Afghanistan, Tajikistan and China meet (Figure 10.1; Khan, 2003). This is one of the greatest watersheds of Central Asia and Pakistan. In the north it separates Pakistan from Tajikistan through a narrow strip of Afghan territory, known as the 'Wakhan Corridor' (Qamer et al. 2012), a source of perennial streams, rivers and ecosystems. The mountain system is nearly 1600 km long and over 300 km wide. It stretches from northeast to southwest (Raza, 2007), and has more than 40 peaks exceeding 7000 m above sea level (Figure 10.2; Rahman, 2010), the highest of which is Tirich Mir (7700 m above sea level), located in the Chitral District of Pakistan. The Hindu Kush–Himalayas (HKH) are the youngest mountains on Earth and they are still tectonically dynamic (Mahmood et al. 2009; Rahman et al. 2011). Because the tectonic uplift is a recent and continuing phenomenon, the area is characterized by fragile bedrock, steep slopes and a high rate of surface erosion (Shams, 2006; Khan et al. 2008). Generally, the height of peaks in the Hindu Kush decreases toward the south (see Figure 10.2).

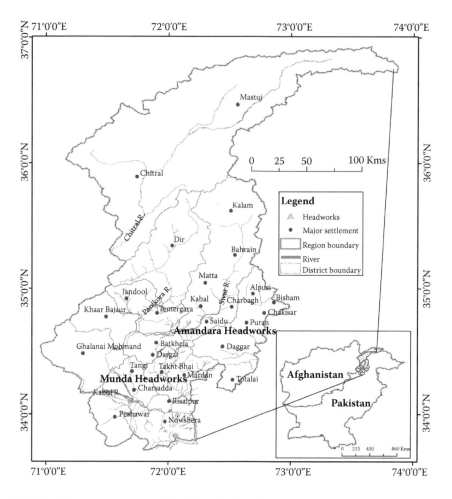

FIGURE 10.1   Location map of the Hindu Kush region.

The Hindu Kush region can broadly be classified into three climatic zones as one proceeds from higher elevation in the north toward the south: humid zone, sub-humid to humid zone, and semi-arid zone. The average annual rainfall varies from more than 2000 mm in the humid to sub-humid areas, to 200 mm in the semi-arid areas. Similarly, there is also a wide variation in the seasonal temperature. During the winter, the mean minimum temperature is in the range –20°C to –50°C at Malam Jabba, Kalam and Dir, whereas in the summer the highest recorded temperatures range from 42°C to 45°C at Saidu and Timergara. Extreme winter temperatures allow snow and glacier ice accumulation, while high summer temperatures acceler-ate the melting of snow and glaciers. This seasonal variation in rainfall and snow poses a variety of disaster risks in the region (Gupta and Sah, 2008; Qamer et al. 2012; Rahman and Dawood, 2017), of which floods are the most frequent and wide-spread, and are the focus of this chapter.

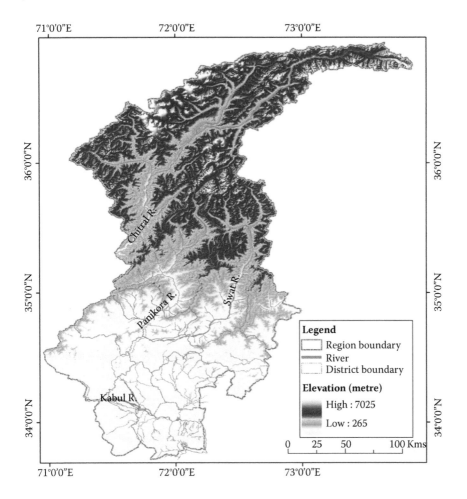

FIGURE 10.2   Digital terrain model of the Hindu Kush region. (Modified after Rahman, A. and Khan, A. N. *Natural Hazards*, 66(2):887–904, 2013.)

10.2   FLOOD HAZARD IN THE HINDU KUSH

As a significant component of water circulation, river runoff is of key significance for sustainable utilization of land and water resources (Dong et al. 2009). In the study area, the channel gradient is steep in the upper reaches while comparatively gentle in the lower reaches. Due to these factors, flash floods are prominent in the upstream areas, and river floods more common in the gently sloping low-lying areas. Fluctuation in river discharge is not only attributed to heavy rainfall and melting of snow/glaciers (Ramos and Reis, 2002) but also to human intervention (Zhang et al. 2006, 2008), while the impacts of climate change and its consequences for the river flow regime is a major determining factor in the region (Rahman and Khan, 2011). Almost every year, flooding causes damage to lives, standing crops and infrastructures: the upper reaches of the rivers are prone to flash flooding while, in the lower reaches, river floods are predominating.

10.3   CASE STUDY: THE 2010 FLOOD

A severe flood hit the eastern Hindu Kush region in Pakistan in 2010. It arose at the end of a 4-day period of intense and prolonged rain, from 27–30 July, together with the substantial melting of snow and glaciers. It resulted in more than 400 fatalities and destroyed a large proportion of physical infrastructure.

10.3.1   THE STUDY AREA

In the Hindu Kush region, there are numerous small and large glaciers (Figure 10.3), and most of the rivers originate from these: The Chitral (Yarkhun) River takes its

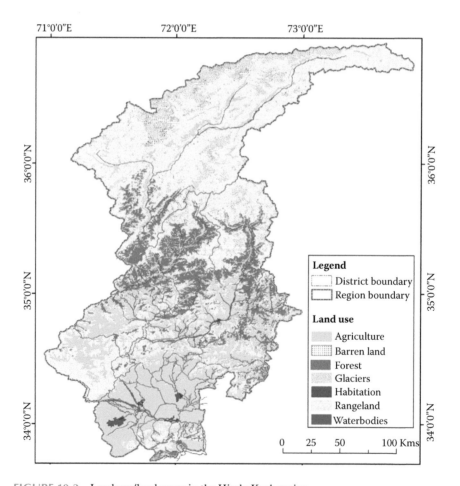

FIGURE 10.3   Land use/land cover in the Hindu Kush region.

origin from the Chiantar glacier, the river Swat from Ushu, Utror and Gabral; while the rivers Karambar, Zhandrai, Panjkora and Bahandra are comparatively small, and fed by numerous small glaciers, snowmelt and springs. Various passes across the Hindu Kush have been of immense historic and military significance (Shams, 2006), of which Baroghil, Lowari, Shandur, Khotgaz, Arandu, Nawa, Bin Shahi and Kankuch are the prominent ones.

The Swat, Chitral, Dir, Shangla and Buner are the main valleys of eastern Hindu Kush, and this is where most of the population lives (GoP, 1999). There are also numerous small valleys in the region. In the Hindu Kush, valleys are narrow and their width doesn't exceed 7 km (Khan, 2003). Agriculture is the main source of livelihood for the rural population, and provides the basis for the region's economy. Due to the rugged topography, agriculture is mainly practiced along the river banks and in low-lying areas. In certain areas, alluvial fans have been terraced for cultivation. Canals taken from rivers, perennial streams and springs are the main sources of irrigation. About 20% of the region is under forest and this is mainly concentrated at higher altitudes (see Figure 10.3; Qamer et al. 2012). Houses for lower-income residents are made of stones and mud, while wealthier residents live in concrete houses.

In the Hindu Kush region, June, July and August are the hottest months. This high temperature leads to intensive glacier and snowmelting, which as a result increases river discharge and multiplies the probability of flood occurrence. Similarly, in summer a low-pressure zone develops over the Hindu Kush mountain system, with high pressure over the Indian Ocean. This drives the summer monsoon, during which time heavy, moisture-laden winds from the Indian Ocean and the Bay of Bengal move north-westward across the subcontinent and into Pakistan (Mirza, 2003). In the study region, monsoon is the major source of summer rainfall in the eastern Hindu Kush region (Khan, 2003), followed by western disturbances (Shams, 2006). Rain is also received from local convections and thunderstorms (Tariq and Van-De Giesen, 2012).

As far as the 2010 flood is concerned, in May 2010 the south-central part of Pakistan was hit by the worst heat wave in a century (Figure 10.4; PACC, 2010; Douglas, 2011). According to the Pakistan Meteorological Department, 11 meteorological stations in Pakistan each recorded their highest-ever temperatures during the last week of May 2010 (PACC, 2010). At Mohenjo-Daro on 26 May 2010, the maximum recorded temperature was 53.3°C. This was the highest-ever recorded temperature in south Asia and the fourth highest in the world (Rahman and Khan, 2013). Because of this heat wave, in July 2010, a historic low-pressure system developed over the Hindu Kush–Himalayas region (Douglas, 2011).

To achieve the study objectives, data were collected both from primary and secondary sources (Figure 10.5) as follows:

Primary data were gathered through interviews with flood-affected people, and officials from flood-related government organizations, focused group discussions (FGDs) with community elders and field observations. During field surveys, flood-affected populations were asked about the possible causes of floods and associated damages. Similarly, officials of the line agencies were interviewed to find the policy gap between community vulnerabilities and strengths. The field observations were recorded in the form of photographs.

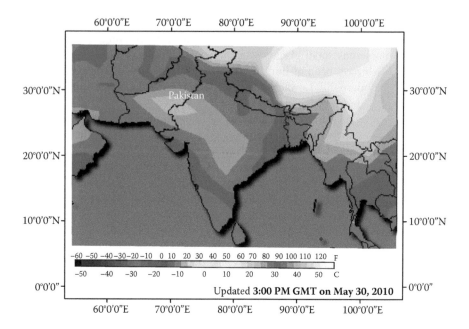

FIGURE 10.4 Heat wave in Pakistan as on May 30, 2010. (Modified after Program Adaptation to Climate Change (PACC). (2010). Ola de calor en Asia que bate su record de temperatura con 53,5°C (in Spanish). http://cambioclimaticoenlosandes.blogspot.com/2010_07_01_archive.html (accessed 30 Mar 2012); Rahman, A. and Khan, A. N. *Natural Hazards*, 66(2):887–904, 2013.)

Secondary data were also obtained from many official sources. In the study area, there are a total of seven meteorological stations: Drosh, Chitral, Dir, Timergara, Kalam, Saidu Sharif and Malam Jabba. Rainfall and temperature data from these stations were collected from the Pakistan Meteorological Department. In addition, temperature and rainfall data from stations in Peshawar, Risalpur and Cherat were also used, since all three stations are located in the Peshawar Valley, where all the rivers of the Hindu Kush region ultimately confluence. The rainfall data were spatially interpolated by applying the inverse distance weighted (IDW) technique in ArcGIS 9.3, to evaluate the spatial variation of rainfall during the 4-day wet spell from 27–30 July 2010. Data pertaining to the monsoon, and to the track of Cyclone Phet, were obtained from the Pakistan Meteorology Department, Islamabad.

Land use data, including forest cover, were obtained from the Pakistan Forest Institute Peshawar. Landsat Images from 2010 were used for generating the land cover map of Hindu Kush region shown in Figure 10.3. Population data were collected from the Population Census Organization of Pakistan.

There are several hydro-gauging stations on the rivers Swat, Chitral and Panjkora, as well as headworks (small barrages) that divert water into the Swat canal system, for use downstream in irrigation and hydro-electricity power generation. Discharge

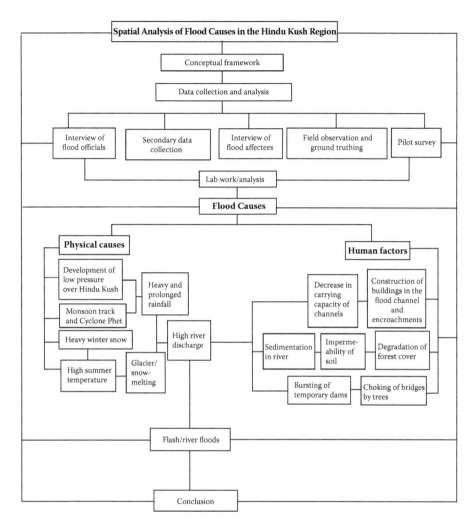

FIGURE 10.5   Research framework and model.

data for the headworks and gauging station at Amandara (located upstream of the Panjkora confluence) for the period 1983–2012, and for the Munda hydrological gauging station (downstream of the confluence of rivers Swat and Panjkora) for the period from 1929 to 2010, were obtained from the hydrology wing of the Irrigation and Drainage Authority in Peshawar. These data revealed that the Rivers Swat and Panjkora have irregular peak flows with high inter-annual variability in run-off. The mean monthly discharge at Amandara rises from March until August, and then it starts gradually falling, while peak runoff occurs during July and August. The data clearly show unusual high discharge during 1983, 1992, 1995, 2001 and 2010 (see Figure 10.6). Maximum discharge was recorded at the Amandara headworks on 9 August 1992 (3588 m^{3}s^{-1}) and 25 July 1995 (3630 m^{3}s^{-1}). In July 2010,

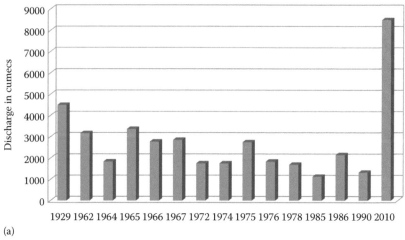

(a)

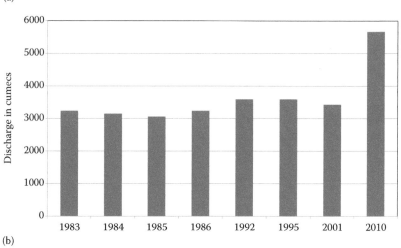

(b)

FIGURE 10.6 (a) River Swat, highest recorded discharge at Munda headworks, 1929–2010, (b) River Swat, highest recorded discharge at Amandara headworks, 1983–2010.

the discharge at Amandara headworks exceeded 5663m³/s (Rahman and Khan, 2013). Similarly, discharge records at Munda show that unusually high runoff was recorded in 1929, 1962 and 1966 (Figure 10.6), but the highest discharge of the past 80 years was recorded at Munda on 29 July 2010, at more than 8495 m³/s. The discharges recorded across all other gauging stations in the area were also at the highest ever levels in 2010, and that year's flood not only broke all previous records, but the volume and force of water involved also destroyed both the Amandara and Munda headworks.

10.3.2   THE CAUSES OF THE FLOOD DISASTER

As has already been noted, floods are a regular occurrence in the Hindu Kush. However, the 2010 event was notable for many reasons, not least of which being the severe cost in terms of loss of life and damage to property and infrastructure. There are a variety of factors that triggered this particular flood, and the causes can be broadly classified into physical and human factors, although in practice the two drivers acted together to amplify and further intensify the floods.

10.3.2.1   Physical Causes

The direction of travel of the 2010 monsoon was a bit different from usual. After its origin in the Bay of Bengal, it reached central India in July. At the same time a high-speed cyclone Phet originated in the Arabian Sea and entered Pakistan (see Figure 10.7; Rahman and Khan, 2013). This also proceeded toward the northwestern part of India, and merged with the active monsoon track. The heavy moisture-laden combined wind systems of the monsoon and Phet then moved toward the Hindu Kush–Himalayas of Pakistan, where a low pressure belt had already developed. This brought four consecutive days of prolonged rainfall, from 27–30 July, and all the meteorological stations in the Hindu Kush region recorded rainfall values

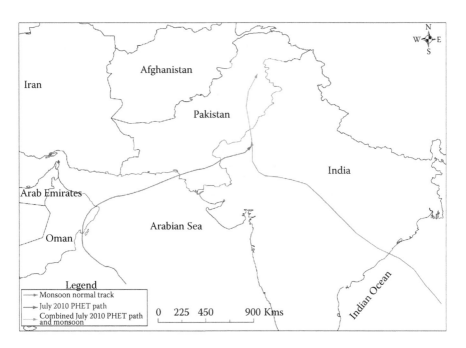

FIGURE 10.7   Monsoon track and path of July 2010 PHET cyclone. (Modified after Rahman, A. and Khan, A. N. *Natural Hazards*, 66(2):887–904, 2013.)

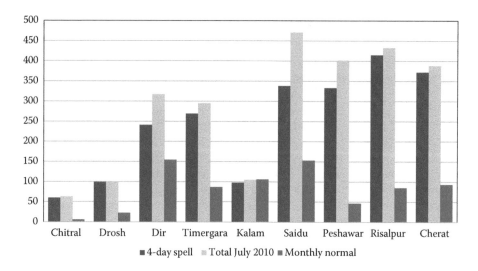

FIGURE 10.8   Four-day wet spell of 27–30 July, 2010, total July 2010 rainfall and monthly rainfall at all the met stations of Hindu Kush region.

above normal as a result (Figure 10.8), including the highest-ever recorded rainfall (Figure 10.9; Rahman and Khan, 2013).

During the four days between 27–30 July, the highest rainfall was recorded at Risalpur (415 mm), followed by Cherat (372 mm), Saidu (338 mm), Peshawar (333 mm), Timergara (269 mm) and Dir (241 mm). However, in the whole of the month of July, Saidu received the highest recorded rainfall (471 mm), followed by Risalpur (433 mm), Peshawar (402 mm), Cherat (388 mm), Dir (317 mm) and Timergara (295 mm). When the 4-day wet spell and total July 2010 rainfall were compared with the normal July rainfall, it was found that the normal rainfall was several times less than the 4-day wet spell (see Figure 10.9). Similarly, more than 150 mm rainfall was recorded in 24 hours in most of the met stations of the Hindu Kush region (Peshawar: 274 mm; Risalpur: 289 mm; Cherat: 257 mm; Dir: 192 mm and Timergara: 150 mm). Such anomaly in rainfall may be linked to climate change. The impact of the rain between 27–30 July was further exacerbated because it fell on ground that was already wet, and whose absorption capacity was therefore severely limited, as a result of previous rainfall that occurred between 19 and 23 July.

In addition to this, heavy snowfall had also occurred in the Hindu Kush region during the previous winter, between November and February, as recorded at the Kalam, Malam Jabba, Dir and Chitral met stations. The 4-day spell of prolonged rain together with the summer period encouraged heavy melting of accumulated snow. Because of these factors, large amounts of water reached the rivers and streams quickly. In addition, dozens of further, localized cloud bursts also contributed to the runoff. Following torrential and prolonged rainfall, the channels were unable to

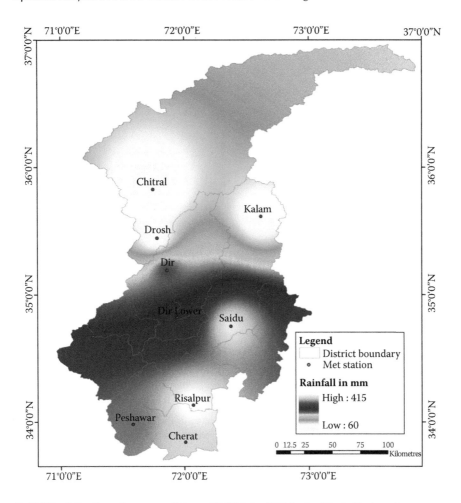

FIGURE 10.9   Four-day wet spell from 27–30 July, 2010 in the Hindu Kush region.

accommodate the high volumes of water (Figure 10.10). As a result, the water over-flowed the levees and inundated the low-lying areas.

In the Hindu Kush region, the area under forest cover varies from district to district. As a whole, about 20% of the area is forest (see Figures 10.3 and 10.9; Qamer et al. 2012). It is recognized that vegetation cover aids the conservation of soil, reduces sediment load in the channels, sustains water flow and maintains the ecosystem (see Figure 10.10; Haeusler et al. 2000; Mirza, 2003). In the study area, due to regional climate variation, there is great disparity in vegetation cover. At the snowline alpine forests dominate, followed by a zone of coniferous forests, while the lower altitudes have semi-arid subtropical forests (Rahman and Khan, 2013).

Generally, thick vegetation covers (especially at higher altitudes) have micro-climatic impacts (UN 2011). When large amounts of water vapour from the axial

(a)   (b)

(c)   (d)

FIGURE 10.10   (a) Deforestation and terraced fields in the Hindu Kush region. (b) After rain, heavy erosion followed by sedimentation even on roads. (c) Flash flood in the upper reaches of Panjkora. (d) Encroachment onto the river.

part of plants enter the atmosphere, it modifies the local climate. The increasing temperature accelerates the rate of transpiration and in turn enhances the chances of convectional rainfall. Therefore, high rainfall in the Hindu Kush region is partly attributed to the local convectional system.

It has been observed that due to environmental degradation and climate change there has been a gradual increase in the magnitude and frequency of floods. Scientists are also of the opinion that due to climate change the occurrence and intensity of rainfall, snowfall and cloud bursting has been multiplying day by day (Dong et al. 2009; Douglas, 2011; Qamer et al. 2012; Rahman and Darwood, 2017). Similarly, climate change experts have also detected fluctuations in river runoff (UN 2011) and subsequent damages. Moreover, it has been determined that the rate of glacier ablation exceeds that of accumulation, which has resulted in glacial retreat. It has been observed that the pace of rapid melting of both glaciers and snow is changing (Shrestha and Shrestha, 2008), perhaps because of climate change, and that has increased the intensity of floods. The analysis further indicates that there is much fluctuation in both rainfall and river runoff, and this is also interpreted as being a clear signal of climate change.

Water is a geomorphic agent (Shams, 2006). It has the potential to pick up material, transport it, and when conditions are favourable, the load is ultimately deposited (Khan, 2003; Mirza, 2003). It has been observed that as the velocity of water increases, the erosional potentials and carrying capacity of the load increases and, wherever the conditions are favourable, siltation also occurs. This process of continuous sedimentation fills the channels and as a result reduces carrying capacity (see Figure 10.10; Ali, 2007). Analysis reveals a rapid deforestation rate over the past two decades (Haeusler et al. 2000; Rahman and Khan, 2013). Therefore, during the 2010 flood incident, all the rivers originating in the Hindu Kush region carried a heavy sediment load from the catchment areas, aggrading the riverbeds. This further intensified the 2010 flood event.

10.3.2.2 Human Factors

Along with physical factors, there are other flood-intensifying factors which accelerate floods and incur damage to the area. In the Hindu Kush region, human encroachments over the active floodplain, population growth, deforestation and overgrazing in the catchment area are some of the key anthropogenic factors.

Forest cover plays a key role in flood risk reduction, whereas deforestation accelerates floods (Haeusler et al. 2000). According to FAO, in the Hindu Kush region 1.8% of forest cover was lost every year over the 10-year period 1990–2000, while during 2000–2010, 2.2% was lost (Rahman and Khan, 2013). Field survey together with the FGDs reveals that the main factors behind deforestation are population pressure, supply of timber to the market, clearing of forests for agriculture and expansion of built-up areas, since growing population also demands new institutions, resource bases and infrastructure. It has been reported that during recent conflict in the province of Khyber Pakhtunkhwa, extensive forest cutting took place in the thickly forested districts of Swat, Shangla, Dir upper and Dir lower (Ali, 2008; Rahman and Khan, 2013), while in certain cases organized crime (the 'timber mafia') have also been blamed for the indiscriminate cutting of forests. In the study area, field surveys, FGDs and interviews with forest department officials revealed that overgrazing and deforestation are the major flood-intensifying factors. This rapid deforestation and overgrazing have seriously affected the fluvial processes of almost all the rivers originating in Hindu Kush region (Khan and Rahman, 2006; Ali, 2008; Rahman, 2010). In particular, deforestation reduces the water absorption capacity of the ground, and contributes much to intensifying the frequency and magnitude of floods in the region.

Population in the Hindu Kush region is increasing at a rapid pace (2.6% per annum), putting tremendous pressure on land resources (GoP 1999). Due to this population growth, both forest and farmland are being converted into built-up areas. Areas supporting impermeable structures such as houses, paved streets, roads, commercial areas and other buildings have increased at the cost of either agricultural or forest land. Because of these developments, the absorption capacity for rainwater has decreased and most of it quickly becomes part of surface runoff. Thus, the increase in built-up areas has a direct relationship with the increasing river runoff and intensifying floods in the Hindu Kush region.

With the rapid pace of population growth, the demand for shelter and other infrastructures has also increased. The lower-income population has no choice but to encroach upon the active floodplains either for agricultural or infrastructure purposes. The river courses narrow down and in effect reduce their carrying capacity; therefore, during intense rainfall the water overflows the levee and ultimately causes floods. During field surveys, it has been observed that human encroachment largely contributed to the occurrence of the 2010 floods. There is therefore a need of policy response for land use regulation, to check encroachments onto channels and watersheds as a flood abatement measure to reduce the risk of future mega-flood events.

There are hundreds of bridges over the streams and rivers of the Hindu Kush region. Torrential rainfall, cloud bursting, lateral erosion and heavy runoff often uproot trees, especially in upstream areas (Figure 10.11; Rahman and Khan, 2013). When carried downstream by the rivers, they often get stuck at the narrow bottlenecks of bridges and thereby block the channel. As a result, a temporary dam develops behind the bridge, putting constant pressure on it which, if it builds up sufficiently, may destroy the bridge and send a surge of stored water to inundate the downstream areas. In the 2010 flood, several bridges were damaged (Figure 10.12)

(a)   (b)

(c)   (d)

FIGURE 10.11   (a) High velocity of floodwater uprooted the trees. (b) Heavy flood passes through a bridge in Dir Lower. (c) The tree trunks struck the pillars of the bridge and choked the river. (d) The breaking of the bridge by water.

FIGURE 10.12   (a) A damaged bridge on the River Swat. (b) Damaged headworks of Amandara. (c) Heavy lateral erosion washed away road. (d) Floodwater in settlement.

and, after the destruction of one of these, the Chakdara Bridge, a major supply line for three districts, was totally cut off from the rest of the country for almost a month. The phenomenon of temporary dams bursting also resulted in peak discharges over hydro-gauging stations. During the 2010 flood, this was observed as a major triggering factor of floods in the entire region.

Therefore, there is a definite need of policy response for land use regulation to check encroachments onto the channel and watershed as a flood abatement measure for reducing the risk of future mega-flood events.

10.4   CONCLUSION

The Hindu Kush region is exposed to numerous natural disasters, but flooding is a repeatedly occurring phenomenon. Flash floods tend to dominate in the upper reaches of these rivers while river floods are more common in the lower reaches. Both types of floods are caused by a combination of physical and human factors. Most floods are triggered initially by climatic events and, in particular, heavy, prolonged rainfall in summer which, as well as contributing additional water to be discharged, also leads to accelerated melting of snow and glacier ice, as does aggradation of the riverbeds by sedimentation. Human factors intensify the probability of floods, and key anthropogenic factors include human encroachments onto the active floodplain, bursting of temporary dams behind the bridges, clearing of forest and agricultural land for infrastructure, increase in the area under impermeable soil, deforestation and overgrazing in the catchment area.

There is a need for stronger, science-based policies to respond to and minimize the impacts of these floods. Areas requiring particular attention in this regard include land use regulation to check encroachments onto the channel, and more effective watershed management plans.

REFERENCES

Ali, A. M. S. (2007). September 2004 flood event in Southwestern Bangladesh: A study of its nature, causes, and human perception and adjustments to a new hazard. *Natural Hazards*, 40:89–111.

Ali, H. B. (2008). Peshawar: Swat forests at the mercy of militants. *Dawn News*, Islamabad, Wednesday 01 October, 2008.

Benestad, R. E. and Haugen, J. E. (2007). On complex extremes: Flood hazards and combined high spring-time precipitation and temperature in Norway. *Clim Change*, 85(3):381–406.

Chang, H., Franczyk, J. and Kim, C. (2009). What is responsible for increasing flood risks? The case of Gangwon Province, Korea. *Natural Hazards*, 48(3):339–354.

Dong, Y., Wang, Z., Chen, Z. and Yin, D. (2009). Centennial fluctuations of flood-season discharge of Upper and Middle Yangtze River Basin, China (1865–1988): Cause and impact. *Front. Earth Sci. China*, 3(4):471–479.

Douglas, L. (2011). Global weather highlights 2010: Flooding, heatwaves, and fires. *Weatherwise*, 64(3):21–28.

Government of NWFP. (2006). *Environmental profile of NWFP*. Environmental Protection Agency, Government of NWFP, Peshawar.

Government of Pakistan (GoP). (1999). *District census report of Swat, 1998*. Population census organization, Islamabad.

Gupta, V. and Sah, M. P. (2008). Impact of the Trans-Himalayan Landslide Lake Outburst Flood (LLOF) in the Satluj catchment, Himachal Pradesh, India. *Natural Hazards*, 45:379–390.

Haeusler, T., Schnurr, J. and Fischer, K. (2000). *Provincial Forest Resource Inventory (PFRI) North West Frontier Province-Pakistan*.

Khan, A. N. and Rahman, A. (2006). Landslide hazards in the mountainous region of Pakistan. *Pakistan Journal of Geography*, 16(1&2):38–51.

Khan, A. N., Rahman, A. and Ali, B. (2008). Adjustment to riverbank erosion hazard: A case study of Village Chalyar, Swat Valley, Pakistan. *Pakistan Journal of Geography*, 18(1&2):46–59.

Khan, F. K. (2003). *Geography of Pakistan: Population, economy and environment*. Oxford University Press, Karachi.

Mahmood, S. A., Shahzad, F. and Gloaguen, R. (2009). Remote sensing analysis of quaternary deformation using river networks in Hindukush region. *Geoscience and Remote Sensing Symposium*, 2009 IEEE International, IGARSS August 2009, Volume: 2. Organized by Remote Sensing Group, Tech. Univ. Bergakad. Freiberg, Germany. DOI:10.1109/IGARSS.2009.5418089

Mirza, M. M. Q. (2003). Three recent extreme floods in Bangladesh: A hydro-meteorological analysis. *Natural Hazards*, 28(1):35–64.

Program Adaptation to Climate Change (PACC). (2010). Ola de calor en Asia que bate su record de temperatura con 53,5°C (in Spanish). http://cambioclimaticoenlosandes.blogspot .com/2010_07_01_archive.html (accessed 30 Mar 2012).

Qamer, F. M., Abbas, S., Saleem, R., Shehzad, K., Ali, H. and Gilani, H. (2012). Forest cover change assessment in conflict-affected areas of northwest Pakistan: The case of Swat and Shangla districts. *Journal of Mountain Science*, 9:297–306.

Rahman, A. (2003). Effectiveness of flood hazard reduction policies: A case study of Kabul-Swat floodplain, Peshawar vale. Unpublished MPhil thesis, submitted to the Department of Geography, Urban and Regional Planning, University of Peshawar.

Rahman, A. (2010). *Disaster risk management: Flood Perspective*. VDM Verlag, Germany.

Rahman, A. and Dawood, M. (2017). Spatio-statistical analysis of temperature fluctuation using Mann-Kendall and Sen's Slope approach. *Climate Dynamics*, 48(3):783–797.

Rahman, A. and Khan, A. N. (2011). Analysis of flood causes and associated socio-economic damages in the Hindu Kush region. *Natural Hazards*, 59(3):1239–1260.

Rahman, A. and Khan, A. N. (2013). Analysis of 2010-flood causes, nature and magnitude in the Khyber Pakhtunkhwa, Pakistan. *Natural Hazards*, 66(2):887–904.

Rahman, A., Khan, A. N., Collins, A. E. and Qazi, F. (2011). Causes and extent of environmental impacts of landslide hazard in the Himalayan region: A case study of Murree, Pakistan. *Natural Hazards*, 57(2):413–434.

Ramos, C. and Reis, E. (2002). The analysis reveals that the flow regimes of Hindu Kush rivers largely depends on the spatio-temporal variation in precipitation. *Mitigation and Adaptation Strategies for Global Change*, 7:267–284.

Raza, M. H. (2007). *Mountains of Pakistan*. Imprint, Rawalpindi.

Shams, F. A. (2006). *Land of Pakistan*. Kitabistan, Lahore.

Shrestha, A. B. and Shrestha, S. D. (2008). Flash flood risk in the Hindu Kush-Himalayas: Causes and Management *Options*, 11.

Tariq, M. A. U. R. and Van De Giesen, N. (2012). Floods and flood management in Pakistan. *J Phys Chem Earth Parts A/B/C*, 47–48:11–20.

United Nations (UN). (2011). *State of the World's Forests 2011*, United Nations, Food and Agriculture Organization, Rome.

Zhang, Q., Liu, C. L., Xu, C. Y. and Jiang, T. (2006). Observed trends of annual maximum water level and stream flow during past 130 years in the Yangtze River basin, China. *Journal of Hydrology*, 324:255–265.

Zhang, Q., Xu, C. Y., Zhang, Z. X., Chen, Y. D., Liu, C. L. and Lin, H. (2008). Spatial and temporal variability of precipitation maxima during 1960–2005 in the Yangtze River basin and possible association with large-scale circulation. *Journal of Hydrology*, 353: 215–227.

11 Integrated Approach for Flood Assessment of Coastal Urban Watersheds

T.I. Eldho, P.E. Zope and Anand T. Kulkarni

CONTENTS

11.1 INTRODUCTION

More than half of the world's population lives in urbanized areas, within 60 km of the shoreline, and it is expected that within the next 30 years, the coastal population will double and much of this growth will be in coastal mega-cities (Li, 2003). Due to large-scale urbanization in recent decades, urban flood disasters have now become common, and are affecting many people, particularly in coastal areas.

From the hydrologic and hydraulic perspective, urbanization is characterized by an increase in impervious areas, and corresponding change in overland flow characteristics and paths due to building blockages. With increased impervious area and limited drainage facilities, increasing rainfall intensity and storm duration in urban areas lead to large peak flows and reduced time to peak (Dewan and Yanaguchi, 2009; Zope et al., 2014). Further, change in land use/land cover (LULC) and climate change effects alter hydrological processes such as evapotranspiration, interception and infiltration, leading to higher runoff generation (Melesse and Shih 2002) that further increases flood risk in coastal areas (Guhathakurta et al., 2011). Generally, severe flood conditions in urban areas occur either if the rainfall exceeds the drainage capacity or if the existing system fails in carrying its designed capacity.

Factors like rapid coastland reclamation, changing rainfall patterns and highly impervious surfaces also contribute to severe flooding. Recently, in many parts of the world, extreme flood events have caused major economic losses to property, industrial production, networks, infrastructure and human lives, usually at the downstream part of the river (Singh and Sharma, 2009). Coastal urban areas are especially vulnerable to flooding when high rainfall coincides with high tidal conditions.

To deal with the coastal urban flooding problem effectively, flood prediction and assessment, based on given or known conditions, is needed to provide appropriate flood warning and remedial measures. Urban hydrology is complex, and development of an appropriate flood model requires sophisticated numerical models integrated with data management tools such as geospatial techniques of remote sensing and GIS. In this chapter, the integrated applications of numerical models coupled with geospatial techniques are discussed and demonstrated with case studies for the coastal urban flood assessment problems.

11.2   COASTAL URBAN DRAINAGE SYSTEM
AND FLOOD PROBLEMS

A typical urban storm water drainage network is a very complex system, which involves various processes of hydrology, hydraulics and transport mechanisms, as well as rainfall events, all of which need to be thoroughly understood for the appropriate design of the drainage system (Parkinson and Mark, 2005). In coastal urban areas, when high-intensity rainfall coincides with high tidal variations, the existing drainage system may lose its carrying capacity, leading to severe flooding in low lying areas (Li, 2003). In many situations, due to nonavailability of free land, it may not be possible to increase the drainage capacities. To avoid flooding in these regions, it is necessary to reduce the size and duration of peak flows. Further, the storm and catchment characteristics influence the hydrological response.

While designing an urban drainage system, the basic elements to be considered are hydrology, hydraulic capability of the system structure and the downstream conditions (Haestad Methods, 2003). Generally, urban storm water drainage systems can be categorized into minor and major systems. The minor drainage systems may consist of roadside drains, culverts and pipe drains, and commonly these are designed for rainfall intensities up to a 10-year return period. The major drainage system may consist of channels or rivers discharging their flow directly into creeks or the sea, and designed for a rainfall intensity with a return period of up to 100 years (ASCE, 1992; Haestad Methods, 2003). In urban coastal cities, where the storm water drainage system discharge at the outlet is influenced by tidal variations, the system design also has to take into account the back-water effect due to tides. In urban areas, as large surface areas are covered by impervious structures, there is less scope for infiltration, which leads to higher runoff and less lag time to peak flow. These also increase the severity and intensity of flooding at the downstream end of the channel or river. In coastal urban areas, detention ponds may be used to reduce the peak discharge during high tides. The detention pond reduces the peak discharge as well as the storm water quality, mitigates downstream drainage capacity problems and recharges the groundwater (ASCE, 1992; Haestad Methods, 2003). Detention ponds can be provided as online or offline detention structures, regional or on-site detention, and independent or interconnected detention structures, depending upon the location and peak discharge.

While designing the storm water drainage system, there are limitations of outlet levels and depths of the channel due to tidal conditions. Many coastal cities, such as Mumbai in India, are partly or wholly constructed on reclaimed land, which changes the hydrological characteristics of the natural watershed and this can lead to saturation of soil due to groundwater retention. If proper reclamation is not carried out with an appropriate drainage system, severe flooding problems may arise. Surges and cyclones can also cause severe flooding problems. For example, many major coastal cities in India have experienced severe flooding problems, witnessing loss of life and property, disruption to traffic and power, among these being Mumbai in 2005, Surat in 2006, Kolkata in 2007, as well as non-coastal cities such as Srinagar in Kashmir (Kulkarni et al., 2014a).

The rainfall-runoff process consists of various components such as rainfall intensity, evaporation/evapotranspiration, infiltration, overland flow, detention or retention storage, channel flow and tidal variations (Chow et al., 1998). The hydrograph generated from the watershed surface runoff is a function of land use, rainfall intensity, duration, shape and size of the area considered and watershed characteristics (Singh, 1996). In urban hydrology, models are now recognized as important tools (Zoppou, 2012) that can give important information about flood hazards and management (Niemczynowicz, 1999). Models can provide an insight into the way rivers and catchments function, a framework for data analysis, the means of investigating problems, and a way of evaluating alternate management strategies (Singh, 1996). Urban storm water models summarize the behaviour of the catchment response as a function of space and time.

11.3   URBAN FLOOD MODELLING APPROACH

Urbanization and city growth lead to a drastic change in LULC, such that the impervious surface area made up of buildings, roads and other structures increases as the built-up areas expand. Accordingly for given rainfall conditions, there is increase in peak discharge, loss of existing drainage capacity and flood frequency. The use of hydrological and hydraulic models can help urban managers and planners achieve better planning and designing of the storm water drainage system. In this section, important hydrologic and hydraulic modelling approaches in flood plain simulation are briefly described.

11.3.1   RAINFALL RUNOFF MODELLING

Simulation of rainfall into runoff is the main objective in a hydrological study (Singh, 1996). In a normal hydrological cycle, various hydrologic processes such as infiltration, evaporation, interception, retention and groundwater seepage take place, as system inputs (precipitation) become converted to throughputs and outputs (Singh, 1996). In urban drainage planning, designing and development, urban storm water models are widely recognized as useful tools (Chen et al., 2009).

11.3.2   URBAN DRAINAGE MODELS

In urban storm water management and planning, typically two types of models are used in combination (ASCE, 1992):

- *Hydrologic models*: Describe quantity of runoff in time and space
- *Hydraulic models*: Provide hydraulic flow parameters in the channel which can be used in the design, analysis and planning

11.3.2.1   Hydrologic Models

In urban drainage designs, hydrologic models simulate the rainfall to runoff process considering various losses and provide the quantity of flow with respect to space and time. The hydrologic models use rainfall information or predictions to provide runoff characteristics including peak flow, time to peak, flood hydrograph and flood frequencies (McCuen, 1989). For urban drainage designs, the selection of appropriate model depends on the objectives, computational and human resources, and availability of appropriate data. The commonly used hydrologic models (Singh, 1998; Zoppou, 2001) can be described briefly as follows.

- *Deterministic model*: Gives specific values for the given input set. Generally, the deterministic models can be based on any of the specific relationships for the specific problem or based on conservation laws of mass, momentum and energy.
- *Stochastic model*: Involves random inputs giving any number of responses for a given set of parameters.
- *Continuous model*: Simulates many storm events over a period of time continuously.
- *Single event model*: Simulates one particular storm event at a time.
- *Empirical model*: Provides specific relationship between variables involved in the process.
- *Lumped model*: Represents a single set of parameters lumped for particular land use or variable and provides a relationship for rainfall to runoff.
- *Distributed model*: Represents all the parameter variations for the problem in a distributed manner and considers complete hydrologic calculations, generally based on the physical laws.

11.3.2.2   Hydraulic Models

The input to the hydraulic model is the known flow amount (typically the output of a hydrologic model) and output results are generally the flow height, location, velocity, direction and pressure at a particular location or time. The hydraulic model type depends on the available input and results required. According to the objectives of the modelling, the hydraulic model can be one dimensional or multidimensional, steady or unsteady, uniform or nonuniform (Singh, 1996). The output from the hydraulic model is used as input for determination of flood plain and flood hazard modelling. Generally, Saint Venant equations in its full form or approximations such as kinematic or diffusion wave forms are used in the hydraulic modelling (Singh, 1996).

11.3.2.3   Integrated Coastal Urban Flood Modelling

Generally, the urban flood model includes many components, which represent various processes based on the geomorphology and other basin characteristics. These are integrated in such a way that the model can predict the hydrologic processes such as runoff and flooding. Figure 11.1 shows a typical structure of an Integrated Coastal Urban Flood Model (ICUFM).

The modelling procedure and components of an ICUFM are briefly described below.

- The model of effective rainfall and land use is input to the overland flow model. The effective rainfall is obtained after deducting the losses such as infiltration, evaporation or interception, if considered depending on the flood model.

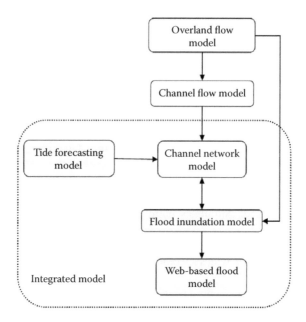

FIGURE 11.1   Typical structure of an Integrated Coastal Urban Flood Model.

- For event-based flood modelling, the overland flow is generally calculated using a physically based approach, with Saint Venant equations, or simplifications based on these, such as a kinematic or diffusion wave approach (Singh, 1996); a simple mass-balance based solution (Singh, 1996); a conceptual/lumped approach such as the Soil Conservation Service–Curve Number (SCS-CN) method; and so on (Zope et al., 2014).
- The output from the overland flow model is input to a stream network model or a sewer model or to the dual drainage model depending on the type of network. In the case of overland flow passes to channel/stream/river, the flow in the channel should be routed using 1D Saint Venant equations or its simplifications such as kinematic or diffusion wave approach.
- In the case of coastal urban cities, the downstream boundary condition will be tidal variation.
- Flood inundation/plain model can give the flood-related details like area, depth, and so on. Flood inundation modelling can be done by direct solution of governing equations such as Saint Venant equations/shallow water equations in two dimensions or using a raster-based approach by coupling with the 1D channel flow model.
- Finally, there can be a management model that may assist the decision makers with regard to flood mitigation measures.

11.3.2.3.1   Distributed Urban Flood Modelling

In urban flood modelling, to capture the spatial and temporal variation of the flood, we need simulation based on physics and the conservation laws of mass, momentum and energy (Singh, 1988). Here the model can simulate the physical processes taking place within the urban watershed, considered with respect to the input data such as rainfall and tidal conditions. The conservation laws give the basic governing partial differential equations, which are then solved using numerical methods. For urban floods, the Saint Venant equations or their approximations can be used for rainfall runoff modelling (Singh, 1988). The effectiveness of the physically based model depends on the approximations and approaches used for a solution, as well as on the accuracy of the data used. Here the physical models used for urban flood simulation are discussed briefly.

11.3.2.4 Overland Flow Modelling

In urban catchments, with rainfall, overland flow occurs on impervious and pervious surfaces such as roofs, driveways, parking lots and lawns (Bedient and Huber, 1988). Here gravitational force is the driving force and runoff occurs in the form of a sheet with very low depths. Generally, for overland flow modelling, the Saint Venant equations of mass conservation and momentum conservation govern the flow behaviour (Singh, 1988). Equations of continuity and momentum adapted from Taylor et al. (Taylor et al., 1974; Singh, 1988, 1996) for a two-dimensional flow are as follows:

The continuity equation is given as:

$$\frac{\partial \bar{u} h}{\partial x} + \frac{\partial \bar{v} h}{\partial y} + \frac{\partial h}{\partial t} = r_e \tag{11.1}$$

where $\bar{u}$ and $\bar{v}$ are spatially averaged velocities in the x and y directions, respectively, h is the depth of flow in the vertical direction and t is the time. Here r_e is the excess rainfall rate and is given as:

$$r_e = r_i - f_i \tag{11.2}$$

where r_i is total rainfall in mm/h and f_i is the infiltration rate. The momentum equations in two dimensions can be written as:

$$\frac{\partial \bar{u}}{\partial t} + \bar{u} \frac{\partial \bar{u}}{\partial x} + \bar{v} \frac{\partial \bar{u}}{\partial y} + g \frac{\partial h}{\partial x} - g(S_{ox} - S_{fx}) + r_e \frac{\bar{u}}{h} = 0 \tag{11.3}$$

$$\frac{\partial \bar{v}}{\partial t} + \bar{u} \frac{\partial \bar{v}}{\partial x} + \bar{v} \frac{\partial \bar{v}}{\partial y} + g \frac{\partial h}{\partial y} - g(S_{oy} - S_{fy}) + r_e \frac{\bar{v}}{h} = 0 \tag{11.4}$$

where g is acceleration due to gravity (m/sec²), S_{ox} is the slope of the watershed element in the x-direction, S_{oy} is the slope of the watershed element in the y-direction, S_{fx} is the frictional slope in the x-direction and S_{fy} is the frictional slope in the y-direction. To find S_{fx} and S_{fy}, the Chezy, Manning or other suitable relationship could be used.

From Manning's equation, the frictional slopes are given (Tisdale et al., 1998) as follows:

$$S_{fx} = \bar{u}(\bar{u}^2 + \bar{v}^2)^{\frac{1}{2}} \frac{n_o^2}{h^{4/3}}; \; S_{fy} = \bar{v}(\bar{u}^2 + \bar{v}^2)^{\frac{1}{2}} \frac{n_o^2}{h^{4/3}} \tag{11.5}$$

where n_o is the overland flow Manning's roughness coefficient in $\text{m}^{-(1/3)}$ sec. Depending on the problem objective, field conditions and data availability, the overland flow may be solved in two dimensions or one dimension. Also, the approximations of the Saint Venant equations such as diffusion or kinematic waveform can be used in the overland flow modelling.

In one dimension, for the diffusion waveform, the governing equations are the continuity equation (Singh, 1996),

$$\frac{\partial q}{\partial x} + \frac{\partial h}{\partial t} = r_e \tag{11.6}$$

and the momentum equation is approximated as

$$\frac{\partial h}{\partial x} = S_o - S_f \tag{11.7}$$

In kinematic waveform, Equation 11.6 is solved with the assumption that $S_0 = S_f$. In Equation 11.6, q is the unit width flow given as $q = \bar{u}h = \sqrt{S_f}/n_o$. The problem is considered to be solved with appropriate initial and boundary conditions. Generally, the initial condition for overland is usually of dry bed condition such as at time $t = 0$, $h = 0$ and $q = 0$ at all specified points. The commonly used boundary conditions include: specified upstream boundary conditions of depth or discharge such as $h = 0$; $q = 0$ at all times t. The downstream boundary condition can be specified or zero depth gradient in case of diffusion wave equation.

In urban flood modelling, the overland flow equations mentioned above are solved numerically using techniques such as finite difference method (FDM), finite element method (FEM), and so on (Vieux, 2001; Desai et al., 2011; Shahapure et al., 2011; Kulkarni et al., 2014a,b and so on), with appropriate initial and boundary conditions.

Further, the overland flow can be simulated also by using a simple mass balance based approximation (Shahapure et al., 2010), for which the equation can be written as:

$$\text{Inflow} - \text{Outflow} = \text{Change in storage} \tag{11.8}$$

For an urban watershed, Equation 11.8 can be represented as

$$r_e \cdot A_c - q \cdot L = \Delta VOL/\Delta t \tag{11.9}$$

where r_e is the excess rainfall which is the inflow for the catchment, that is, rainfall minus the infiltration, q is the overland flow from the catchment into the stream element. The unit of q is taken to be catchment flow per unit length of the stream. Equation (11.9) can be solved iteratively as explained in Kulkarni et al. (2014) and Shahapure et al. (2010).

11.3.2.5   Channel Flow Modelling

In urban flood modelling, channel flow is the main form of surface water flow and all other surface flow processes contribute to it. Generally, the channel flow is solved using one-dimensional Saint Venant equations of mass and momentum conservation (Singh, 1988). The full form of one-dimensional (1D) Saint Venant equations for a channel is (Singh, 1989) the continuity equation:

$$\frac{\partial Q}{\partial x} + \frac{\partial A}{\partial t} - q = 0 \qquad (11.10)$$

The momentum equation can be written as

$$\frac{\partial Q}{\partial t} + \frac{\partial}{\partial x}\left(\frac{Q^2}{A}\right) + gA\left(\frac{\partial h_c}{\partial x} + S_{fc} - S_c\right) - qv_x = 0 \qquad (11.11)$$

where Q is the discharge in the channel (m³/sec), q is lateral inflow per unit width of flow plane (overland flow), A is the area of flow in the channel (m²), S_c is the bed slope of channel, S_{fc} is the friction slope of channel, g is the acceleration due to gravity, h_c is the depth of flow in channel, x, t are the spatial and temporal coordinates and v_x is the velocity component of lateral discharge in x direction. Further, Equation (11.11) can be approximated to the diffusion waveform by putting $\frac{\partial h_c}{\partial x} = S_c - S_{fc}$ and to the kinematic waveform by putting $S_c = S_{fc}$ (Singh, 1996).

In urban flood modelling, the solution of overland flow equations provides the lateral inflow. Further, the channel flow equations are solved numerically using techniques such as FDM, FEM, and so on (Vieux, 2001; Desai et al., 2011; Shahapure et al., 2011; Kulkarni et al., 2014), with appropriate initial and boundary conditions, to get the discharge or flow depth at any location of the channel at the time required. In urban flood simulation, generally the overland flow model and channel flow model are coupled so that at the given time step, the flow depth or discharge can be obtained at any location.

11.3.2.6   Tidal Flow Modelling

Most coastal urban watersheds will be affected by tidal flows. The tidal flow will be directly affecting the drainage channels and can be considered as boundary condition for the channel flow. For example, the tidal flows in the case study areas presented in this chapter are semidiurnal in nature, and the tidal stage at any time t is obtained using the following equation (Kulkarni et al., 2014):

$$S_{ts} = S_{tsm} + S_{tsr} \sin(2\pi \times t / T_p) \qquad (11.12)$$

where S_{ts} is the tidal stage in m at any time t in sec; S_{tsm} is the mean tidal stage in m; S_{tsr} is half the oscillation range in m and T_p is the time period for one tidal cycle.

11.3.2.7 Holding Pond Simulation

In many of the coastal urban areas, to control the flood, holding ponds are used to store the runoff from excess rainfall during the high tides. Holding ponds can be located either at the downstream end of the channel or adjacent to the channel at a suitable location as a flood mitigation measure. The holding pond is normally fitted with a one-way flow valve (i.e. outflow from pond into channel and not vice-versa). In the urban catchment, the pond receives the inflow from its up-stream overland flow elements and releases downstream into the channel during the low tide condition. The hydrologic storage equation for the holding pond is given as (Akan, 1990):

$$\frac{dV_p}{dt} = (Q_{in} - Q_{out}) \cdot dt \text{ where } Q_{in} = \sum q_i + (A_p \cdot r) \qquad (11.13)$$

where Q_{in} = inflow; Q_{out} = outflow; V_p = volume of storage in pond; $\sum q_i$ = sum of all discharges connected to any holding pond; A_p = area of holding pond and r = rainfall intensity. Generally, the holding pond is assumed to have vertical walls with a storage elevation relationship as:

$$V_p = A_p \cdot (h_p - h_{ip}) \qquad (11.14)$$

In Equation (11.13), h_p = water level in the pond; h_{ip} = invert level of outlet discharge of pond. The outflow from the pond will depend upon the discharge head (h_d) above the outlet. If h_{tl} = tail water level in the pond, then $h_d = h_p - h_{tl}$ if $h_p > h_{tl}$ and $h_{tl} > h_{ip}$; also $h_d = h_p - h_{ip}$ if $h_{ip} > h_{tl}$. The outflow from the pond is approximated as $Q_{out} = C_d a_p \sqrt{2gh_d}$ if $h_p > h_{tl}$; $Q_{out} = 0$ if $h_p < h_{tl}$. C_d is coefficient of discharge; a_p is the total area of pipes. The tidal condition occurs at the outlet of the pond while the water level in the pond acts as a downstream boundary condition for the channel flow (Shahapure et al., 2010, 2011; Kulkarni et al., 2014).

11.3.2.8 Floodplain Simulation

In urban flood modelling, in case of flooding, we need to estimate the flood depth at any location for appropriate remedial measures. In case of major flooding due to tidal flows and heavy rainfall conditions, the channel flow raises and spills over to the banks causing the floods. In a channel, the flood essentially consists of a large flow amplitude wave propagating down the valley. When a channel reaches its bankful stage, water ceases to be contained in the channel and spills over into the floodplains. In two-dimensional flood modelling, the floodplain flow can be directly captured. Two-dimensional flow simulation over inundated floodplains is often a slow and shallow phenomenon where local free surface slopes are very small. Hence, floodplain flows are primarily influenced by bed roughness than topographically induced velocity gradients, thus allowing inertial terms to be neglected from the dynamic governing equations of flow (Hunter et al., 2005).

However, if overland flow is modelled in one dimension, we need a separate model to capture the floodplain flow depth. One of the approaches to simulate floodplain is to consider a digital elevation model (DEM) of a sufficient resolution and accuracy as the basic grid over which flood level computations are performed (Bates et al., 2010). When the flood level in the channel/river cell exceeds the bank level (with head causing flow), it spills into an adjacent plain (Figure 11.2) (Bates et al., 2005). Using DEM, each floodplain cell is identified by its elevation. In flood plain modelling, the simplest way to achieve distributed routing of water over the floodplain is to treat each cell as storage volume and solve the continuity equation. In the 1D-2D urban flood model, the channel flow model is integrated within the simplified urban flood model to simulate the channel bank overtopping and subsequent inundation into the floodplains. The basic governing equation under this approach is the continuity equation. The mass continuity equation can be written as (Bates et al., 2005; Kulkarni et al., 2014):

$$\frac{V_{i,j}^{t+\Delta t} - V_{i,j}^{t}}{\Delta t} = Q_{up} + Q_{down} + Q_{left} + Q_{right} \tag{11.15}$$

Here, $V_{i,j}^{t}$ is the volume of water in the cell of ith row and j^{th} column at time t, Q_{up}, Q_{down}, Q_{left} and Q_{right} are the flow rates (here flux entering into cell is considered as positive) from the up, down, left and right adjacent cells, respectively. The change in the cell volume over the time is equal to the fluxes into and out of it during the time step as shown in Figure 11.2. Here, the water level in the central cell is determined based on the water levels in the adjacent cells and their corresponding discharges.

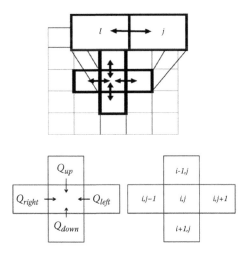

FIGURE 11.2   Representation of typical flow between cells in a floodplain model.

The interflow between two neighbouring cells can be simulated using uniform or diffusive flow equations (Bates et al., 2010). In this study, uniform flow equation has been used which can be written as (Bates and De Roo, 2000; Kulkarni et al., 2014):

$$Q_{up} = \frac{h_{flow}^{5/3}}{n_{fp}} \left(\frac{h_{i-1,j} - h_{i,j}}{\Delta x} \right)^{1/2} \times \Delta x \qquad (11.16)$$

where $h_{i,j}$ is the free water surface level (with reference common datum) at cell (i,j) and Δx is the cell in linear dimension, n_{fp} is the Manning's roughness coefficient. The flow depth, h_{flow}, is the difference between the highest free water surface between the two cells and the highest bed elevation. Mathematically flow depth is represented as

$$h_{flow} = \max(h_{i-1,j}, h_{i,j}) - \max(z_{i-1,j}, z_{i,j}) \qquad (11.17)$$

Here $z_{i,j}$ indicates the bed level for cell (i,j) with respect to a common datum.

11.3.3 NUMERICAL FLOOD MODELLING

As discussed above, in urban flood modelling using physically based approaches, we have to solve Saint Venant's partial differential equation for mass and momentum which describes surface and channel flow. For the solution of these partial differential equations, in the last few decades, a variety of numerical methods such as FDM, method of characteristics, finite volume method (FVM), FEM, boundary element method (BEM), mesh free method, and so on, have been developed (Vieux, 2001; Desai et al., 2011; Shahapure et al., 2011; Kulkarni et al., 2014; etc.). In the numerical methods, the partial differential equations are converted to a set of algebraic equations by particular numerical techniques and solved for unknown discharge or flow depth after application of appropriate boundary conditions. In a finite difference scheme, the single point values of key variables are used to estimate the gradient terms in the governing equations (Hsu et al., 2000; Fang and Su, 2006; etc.). FD-based models have been extensively used for semi-urban and rural flooding applications. In the case of a finite element scheme, the domain is subdivided into elements and a smooth and continuous solution is obtained for each element by integrating the governing equation over each element while ensuring matching values at the common nodes between connecting elements (Bradford and Sanders, 2002; Desai et al., 2011; Babister and Team, 2012). In finite volume methods, the domain is divided into control volume and represents volume averaged values of the variables. Since the formulations are based on an integral form of the shallow water equations, finite volume schemes are better equipped to handle shocks (Bradford and Sanders, 2002; Kadi et al., 2008). Reduced complexity models can be said to use some form of finite difference scheme (Yu and Lane, 2006; Bates et al., 2010; Kulkarni et al., 2014).

11.4   RECENT ADVANCES IN URBAN FLOODPLAIN MODELLING

11.4.1   INTEGRATED USE OF GIS, REMOTE SENSING AND NUMERICAL MODELS

A geographical information system (GIS) may be considered as a database management tool for collecting, storing, retrieving, transforming and manipulating spatial data from the real world for a particular set of purposes (Ogden et al., 2001), and for displaying the results of analysis in map form. GIS has capabilities for data query, input, analysis, manipulation, output, measurements, and so on. GIS platforms have become almost indispensable for preprocessing inputs and postprocessing outputs in urban flood modelling, and may be used to provide spatial information products such as digital elevation models, slope maps and LULC maps, all of which are useful for flood analysis. Remote sensing (RS) is the technique used for collecting information about an object and its surroundings from a distance, and without physical contact (Lillesand and Kiefer, 2000). For example, information required for flood modelling such as LULC and other land features can be obtained by sensor systems located on a satellite or aircraft which measures electromagnetic radiation received from the ground. The data products derived from RS technology include synthetic aperture radar (SAR), LIDAR data and stereo images, as well as other optical images (Lillesand and Kiefer, 2000).

In integrated coastal urban flood modelling, the database required for the flood modelling is prepared from available maps, GIS and RS (Fortin et al., 2001). GIS is also used for preprocessing of various modelling requirements including grid preparation and visualization. A numerical model is the simulator of a hydrologic/hydraulic model. Once the simulation is done, the results can be postprocessed with the help of GIS to prepare plots, graphs, contours and animations. A general framework for an integrated urban watershed modelling consists of selection of a watershed with collection of necessary field data like channel geometry, rainfall, discharge, and so on. The terrain data can be obtained from secondary sources like digitizing topo sheet, downloading available DEM and optical images and preparation of thematic maps. These form the input to the chosen hydrologic/hydraulic model. The ability of GIS software to integrate the different geospatial data sources and extract information across layers has led to efficient modelling techniques (Ogden et al., 2001). The desired geometric data in the overland flow grid can be obtained from DEM and LU/LC layers through geospatial manipulation in GIS software. Thus, GIS and RS have become integral parts of coastal urban flood modelling.

11.4.2   SUB-GRID SCALE POROSITY MODEL

In urban flood modelling, it is essential to understand the spatial and temporal variation of the flooding process to take care of various remedial measures and hence high-resolution modelling is required. However, high-resolution flood modelling still remains computationally expensive, when applied to large areas and generally coarse resolution modelling is adopted. Every flood model domain usually contains important features that cannot be presented at chosen resolution due to their relative sub-grid features, typically consisting of either localized obstructions and blockages or conveyance features like small drains, gutters, and so on (Babister and Team, 2012). To account for such complexities at coarser spatial resolution, Yu and Lane (2006) proposed a methodology for sub-grid scale treatment, which improved the inundation prediction in urban areas over traditional calibration techniques. McMillian and Brasington (2007) in their paper proposed a sub-grid scale porosity treatment to represent the effect of buildings and micro-relief on flow path and floodplain storage at coarse spatial resolution, thereby retaining the effect of topographic features in the inundation results. Fewtrell (2008) in his study reviewed various sub-grid scale porosity techniques, namely, areal-based porosity approaches (simple porosity scaling and water height dependent) and boundary-based porosity approaches. Application of these techniques to field studies indicated that simple areal porosity significantly improved the results at coarser resolution while other techniques marginally improved the results further.

11.4.3 Web GIS-Based Urban Flood Models

In urban flood modelling, the developed flood model for a particular area may be used by many users, including concerned engineers, decision makers and the general public. Hence, if the developed model is integrated within a web GIS (i.e. integrated product of GIS and Internet technologies), it will be very useful and there will be many advantages including ease of access, ready delivery of information to the public, data transparency, platform independence, absence of additional hardware/software requirements, better visualization, cost effectiveness, and so on. If access to web GIS-based environmental solutions can be made available, the stakeholders or local community can participate in resolving the particular issues such as flooding/drainage designs that may directly affect them (Al-Sabhan et al., 2003). In recent times, many environmental applications have been made within web GIS. The Hydrological Simulation Program Fortran (HSPF), a general purpose model, has been integrated within web GIS by Lohani et al. (2002) to assess the impact of land use change on hydrological processes of any urban catchment. This model has capabilities to simulate hydrologic and water quality processes in urban watersheds. Engel et al. (2003) presented a web-based decision support system (DSS) with a distributed conceptual model for hydrologic impact evaluation of small watersheds on land use changes. By integrating meteorology, hydrology and hydraulic models using web based grid computing techniques, Hluchy et al. (2004) presented flood forecasting for a river basin. Jia et al. (2009) developed a web GIS-based rainfall runoff prediction system using a distributed conceptual model. Thus, researchers have developed many general purpose hydrological tools, applicable to most of the urban catchments for decision making, by harnessing the power of the Internet and GIS. Recently, web GIS-based computing techniques have further demonstrated the effectiveness in computation of flood simulation, data accessibility and visualization (Kulkarni et al., 2014a).

For better flood forecasting, distributed physics-based models based on Saint Venant equations/Navier Stokes equations bring out the actual hydrodynamic behavior for the urban catchments. However, these models are computationally intensive and more complex when running on web servers. The reliable flood estimates and model computational time requirements are some of the challenges that are being addressed for real time flood forecasting systems based on distributed models (Hénonin et al., 2010). Kulkarni et al. (2014b) presented an Integrated Flood Assessment Model (IFAM) wherein a 1D/2D coastal urban model was integrated within a web GIS framework for flood simulation and database visualization. IFAM is a general-purpose model applicable to coastal urban watersheds for flood simulation.

11.5   SOME IMPORTANT URBAN FLOOD MODELS

In the last few decades, a number of urban flood models have been developed by researchers, and many of these are used by field engineers for flood simulations. Some of the common hydrologic and hydraulic models being used for urban flood simulation all over the world are briefly described here with their applications, advantages and limitations.

11.5.1   HEC-HMS MODEL

The Hydrologic Engineering Center's, Hydrologic Modelling System (HEC-HMS) is a hydrologic simulation model, with a graphical user interface software developed by U.S Army Corps of Engineers with a number of upgrades. The first version of the model in the form of HEC-1 was developed in 1998. It can be used to simulate the rainfall-runoff process for dendrite watershed systems, and can be used as a distributed, semidistributed or lumped model, event-based or continuous simulation model (www.hec.usace.army.mil/software/hec-hms/). The software contains a basin model, a meteorological model, control specifications and time series data manager. Input data required for the basin model is hydrological soil group, and curve number for each subwatershed, infiltration data and time of concentration. The control specifications component has inputs such as duration and time step (Suriya and Mudgal, 2012). Background files required for the basin model are generated using the HEC-GeoHMS model in ArcGIS. For simulation of flow in open channel, hydrologic routing methods can be used as kinematic wave, lag, Muskingum, Muskingm–Cunge, lag modified plus and studded stagger methods (Zope, 2016). Some of the limitations of this model are: 2D/3D simulations not possible, hydraulic routing in channel to be done separately and incapability to deal with rapidly varied flows. The output generated from this model is in the form of hydrographs and peak discharge is being used as input to hydraulic models such as the HEC-RAS model.

11.5.2   HEC-RAS MODEL

Hydrologic Engineering Center's (HEC) River Analysis Software (RAS) developed by U.S Army Corps of Engineers is a one-dimensional graphic user interface hydrodynamic model having capabilities to perform steady and unsteady river flow analysis (www.hec.usace.army.mil/.../hec-ras/). The first version of the model was developed in July 1995. Water surface profiles are generated by solving Saint Venant equations for unidirectional flow conditions and by solving energy equations

(Lastra et al., 2008). In HEC-RAS, the implicit finite difference scheme is applied for solving unsteady flow equations. HEC-RAS is designed for steady, unsteady flow simulations, movable boundary sediment transport computation and for water quality analysis. A geometry file, as the main input for this model, is generated using the HEC-GeoRAS software model. Geometric data preparation includes stream centreline, banks, flow path centrelines, cross-section cut lines referred to as RAS layers, and extraction of their attributes are being done by executing the preprocessing process in HEC-GeoRAS with integration of ArcGIS (Zope, 2016). Output from the HEC-HMS model in the form of peak discharge for the steady flow data and flow hydrograph for unsteady flow data is the main input to the HEC-RAS model for generation of water depth and extent. Manning's 'n' is also the main input. The water surface profiles and water surface extents generated in HEC-RAS are used as input for generation of flood plain mapping in ArcGIS using postprocessing with HEC-GeoRAS. One of the major limitations of this model is that the simulation is possible only in one-dimensional form.

11.5.3   SWMM Model

The U.S. Environmental Protection Agency (USEPA) developed an urban dynamic rainfall-runoff model called Storm Water Management Model (SWMM) in 1971 and afterward several modifications have been done. The model has capabilities of simulation of single as well as continuous events of urban runoff quantity, as well as quality in storm water and sewered, combined drainage systems. The model has hydrologic as well as hydraulic features. Screening, planning, designing and operation activities are being designed using this model. SWMM can be used for drainage design, flood control strategies, flood plain mapping and for non-point source loading purposes. Outputs generated from this model are in the form of runoff hydrographs, pollutographs, storage volumes, flow stages and depths. (http://www.epa.gov/ednnrmrl/models/swmm).

11.5.4   Storm

The Storage, Treatment, Overflow, Runoff Model (STORM) was developed by the U.S. Corps of Engineers in January 1973. Runoff and pollutant load from watersheds for a single event with hourly precipitation from urban and nonurban watersheds are simulated using this model. The coefficient method, soil complex cover method and unit hydrograph method are used in this model for simulation of runoff (Zoppou, 2001). The main purpose of the model is to control the quantity and quality of storm water runoff and land surface erosions. The major steps involved in this model are runoff quantity and quality estimation, computation of treatment, storage, overflow and land surface erosion (USACE, 1977).

11.5.5   PRMS

The precipitation-runoff modelling system (PRMS) was developed by the U.S. Geological Survey (USGS). The first version of the program was released in 1983 as a single FORTRAN 77 program. Subsequently, it was modified in 1996 and now the latest version PRMS-IV has been released in 2014 as a stand-alone program which can be executed on Linux or Microsoft Windows platforms. This model is deterministic and used to simulate the watershed response toward climate change and land use (USGS, 2015). The model has capabilities of simulation both for daily and for very small time intervals, using variable time steps (Wurbs, 1995). The kinematic flood routing method is used in this model.

11.5.6 MIKE 11

The model developed by the Danish Hydraulic Institute (DHI) Water and Environment can be used for simulation of river and channel flows, water quality and sediment transport system of all types of water bodies. The model provides dynamic, diffusion and kinematic wave approaches, depending on the situation of the flow regime. It has the capabilities of a floodplain, flood encroachment, and flood control design strategies. This model also can be used for flood forecasting and resource operation, operation of irrigation and drainage system and tidal and storm surge studies in rivers and estuaries (www.mikebydhi.com).

11.5.7 HYDROLOGIC SIMULATION PROGRAM–FORTRAN

The hydrologic simulation program–Fortran (HSPF) was developed in the mid-1970s by the U.S. Environmental Protection Agency to model a broad range of hydrologic and water quality processes in watersheds. Urban watersheds can also be simulated using HSPF. This model is being used for the continuous or single event simulation of runoff quantity and quality from watershed. It has capabilities to simulate conventional and toxic organic pollutants from urban as well as agricultural watersheds. Outputs generated are in the form of time series information for water quality and quantity, flow rates, sediment loads and nutrient and pesticide concentrations (Zoppou, 2001).

11.5.8 DR$_3$M-QUAL

The Distributed Routing Rainfall–Runoff Model (DR$_3$M) developed by the U.S. Geological Survey was first developed in 1972 for small nonurban watersheds. The latest version was released in 1991. This model is an urban drainage model. In this model, the basin is represented by an overland flow, channel, pipe and reservoir elements (Zoppou, 2001). The main input is rainfall and soil moisture accounting between storms. This model cannot be used to simulate snow accumulation, interflow and base flow (https://water.usgs.gov/cgi-bin/man_wrdapp?dr3m).

11.5.9 PENN STATE URBAN RUNOFF MODEL

The Penn State Urban Runoff Model (PRSM) is available at the Department of Civil Engineering, Penn State University. PRSM can simulate a single event, infiltration is computed using Soil Conservation Service (SCS) methods and overland flow is computed using the nonlinear reservoir method (Bedient and Huber, 1988).

11.5.10   IFAM MODEL

The Integrated Flood Assessment Model (IFAM) is a web GIS-based integrated flood model with capabilities of 1D-1D and 1D-2D floodplain models (Kulkarni et al., 2014a,b). Both the web GIS server and the associated hydrological model have been indigenously built at IIT Bombay. The web GIS server has been built using Java, Java Servlet Page, JQuery, HTML and XML technologies, while the associated hydrological model has been built in MATLAB® language and both are stored on the server side. The data input to the model is from the client side through the web browser. The model is capable of simulating 1D overland flow using mass balance approach, 1D diffusion wave based channel flow model and quasi-2D raster based floodplain model. The three main outputs from the IFAM tool are: (1) generation discharge and stage hydrographs at any point along the channel; (2) water level profile plot at any hour of the simulation and (3) flood map animation in case of flooding in the channel. The IFAM model has been used for the coastal urban flood simulation of many catchments (Kulkarni et al., 2014a). The model has been applied for two coastal urban watersheds in the Navi Mumbai area of Maharashtra in India. The results of the model applications indicate that the model can be used as an effective coastal urban flood simulation tool (Kulkarni et al., 2014b).

11.6   CASE STUDIES

To demonstrate the applications of the integrated coastal urban flood modelling approach, here two case studies are considered for flood simulation. Both the case study areas are from Navi Mumbai coastal watersheds, in Greater Mumbai, India. The first case study is the application of the Integrated Flood Assessment Model (IFAM), developed at IIT Bombay (Kulkarni et al., 2014a,b). The second case study is the integrated application of the HEC-GeoHMS-RAS model.

Mumbai, the capital of Maharashtra as well as the financial capital of India, is an island city with limitations on its horizontal expansion. To decongest the city, a new urban planned township, Navi Mumbai, was developed in 1972 adjacent to Mumbai. Being the newly developed planned city, the infrastructure takes into consideration future urban growth needs (Shaw, 1999). Navi Mumbai Township is 344 km^2 and the creek line is 150 km out of 720 km of the Konkan coast (City and Industrial Development Corporation, 2003). The area receives an average rainfall of 2250 mm during the monsoon period. As the area is surrounded by creeks on the western side and hills on the eastern side, when high intensity rainfall coincides with tides, severe flooding takes place in many low-lying parts of the city (Kulkarni et al., 2014a). Here two coastal urban watersheds of Navi Mumbai are selected as case studies to demonstrate the integrated coastal urban flood modelling. The first case study area is Belapur-3, an urban coastal catchment, and flood plain analysis is carried out using the IFAM model. The second case study area is Koparkhairane catchment discharging the storm water flow into Thane Creek. The floodplain and flood hazard analysis is carried out using HEC-HMS, HEC-GeoHMS, HEC-GeoRAS and HEC-RAS software with integration of GIS and remote sensing.

11.6.1   Case Study 11.1: Belapur-3 Coastal Urban Catchment

Belapur-3 is located between North latitudes of 19° 1' 2.954" and 19° 3' 5.29", and East longitudes of 73° 2' 58.548" and 73° 2' 12.241" with a watershed area of 3.10 km² as shown in Figure 11.3. The dominant flow direction is toward the south joining the Kharghar/Panvel Creek which is subjected to oscillating tides. The northern boundary is shared with Juinagar and Kharghar watersheds while it is flanked by Belapur-2 watershed on the Western side. The ground elevation varies from 3 m to 245 m above msl. The watershed has been divided into 173 overland flow grids with average intercepting length of grid being 130 m, as shown in Figure 11.3. The length of the main channel is 2.28 km. The land use/land cover classes observed in this watershed are: built up (12.27%), marshy land (10.25%), open/barren land (6.33%) and light vegetation (71.15%).

An extreme rainfall event of 26 July 2005 (see Figure 11.4) has been simulated using IFAM (Kulkarni et al., 2014a). For this event, the rainfall started at 03:00 h of the day during which the tidal stage was in high water condition at 1.72 m (above msl). The event has peak rainfall intensity of 76 mm/h occurring for over 1 h duration. The simulated discharge and stage hydrographs at chainage 2120.0 m are shown in Figure 11.4. The discharge in the channel gradually increases to around 55 m³/s at the 5th hour and reaches the peak value of 63.55 m³/s at the 10th hour of the simulation. Stage hydrograph at chainage (ch) 2120.0 m in Figure 11.4 indicates that channel overtopping has not occurred during the event except at around 12.5 hour due to rising tides. The longitudinal channel profile and water levels shown in Figure 11.5 indicates that water level at the 5th and 10th hour has been overtopped for more than 70% of the channel length (i.e. between ch 0 ~ 1600 m). The simulated flood extent at the 4th, 10th and 24th hour of the simulation is shown in Figure 11.6, which indicates wetting and drying of the floodplain at the 4th and 24th hour. The inundated area at

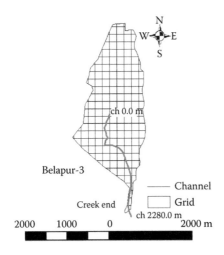

FIGURE 11.3   Coastal urban watershed of Belapur-3 with grid map and channel.

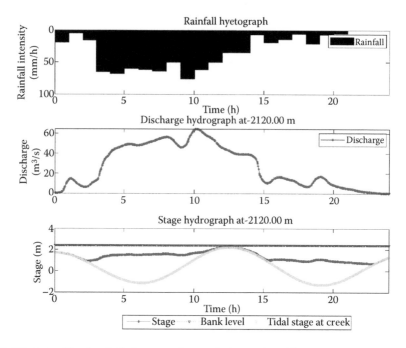

FIGURE 11.4   Simulated discharge and stage hydrographs of Belapur-3 using IFAM (26 July 2005).

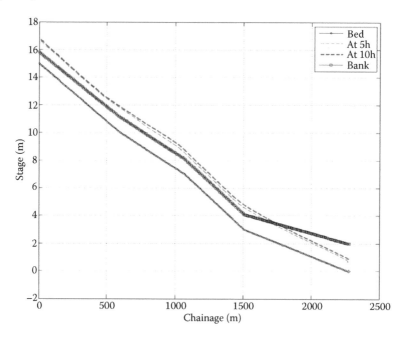

FIGURE 11.5   Longitudinal channel profile and water levels of Belapu-3 watershed (26 July 2005).

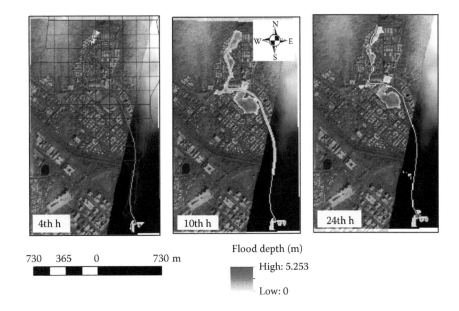

FIGURE 11.6   Simulated flood depth of Belapu-3 watershed (26 July 2005).

the 10th hour of the simulation is around 5% of the area. Being an extreme event, no observed data were available for flooding in the area and hence simulated results could not be verified. However, an enquiry with the municipal authorities confirmed flooding in the areas as shown in Figure 11.6. Moreover, the model used in this study, IFAM, is well validated for a number of coastal urban watersheds in this region as reported by Kulkarni et al. (2014a,b).

The Belapur-3 watershed is leaf shaped, with channel origination at almost the geometric centre of the watershed. The majority of the overland flow contribution occurs in two-thirds of the channel reach. Around 18% of the area is impervious (including open/barren class) while a major portion is degraded forest. The watershed has high overland flow slopes leading to quick runoff. The flood modelling of the watershed using the IFAM model for the rainfall event of 26 July 2005 indicated that the upstream reach of the channel is vulnerable to flooding, which is also physically intuitive as half of the watershed areas join there to form channel flow. The playground/park located near the middle reach of the channel seems to be acting as a natural storage basin, absorbing the surface runoff from the hills and thereby preventing downstream flooding. This case study shows the effectiveness of the IFAM.

11.6.2   CASE STUDY 11.2: KOPARKHAIRANE CATCHMENT

Koparkhairane watershed is located between north latitudes of 19° 4' 24.964" and 19° 7' 56.285", and east longitudes of 73° 0' 15.758" and 73° 2' 48.059" with a catchment area of 14.8 km^2 (Figure 11.7). The dominant flow direction is toward the west, joining the Thane Creek which is subjected to oscillating tides. The eastern boundary is governed by the Parsik Hills and the southern boundary is surrounded by the

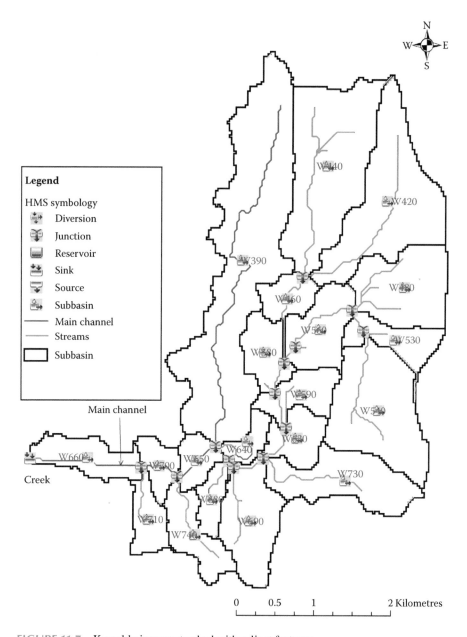

FIGURE 11.7   Koparkhairane watershed with salient features.

Juinagar watershed. The ground elevation varies from 5 m to just over 390 m with respect to msl. The hydrological modelling has been carried out using the HEC-HMS model described above.

11.6.2.1 Hydrologic Modelling Using HEC-HMS

The main components of the HEC-HMS are basin model, meteorological model, control specification and time series data manager. The basin model components such as streams, subbasins, reach, source, junctions and sinks has been generated in the ArcGIS with integration of HEC-GeoHMS software. The delineated subbasins are 20 numbers. The main inputs for the HEC-HMS model are Manning's roughness coefficient, curve number, area of subbasin, stream length, slope of the stream, slope of the watershed, time of concentration and infiltration. To prepare a curve number map, soil map and land use map, a coupled hydrological soil group map was prepared and a composite curve number for each sub-watershed is derived. In this study, the Soil Conservation Service–Curve Number (SCS-CN) method has been used for loss estimation, the SCS-unit hydrograph method has been used for transformation and the kinematic wave method has been used for channel routing. The rainfall event considered is the 26 July 2005 event with simulation time of 24 h. Peak discharge as well as flood hydrographs at each junction node and at outlet were generated. The generated flood hydrograph at each junction is the main input in HEC-RAS modelling. The simulated discharge hydrograph at the channel end is shown in Figure 11.8.

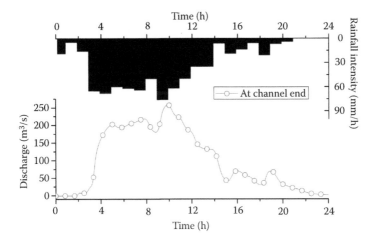

FIGURE 11.8 Simulated discharge hydrograph at channel end of the Koparkhairane watershed (26 July 2005).

11.6.2.2 Hydraulic Modelling Using HEC-RAS and Flood Plain Map

For hydraulic modelling using HEC-RAS, the one-dimensional graphical user interface hydrodynamic model developed by the U.S. Army Corps of Engineers has been used. Water surface profiles are generated by solving Saint Venant equations for unidirectional flow conditions and by solving energy equations (Lastra et al., 2008). Water surface profiles and water surface extents can be generated for steady flow as well as unsteady flow conditions. When the velocity head at the downstream is greater than the velocity head at upstream, it is assumed that contraction is occurring and contraction loss is taken into consideration (USACE, 2010). The main components of the HEC-RAS model are geometric data, flow data and boundary conditions at upstream and downstream. The geometric data components such as channel centreline, bank lines, flow paths and cut lines are generated using HEC-GeoRAS plugin into ArcGIS software (USACE, 2011). In this case, the cutlines are taken at 50-metre intervals. The main inputs to the HEC-RAS model are Manning's roughness coefficient, land use map and boundary conditions. Here, the initial condition is the initial flow hydrograph and downstream boundary condition is the tidal wave and flow hydrograph at outlet. The generated flood hydrograph at each junction from HEC-HMS is the main input in HEC-RAS modelling. A triangulated irregular network (TIN) data model is used for extraction of attributes of spatial geometric data in the form of flood depth and extent (USACE, 2011). The water surface profiles generated in HEC-RAS are exported to Arc-GIS and used as input for generation of flood plain polygon by using postprocessing in HEC-GeoRAS software. The floodplain maps are generated using HEC-GeoRAS tool as shown in Figure 11.9. Being an extreme event, no observed data were available for the flooding in the area and hence simulated results could not be verified. However, as per the personal discussions held with the municipal authorities, they confirmed flooding in the areas as shown in Figure 11.9.

A flood plain map for maximum water depth profile was prepared from the extent of the flood plain polygon whereas the flood depth denotes the flood risk. The flood depth and flood plain map are used as one of the inputs for preparation of a flood hazard map, and can also be used for flood mitigation, flood forecasting and evacuation system. Based on the flood depth derived, the design criteria of the drainage system can be decided.

11.6.2.3 Flood Hazard Map

Flood hazard maps can be used effectively for flood disaster management planning, effective flood warning and flood insurance and risk assessment systems. Flood plain maps are used as the main input for the generation of flood hazard maps. For generation of the flood hazard map, a distance grid raster from the stream channel, a cost grid raster, a distance grid raster from the sea or creek and a flood depth raster are used. Weighted overlay analysis has been undertaken in Arc GIS, using the spatial analysis tool (Zope et al., 2014). Distance grid raster from stream channel calculates, for each cell, the Euclidean distance to the closest source. In this case, the stream centreline is used as input, and distance from the centreline of the channel to the flood polygon boundary is calculated and the raster map is prepared. For

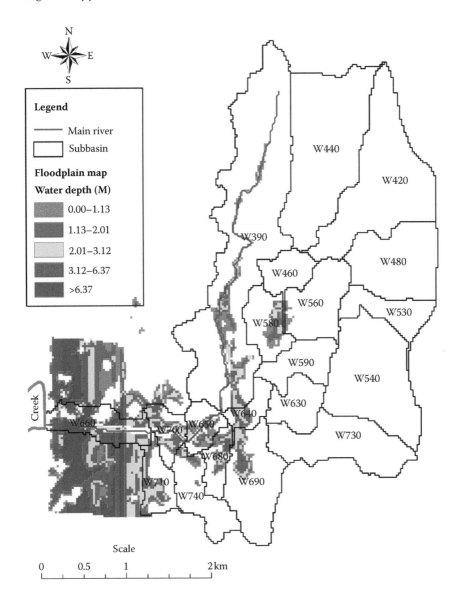

FIGURE 11.9   Simulated flood extent of Koparkhairane watershed for maximum water depth profile (26 July 2005) using the HEC-RAS and HEC-GeoRAS modelling tools.

the cost grid, a raster stream channel and a basin slope layer are used as input. In weighted overlay analysis, the weightage given are: 0.3 to distance grid raster, 0.25 to cost grid raster, 0.4 to flood depth raster and 0.05 to distance grid raster from sea. The generated flood hazard map for the rainfall event of 26 July 2005 is shown in Figure 11.10. The flood hazard categories have been classified as very high, high, medium, low and very low hazards. Flood preparedness and mitigation plans can

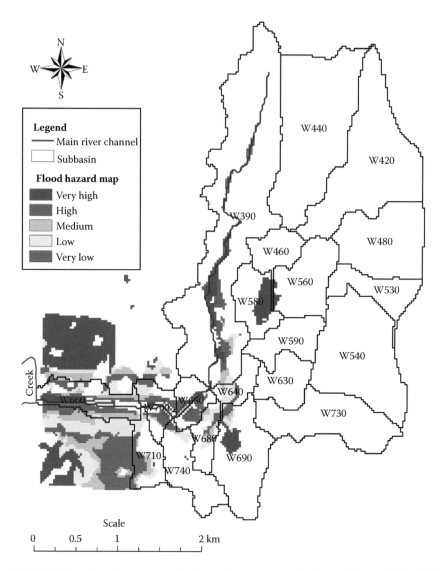

FIGURE 11.10 Flood hazard map for Koparkhairane watershed for 26 July 2005 rainfall event.

be prepared by using the flood hazard maps. This case study shows the effectiveness of the Integrated Coastal Urban Flood Model approach using HEC-HMS and HEC-GeoRAS.

11.7   CONCLUDING REMARKS

In many parts of the world, coastal urban flooding has become a serious issue in recent times. The serious flooding problem will influence the way in which the existing and developing coastal cities grow in the future. The urban flooding in coastal areas is a complex phenomenon which can occur in various forms, namely, flooding due to high tides; surface flooding due to high intensity rainfall; flooding due to inadequate drainage and flooding caused by overtopping in the channels or rivers; basement flooding due to sewer running full, and so on. In most of the serious flood scenarios, flooding can be a combination of tidal flooding, surface flooding, channel overtopping and sewer flooding rather than isolated processes. For example, in coastal cities like Mumbai, which are affected by tides, when there is a marginal increase in flood peak coinciding with high tide, the existing drainage system loses its capacity and severe flooding takes place. For effective management of coastal urban floods, we need to simulate the flooding for extreme rainfall events and other design scenarios.

In this chapter, an integrated approach of coastal urban flood assessment using simulation models is presented. Integration of hydrologic and hydraulic models with GIS and remote sensing can be effectively used in coastal urban flood simulation. In this chapter, various theoretical backgrounds of coastal urban flooding with various modelling approaches are discussed in detail. The commonly used integrated flood models are briefly described. Further, the applications of two integrated urban flood models are explained with two case studies. For the case studies presented, for an extreme rainfall event, the flooded areas and floodplain maps/flood hazard maps are developed. The floodplain and flood hazard maps can be used as effective tools for preparing the flood mitigation/management plan by decision makers or local self-government authorities. The integrated coastal urban flood models can be effectively used as decision making tools for designing the proper drainage system and to generate floodplain maps for initiating adequate flood forecasting. Further, these models can be used in flood mitigation measures, flood warning and evacuation systems under a comprehensive disaster management control plan by city authorities.

ACKNOWLEDGEMENTS

The first author acknowledges the contributions of his former PhD students, Dr. Venkata K. Reddy and Dr. S.S. Shahapure. The authors would like to thank the Department of Science and Technology, Government of India for sponsoring Project No. 09DST033. We would also like to thank the staff at City Industrial Development Corporation (CIDCO) and Navi Mumbai Municipal Corporation for providing the necessary data for the case study areas.

REFERENCES

Akan, A. O. (1990). Single outlet detention pond analysis and design. *Journal of Irrigation and Drainage Engineering.* 116(4), 527–536.

Al-Sabhan, W., Mulligan, M., Blackburn, G. (2003). A real-time hydrological model for flood prediction using GIS and the WWW. *Computers, Environment and Urban Systems.* 27(1), 9–32.

ASCE (American Society of Civil Engineers). (1992). *Design and construction of urban storm water management systems*, ASCE, New York.

Babister, M. and Team (2012). Two dimensional modelling in urban and rural floodplains Stage 1 & 2 report. Barton, Australia, 202.

Bates, P. D., Dawson, R. J., Hall, J. W. et al. (2005). Simplified two-dimensional numerical modelling of coastal flooding and example applications. *Coastal Engineering.* 52(9), 793–810.

Bates, P. D. and De Roo, A. (2000). A simple raster-based model for flood inundation simulation. *Journal of Hydrology.* 236(1–2), 54–77.

Bates, P. D., Horritt, M. S., Fewtrell, T. J. (2010). A simple inertial formulation of the shallow water equations for efficient two-dimensional flood inundation modelling. *Journal of Hydrology.* 387(1–2), 33–45.

Bedient, P. B. and Huber, W. C. (1988). *Hydrology and flood plain analysis.* Addison-Wesley Publishing Company, London.

Bradford, S. F. and Sanders, B. F. (2002). Finite-volume model for shallow-water flooding of arbitrary topography. *Journal of Hydraulic Engineering.* 128(3), 289–298.

Chen, J., Hill, A. A., and Urbano, L. D. (2009). A GIS-based model for urban flood inundation. *Journal of Hydrology.* 373, 184–192.

Chow, V., Maidment, D. R., and Mays, L. W. (1998). *Applied hydrology.* McGraw-Hill Book Company.

City and Industrial Development Corporation. (2003). Some studies on the Nerul node drainage system of New Bombay. (Technical Report, unpublished). Navi Mumbai, India.

Desai, Y. M., Eldho, T. I., and Shah, A. H. (2011). *Finite element method with applications in engineering.* Pearson Education, New Delhi.

Dewan, A. M. and Yamaguchi, Y. (2009). Land use and land cover change in Greater Dhaka, Bangladesh: Using remote sensing to promote sustainable urbanization, *Applied Geography.* 29, 390–401.

Engel, B. A., Choi, J.-Y., Harbor, J. and Pandey, S. (2003). Web-based DSS for hydrologic impact evaluation of small watershed land use changes. *Computers and Electronics in Agriculture.* 39(3), 241–249.

Fang, X. and Su, D. (2006). An integrated one-dimensional and two-dimensional urban stormwater flood simulation model. *Journal of the American Water Resources Association.* 42(3), 713–724.

Fewtrell, T. J. (2008). Development of simple numerical methods for improving twodimensional hydraulic models of urban flooding. Unpublished PhD. Thesis, University of Bristol.

Fortin, J. P., Turcotte, R., Massicotte, S. et al. (2001). Distributed watershed model compatible with remote sensing and GIS data, I. Description of model. *Journal of Hydrologic Engineering.* ASCE. 6(2), 91–99.

Guhathakurta, P., Sreejith, O. P. and Menon, P. A. (2011). Impact of climate change on extreme rainfall event and flood risk in India. *Journal of Earth System Science.* 120, 359–373.

Haestad Methods and Durans, S. R. (2003). *Storm water conveyance modelling and design,* Haestad Press, Waterbury, CT.

Hénonin, J., Russo, B., Roqueta, D. S. et al. (2010). Urban flood real-time forecasting and modelling: A state of the art review. MIKE by DHI Conference, Copenhagen, 6–8.

Hluchy, L., Tran, V. D., Habala, O., Simo, B., Gatial, E., Astalos, J., and Dobrucky, M. (2004). Flood forecasting in crossgrid project. Second European AcrossGrids conference, AxGrids 2004, Nicosia, Cyprus, January 28–30, LNCS 3165, M. Dikaiakos, Ed., Springer-Verlag, Berlin, 51–60.

Hsu, M., Chen, S. and Chang, T. (2000). Inundation simulation for urban drainage basin with storm sewer system. *Journal of Hydrology.* 234(1–2), 21–37.

Hunter, N. M., Horritt, M. S., Bates, B. P. D. et al. (2005). An adaptive time step solution for raster-based storage cell modelling of floodplain inundation. *Advances in Water Resources.* 28(9), 975–991.

Jia, Y., Zhao, H., Niu, C. et al. (2009). A Web GIS-based system for rainfall-runoff prediction and real-time water resources assessment for Beijing. *Computers & Geosciences.* 35(7), 1517–1528.

Kadi, A., Paquier, A. and Mignot, E. (2008). Modelling flash flood propagation in urban areas using a two-dimensional numerical model. *Natural Hazards.* 50(3), 433–460.

Kulkarni, A. T., Eldho, T. I., Rao, E. P. et al. (2014b). An integrated flood inundation model for coastal urban watershed of Navi Mumbai, India. *Natural Hazards.* 73(2), 403–425.

Kulkarni, A. T., Mohanty, J., Eldho, T. I. et al. (2014a). A web GIS based integrated flood assessment modelling tool for coastal urban watersheds. *Computers & Geosciences.* 64, 7–14.

Lastra, J., Fernandez, E., Diez-herrero, A. et al. (2008). Flood hazard delineation combining geomorphological and hydrological methods: An example in the Northern Iberian Peninsula. *Natural Hazards.* 45, 277–293.

Li, H. (2003). Management of coastal mega-cities—A new challenge in the 21st century. *Marine Policy.* 27(4), 333–337.

Lillesand, T. and Kiefer, R. (2000). *Remote sensing and image interpretation.* John Wiley, New York.

Lohani, V., Kibler, D. F. and Chanat, J. (2002). Constructing a problem solving environment tool for hydrologic assessment of land use change. *Journal of the American Water Resources Association.* 38(2), 439–452.

McCuen, R. H. (1989). *Hydrologic analysis and design,* Prentice Hall, Englewood Cliffs, NJ.

McMillan, H. K. and Brasington, J. (2007). Reduced complexity strategies for modelling urban floodplain inundation. *Geomorphology.* 90(3–4), 226–243.

Melesse, A. M. and Shih, S. F. (2002). Spatially distributed storm runoff depth estimation using Landsat images and GIS. *Computers and Electronics in Agriculture.* 37, 173–183.

Niemczynowicz, J. (1999). Urban hydrology and water management – Present and future challenges. *Urban Water.* 1(1), 1–14.

Ogden, F. L., Garbrecht, J., DeBarry, P. A. et al. (2001). GIS and distributed watershed models. II: Modules, interfaces, and models. *Journal of Hydrologic Engineering.* 6(6), 515–523.

Parkinson, J. and Mark, O. (2005). *Urban stormwater management in developing countries,* IWA Publishing, London.

Shahapure, S. S., Eldho, T. I., and Rao, E. P. (2010). Coastal urban flood simulation using FEM, GIS and remote sensing. *Water Resources Management.* 24(13), 3615–3640.

Shahapure, S. S., Eldho, T. I. and Rao, E. P. (2011). Flood simulation in an urban catchment of Navi Mumbai City with detention pond and tidal effects using FEM, GIS, and remote sensing. *Journal of Waterway, Port, Coastal, and Ocean Engineering.* 137(6), 286–299.

Shaw, A. (1999). The planning and development of New Bombay. *Modern Asian Studies.* 33(4), 951–988.

Singh A. K. and Sharma Arun, K. (2009). GIS and a remote sensing based approach for urban flood-plain mapping for the Tapi catchment, India, Proc. of Symposium JS 4 at the Joint IAHS and IAH Convection, IAH Publ. 331, 389–394.

Singh, V. P. (1988). *Hydrologic systems, rainfall-runoff modelling, Vol-I.* Prentice Hall, Englewood Cliffs, NJ.

Singh, V. P. (1989). *Hydrologic systems, watershed modelling, Vol-II.* Prentice Hall, Englewood Cliffs, NJ.

Singh, V. P. (1996). *Kinematic wave modelling in water resources.* Wiley-Interscience, New York.

Suriya, S. and Mudgal, B. V. (2012). Impact of urbanization on flooding: The Thirusoolam sub watershed – A case study. *Journal of Hydrol.* 412, 210–219.

Taylor, C., Al-Mashidani, G., and Davis, J. M. (1974). A finite element approach to watershed runoff. *Journal of Hydrology.* 21(3), 231–246.

Tisdale, T. S., Scarlatos, P. D., and Hamrick, J. M. (1998). Streamline upwind finite-element method for overland flow. *Journal of Hydraulic Engineering*. 124(4), 350–357.

USACE. (1977). *STORM—Storage, Treatment, Overflow, Runoff Model. Users manual*, USACE, Davis, CA.

USACE. (2011). *GIS tools for support of HEC-RAS USING ArcGIS, HEC-GeoRAS: Users manual, Version 4.3.93*, USACE, Davis, CA.

USACE. (2010). *HEC-RAS River analysis system, Hydraulic reference manual, Version 4.1*, USACE, Davis, CA.

USGS. (2015). *PRMS-IV, The precipitation runoff modeling system, Users Manual, USGS, Version 4*, Reston, VA.

Vieux, B. E. (2001). *Distributed hydrologic modeling using GIS*. Kluwer Academic Publishers.

Wurbs, R. A. (1995). *Water management models – A guide to software*. Prentice Hall, New York, 29.

Yu, D. and Lane, S. N. (2006). Urban fluvial flood modelling using a two-dimensional diffusion-wave treatment, part 2: Development of a sub-grid-scale treatment. *Hydrological Processes*. 20(7), 1567–1583.

Zope, P. E. (2016). Integrated urban flood management with flood models, hazard, vulnerability and risk assessment, Unpublished PhD Thesis, submitted to Department of Civil Engineering, I.I.T. Bombay, Mumbai, India.

Zope, P. E., Eldho, T. I. and Jothiprakash, V. (2014). Impacts of urbanization on flooding of a coastal urban catchment: A case study of Mumbai City, India. *Natural Hazards*. DOI 10.1007/s11069-014-1356-4.

Zoppou, C. (2001). Review of urban storm water models. *Environmental Modeling & Software*. 16, 195–231.

12 Lightning

Colin Price

CONTENTS

12.1 INTRODUCTION

Lightning is one of nature's most beautiful and awesome sights (Figure 12.1). Yet it can also be extremely dangerous, presenting a major natural hazard in many different environments, from power utility companies, to civil aviation, to golfers and more. Thousands of people are killed every year by lightning bolts, while tens of thousands are injured as well (Cooray et al., 2007). Lightning impacts both our daily commercial and recreational activities. In the United States alone damages due to lightning strikes amount to tens of millions of dollars annually (Curran et al., 2000). In recent years, with increasing interest in renewal energy, wind turbines have become extremely vulnerable to lightning damage (Glushakow, 2007). Furthermore, most commercial airliners are struck about once a year by lightning; however, due to the protective metal skin, generally little damage is incurred. Tens of thousands of fires are also ignited by lightning every year, generally in temperate or high latitudes (e.g. Canada, Siberia, etc.) (Stocks et al., 2002). In such cases, tens of fires can be ignited locally on the same day as a storm passes through, causing major problems for fire crews and fire management. Hence, knowledge of lightning occurrences, distributions, lightning physics, and the characteristics of lightning can help us develop ways to protect ourselves, our businesses and our homes from this natural hazard. Furthermore, due to the connection between lightning and other natural hazards (such as flash floods, tornadoes, hail storm and even hurricanes, many of which are discussed in other chapters of this volume), lightning can also be used as a tool to reduce our vulnerability to other meteorological hazards impacting our lives.

Every second of the day there are approximately 50 lightning flashes in total around the planet (Christian et al., 2003). The distribution of these is not random, but follows the general circulation patterns of the atmosphere that are driven by solar heating (Price, 2006). Solar heating in the tropics results in the creation of warm moist air that initiates vertical convection and mixing of the atmosphere. Depending on the atmospheric stability, the convection can lead to the development of thunderstorm activity. However, this region of tropical convection decreases with increasing

FIGURE 12.1   Image of lightning discharges at night.

latitude, due to the circulation patterns in the atmosphere that result in sinking air (subsidence) in the subtropical regions around 30° N and 30° S. The rising air in the tropics, and sinking air in the subtropics, produces the meridional Hadley cell (Figure 12.2), where the region of subsidence and high barometric pressure over the global deserts is primarily determined by the amplitude of the Coriolis force,

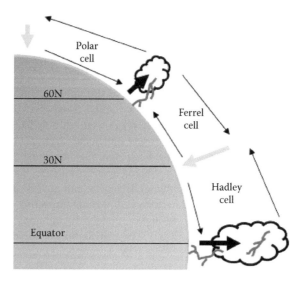

FIGURE 12.2   Schematic representation of the general circulation cells in the atmosphere, between the equator and the pole, showing the thunderstorms in the equatorial regions and in mid-latitudes (60N), while in areas of subsidence few thunderstorms are found (30N and polar regions).

which is directly linked to the Earth's rate of rotation. The Coriolis force is an apparent force resulting from the different speed of rotation of different latitude belts. While every latitude band rotates once every 24 hours, the distance covered in that 24 hours is very different at different latitudes. At the poles the distance is basically zero, resulting in zero velocity. At the equator with a circumference of 40,000 km, the velocity of an object sitting on the surface of the Earth is 1667 km/h in order to complete one revolution in 24 hours. This difference in latitudinal velocity results in the deviation of winds to the 'right' of their direction of motion in the Northern Hemisphere, and to the 'left' of their direction of motion in the Southern Hemisphere. And if the planet was to rotate slower or faster, the tropical zonal band of thunderstorms would expand or contract due to changes in the Coriolis force.

The descending air at 30° will diverge at the surface and spread both toward the equator (resulting in the easterly trade winds) and also poleward (resulting in the westerly mid-latitude winds). These poleward winds eventually meet cold dry polar air around 50–60°, along the polar front. This area of convergence between the cold polar air and the warmer subtropical air results in an additional region of thunderstorms in mid-latitudes, often associated with cold and warm fronts, and other severe weather. Thus, this north-south meridional circulation results in three circulation cells: the Hadley cell between the equator and 30° latitude; the Ferrel cell between 30 and 60° latitude; and the Polar cell between 60° and the pole (Figure 11.2). These cells occur in both the northern and Southern Hemispheres, and shift with the seasons.

In the longitudinal direction, the Earth is separated into tropical continental regions (Americas, Africa and Asia) separated by the various oceans (Atlantic, Indian and Pacific). While solar radiation during the day is absorbed in only a few centimetres of soil, the same radiation is absorbed in a few tens of metres of water in the ocean. This, together with the different heat capacity of water and soil, results in the continents heating much more rapidly in the daytime compared with the oceans. Hence, atmospheric convection and thunderstorms are much more common over the tropical continents compared with the tropical oceans. In addition, water vapour, and particularly the release of latent heat during condensation and freezing, plays a vital role in thunderstorm development. Since the saturation water vapour concentrations increase ~7% for every 1°C increase in temperature (Clasius-Clapeyron relationship), the tropical atmosphere has more than 10 times more water vapor in its atmosphere compared with the polar atmospheres. This water vapor can condense into water and ice, impacting the latent heat release in the tropical atmosphere. For the above reasons, the vast majority of lightning occurs in the tropics between 30° N–30° S, while 90% of all lightning occurs over continental regions, and in the summer hemisphere (Christian et al., 2003) (Figure 12.3).

In recent decades a new thunderstorm-related phenomenon has been discovered, related to upper atmospheric discharges occurring above thunderstorms in the stratosphere and mesosphere. These discharges occur at altitudes from 20–100 km above the Earth's surface, but always above thunderstorms, and are related to the electrical activity in the thunderstorms themselves. Collectively, these discharges have been named transient luminous events (TLEs), and cover a host of phenomena given names such as 'red sprites', 'blue jets', 'ELVES', 'gigantic jets' and more (Figure 12.4). The physical phenomenon for each TLE is slightly different, and for

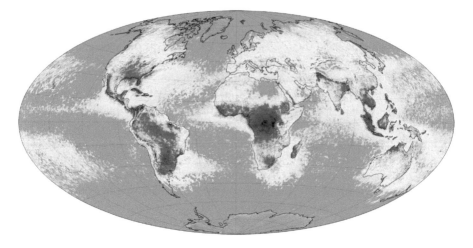

FIGURE 12.3   Global distribution of lightning from the NASA OTD/LIS satellite (Christian et al., 2003). (Courtesy of NASA Marshall Space Flight Center, http://thunder.msfc.nasa.gov/.)

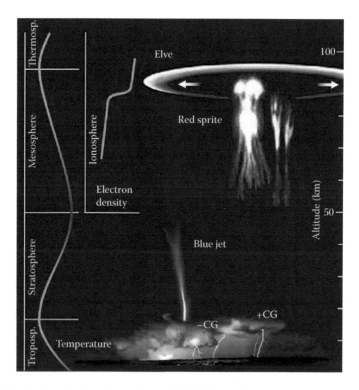

FIGURE 12.4   Transient Luminous Events (TLEs) above thunderstorms.

further details the reader is directed to a review of the topic (Pasko et al., 2011). While most commercial aircraft try to avoid flying near thunderstorms, and none fly above thunderstorms, TLEs may be a hazard to high-altitude aircraft, and space vehicles during launch or descent.

Finally, another type of lightning that poses a hazard to the public is ball lightning. While little is known about ball lightning, there have been many documented sightings, often within buildings and homes. All ball lightning appears to be associated with thunderstorms, although the ball of light can last for many seconds as it hovers at a fixed height above the ground. Ball lightning ranges in size from pea-size to several metres in diameter, and can have different colours (Abrahamson and Dinniss, 2000).

12.2   CLOUD MICROPHYSICS AND DYNAMICS

Lightning activity in thunderstorms depends both on the microphysics and dynamics of the clouds. Some of the very initial studies of clouds and ions in the atmosphere were performed by C.T.R. Wilson at the end of the nineteenth century (Wilson, 1921, 1924). Wilson eventually received the Nobel Prize in Physics for developing his 'cloud chamber' that allowed scientists to study cloud microphysics in laboratory settings.

It is now well known that the electrification process in thunderstorms is related to the existence of hydrometeors (drops, ice crystals, hail, etc.) in different phases and sizes interacting with each other through collisions, freezing, melting, coalescence and breakup (Williams et al., 1991). In the thunderstorm there is a layer where we can find liquid water (supercooled), ice crystals, snow, hail and graupel (soft hail) all existing together. This is at altitudes between the $0°C$ isotherm, and the $-40°C$ isotherm. At temperatures higher than $0°C$, all ice will start melting and turn to water drops. At temperatures below $-40°C$, all hydrometeors will be frozen solid. However, water can exist in the liquid form at temperatures below $0°C$ and above $-40°C$, called supercooled water. It has been shown in laboratory studies that collisions between all these particles (especially ice and graupel), in this mixed-phase region of clouds, is key for the charge transfer between cloud particles (Takahashi, 1978; Saunders et al., 1991). Cloud particle collisions are thought to be the main mechanism for cloud electrification. Rebounding particles carry away equal and opposite charges. Observations show that clouds having predominantly ice crystals in the mixed phase region, and small amounts of supercooled water and graupel, show little electrification (Takahashi, 2006), and little lightning. However, it should be noted that in thunderstorm anvils with no liquid water or graupel, in situ charging has been documented (Kuhlman et al., 2009).

What determines which clouds have hail, graupel and supercooled water in the mixed phase region of convective clouds? This region in summer thunderstorms can extend from around 2–10 km altitude, and therefore we need significant updrafts in these clouds to carry the heavier particles up above the freezing level. It turns out that the threshold updraft speed is around 10 m/s. Clouds with little lightning activity (maritime clouds) often have maximum updrafts less than 10 m/s (Deierling and

Petersen, 2008), while electrically active storms have updrafts reaching up to 50 m/s. In addition to the transport of larger hydrometeors into the mixed phase region of clouds, the stronger updrafts also enhance the collisions between different sized particles. Increased collisions result in increased charge transfer between particles, leading to rapid charge buildup in clouds.

It should be noted that weak updraft speeds may influence the production of rainfall, but not necessarily the production of lightning. While weak updrafts cannot carry supercooled drops above the freezing level, the collision and coalescence between drops of different sizes can very efficiently result in 'warm' rain production, with no involvement of the ice phase (Johnson, 1981; Beard and Ochs, 1993). Such heavy rainfall is seen in many monsoon regions, as well as over the equatorial oceans, but with little lightning activity. In fact, the rainiest regions of the globe paradoxically have the least lightning activity (Price, 2009).

Charge transfer between particles is not a sufficient condition alone to produce lightning. To build up large-scale electric fields in convective clouds, we need to separate the positively charged particles from the negatively charged particles in the cloud. If all particles in clouds randomly received either a positive or a negative charge, we would have a cloud filled with charged particles, but with no net electric field on the large (kilometres) scale. However, it has been established that the smaller ice crystals in clouds, in general, acquire a net positive charge due to all the collisions, while the larger graupel particles acquire a net negative charge (Saunders et al., 2006). Due to their different sizes (and terminal velocities), the smaller positive crystals are carried aloft to the top of the cloud by the updrafts, while the larger negative hail stones drift downward to the base of the cloud. In this way, within 20 minutes of the formation of the cloud, regions of positive and negative net charge can be built up, with electric fields approaching the breakdown electric field in air. The maximum observed electric fields in thunderstorms can reach 400 kV/m. When that threshold is passed, lightning occurs.

These natural electric discharges in the atmosphere can span tens of kilometres vertically and horizontally, and transfer electric charge (or current) between different regions of the clouds, or between clouds and the ground. Approximately 75% of all lightning remains within clouds, called intracloud (IC) lightning, while the remaining 25% occurs between the cloud and the ground, known as cloud-to-ground (CG) flashes. Of these 25% CG flashes, more than 90% have a negative polarity (–CG), transferring negative charge to Earth, while the remaining CG flashes have positive polarity (+CG) (Figure 12.5).

While lightning activity in thunderstorms is directly related to the microphysics and dynamics of thunderstorms, the dynamics impacts the microphysics and vice versa, due to latent heat release as the storm develops. However, thunderstorm cells often follow a regular cycle of birth, development, maturity, decay and dissipation (Figure 12.6). This cycle can take around 1 hour from initiation to completion, with new thunderstorm cells forming on the cold outflow boundary of previous cells. From the birth of the convective cloud to its dissipation, the cloud is electrified at different levels, while we only see the lightning activity during the developing, mature and decaying stage. The lightning activity itself follows a specific pattern, with the IC lightning normally appearing first (during developing stage), followed by the CG

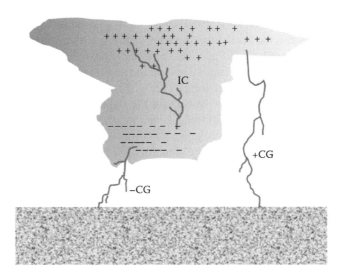

FIGURE 12.5   Main types of lightning in thunderstorms, showing the dipole structure with positive charge in the top of thunderstorms, and negative charge near the base.

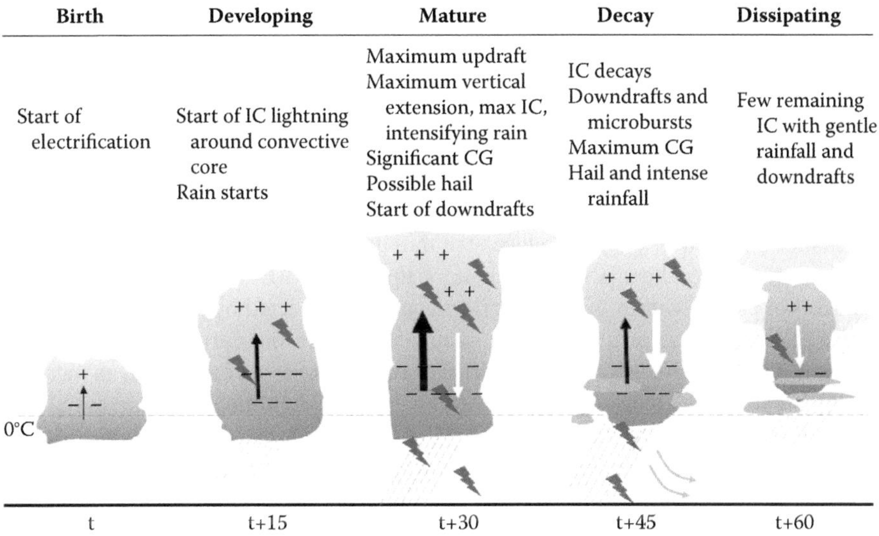

Birth	Developing	Mature	Decay	Dissipating
Start of electrification	Start of IC lightning around convective core Rain starts	Maximum updraft Maximum vertical extension, max IC, intensifying rain Significant CG Possible hail Start of downdrafts	IC decays Downdrafts and microbursts Maximum CG Hail and intense rainfall	Few remaining IC with gentle rainfall and downdrafts
t	t+15	t+30	t+45	t+60

FIGURE 12.6   The different stages of development of a thunderstorm from birth to dissipation, with the associated lightning and severe weather associated with each stage.

lightning starting during the mature stage (Williams et al., 1989). Both types of lightning can occur during the decaying stage. In addition to the lightning changes, the mature stage can be associated with heavy rainfall, hail and tornadoes, while the dissipating stage is known to be associated with downdrafts/microbursts and wind shear.

The lightning discharge starts with the preliminary breakdown process, when the local field rises above the breakdown threshold. This may occur due to the enhanced electric fields produced around pointed sharp ice crystals or graupel in the cloud. When the fields are intense enough to initiate the preliminary breakdown of the air, the ionized channel propagates in jumps of 50–100 m toward the ground, called the stepped leader. The leader leads the way for the lightning discharge, and prepares the channel to be used by the lightning discharge when it connects to the ground. The leader can develop between negative and positive charged regions in the cloud, or between the charged regions in the cloud and the ground. As the stepped leader propagates in space and time, it can split into branches, searching for the easiest route to travel through the cloud or to the ground. These branches give the typical morphology of the lightning discharge, with the branches always pointing in the direction of propagation. Hence, downward-pointing branches occur when the lightning discharges propagate from the cloud to the ground (see Figures 12.1 and 12.5), while upward-pointing branches imply the leader started at the ground and moved upward to the cloud.

When the leader reaches approximately 100 m from the ground, a second leader or 'streamer' starts to rise from sharp objects at the surface (trees, rooftops, utility poles, etc.) to meet the stepped leader. This leader process takes on the order of 1 msec. When the leader meets the up-going streamer, we get the electrical short circuiting of the cloud to ground, and the charge deposited along the leader flows rapidly to the Earth (for a −CG). This stage of the lightning flash is called the return stroke, and the current rises rapidly to values of 20–30 kilo Amperes in a few microseconds. Since air has a finite conductivity (or resistance), the air in the lightning channel heats up to temperatures of 30,000 K, which also results in the bright optical emissions – the luminous 'flash' of the lightning – during this stage, that illuminates the lightning channel. The high temperatures also result in the production of an acoustic shock wave, which produces the thunder we hear after lightning. After the first return stroke, there is often a break of up to 40×10^{-3} sec, during which other pockets of charge within the thunderstorm can be connected by a new leader (called a dart leader) to the original lightning channel. When the dart leader reaches the ground, a second return stroke will occur, again causing a rise in current, an increase in temperature, and a brightening of the channel. One lightning flash can have up to tens of return strokes, each separated by a few tens of msec, resulting in the flickering effect we notice when we observe lightning in a thunderstorm. The high temperatures and high currents in the return stroke are what makes lightning such a great natural hazard to people and electrical equipment of all kinds, as well as having the potential to ignite forest and other fires.

12.3 LIGHTNING AND SEVERE WEATHER

As mentioned above, lightning in thunderstorms is strongly linked to the microphysics and dynamics of thunderstorms, and hence changes in the lightning activity can tell us about changes in the internal processes within the thunderstorms. Both the amount of lightning in thunderstorms, as well as the polarity of the lightning discharges, are found to be associated with specific severe weather phenomena

(Williams, 2001; Dotzek and Price, 2009). The amount of lightning can be related to the intensity of the updrafts, which affects the rate of charge transfer and charge separation. As to the polarity of CG lightning, this can change by varying either the temperature or the liquid water content (LWC) in the charging zone (Takahashi, 1978; Saunders et al., 1991).

Hailstorms have been studied in many countries due to the damage caused both to agricultural harvests, as well as the damage to property, cars, and so on. Hail size is directly related to the updraft speed, with pea-sized hail needing updrafts of 35 km/h while grapefruit-sized hail needs updrafts reaching 160 km/h, vertically upward! Hail forms by multiple ascents and descents within the thunderstorm, ascending through the mixed-layer region, where any supercooled water colliding with the hailstone will freeze and build a new layer (like an onion) on the hailstone. While descending from the top of the mixed-phase region, the hail will collect another layer of water that will add to the mass of the hailstone. Eventually, the hailstone will be too heavy for the updrafts to support the weight, or the hailstone will exit the updraft and fall out of the cloud. Many studies have shown a link between lightning activity and hail occurrence on the ground (McGorman and Burgess, 1994; Carey and Rutledge, 1998; Emersic et al., 2011). Chagnon (1992) and Montanya et al. (2009) have shown that lightning activity rapidly increases at the time of hail occurrence on the ground. Liu et al. (2009) have shown that the charge in the CG lightning shifts to being primarily positive (+CG) during the hail portion of the storm. Generally, nonsevere thunderstorms have CG lightning that is predominantly negative charged (carries negative charge to ground), and thunderstorms with a large fraction of +CG are quite rare.

Tornadoes are also associated with certain lightning signatures. Two distinct signals have been observed. The first is what is called the lightning 'jump' in total lightning (both IC and CG flashes) 10–20 minutes before the touchdown of a tornado (Kane, 1991; Perez et al., 1997; Weber et al., 1998; Williams et al., 1999; Schultz et al., 2009; Gatlin and Goodman, 2010). In addition, numerous studies also show a shift in the CG charge (polarity) to positive lightning around the time of tornado sightings. Carey et al. (2003) showed that during an episode of five tornadoes within one hour, the +CG fraction increased to ~60% of all CG lightning. Another type of severe windstorm, but with straight-line rather than circular winds, is called a derecho. Such severe storms can cause tremendous damage over hundreds of kilometres. Price and Murphy (2002) studied a derecho along the U.S./Canada border that exhibited predominantly positive CG lightning activity during the most intense part of the storm. As mentioned above, polarity changes in the laboratory have been linked with changes in liquid water content or temperature in clouds. Hence, such changes in lightning may tell us something about the inner workings of these severe storms.

Due to the strong link between lightning, cloud microphysics and cloud dynamics, lightning in individual storms is generally positively correlated with rainfall amounts. However, this relationship is very variable based on location and season (Piepgrass et al., 1982; Petersen and Rutledge, 1998; Gungle and Krider, 2006; Price and Federmesser, 2006). In particular, rapidly developing storms can produce heavy precipitation in continental regions, which can result in flash floods due to heavy rainfall in short periods of time. Furthermore, lightning activity is generally observed to precede the rainfall in thunderstorms by 10–20 minutes, allowing for some short-term forecasting skill. For this reason, many groups have attempted to use lightning data to estimate regions of heavy rainfall, and possibly flash flooding (Kohn et al., 2010; Price et al., 2011).

Tropical storms, hurricanes and typhoons have embedded within them thunderstorm cells that influence the development and intensification of these monster storms (Molinari et al., 1994; Samsury and Orville, 1994; Black and Hallett, 1999). In recent years, we have started to study these oceanic storms using global lightning networks (Black and Hallett, 1999). Since these storms have lifetimes of 1–2 weeks, and migrate over thousands of kilometres, we need global monitoring capabilities to track and study them. It has been shown that lightning activity in hurricanes peaks 24 hours before the peak intensity of the storm (maximum winds) (Price et al., 2009). The lightning activity may allow us to better forecast the intensification of these killer storms.

Finally, lightning is also a major cause of forest fires in temperate latitudes, burning more than 1.5 million hectares of wilderness in Canada alone every year (Stocks et al., 2002). Here, too, +CG lightning appears to be a key in the ignition of fires, since the CG flashes generally are characterized by the presence of 'continuing current' that allows enough time for the biomass to ignite (Latham and Schleiter, 1989). When the current pulse is too short (<100 microseconds) the lightning flash is less likely to start a fire. Hence, tracking the polarity in lightning across regions may supply important information about the likelihood of forest fires due to lightning.

Due to the radio waves emitted by the lightning discharges, we can monitor lightning activity in storms remotely, and over great distances, with information about the time, location, polarity, peak current and multiplicity available from ground detection networks (Latham and Schleiter, 1989; Lagouvardos et al., 2009; Abreu et al., 2010). By 2020, both the Europeans and the Americans will have launched lightning sensors into geostationary orbit, allowing us hemispheric observations of total lightning continuously in time and space (Stuhlman et al., 2005; Goodman et al., 2010). This will allow us to better monitor severe weather and perhaps to provide better information about the storm severity, intensification, precipitation, and so on, allowing us to better warn the public and other stakeholders in real time.

12.4 LIGHTNING AND CLIMATE CHANGE

The Earth's climate may significantly change in the near future due to increasing greenhouse gases in the atmosphere (IPCC, 2013), and it is important to know how lightning and thunderstorms are likely to change due to these climate changes.

It has been shown by many studies that lightning activity, thunderstorm days, or indices linked to global lightning activity (ionospheric potential, Schumann resonances, etc.) are sensitive indicators of surface temperature changes (Williams, 1992, 2005, 2009; Price, 1993; Markson and Price, 1999; Reeve and Toumi, 1999; Price and Asfur, 2006; Markson, 2007). These studies show that on different temporal and spatial scales, small increases in surface temperature result in large increases in thunderstorm and lightning activity (a nonlinear link).

The IPCC report (IPCC, 2013) predicts a global warming of 1–5°C by the end of this century, depending on the scenario we use for future uses of energy and land use. However, one of the weaknesses of the models used for these predictions is the simulation of convective clouds, which are a sub-grid scale process that needs to be parameterized in climate models (Del Genio et al., 2007; Siingh et al., 2011). While climate models have problems accurately modelling convective clouds, they are even more problematic modelling lightning activity, although this has been attempted (Price and Rind, 1994a,b; Shindell et al., 2006; Futyan and Del Genio, 2007).

For understanding lightning changes in a warmer world, we need to not only look at surface temperatures, but also the temperature profile (lapse rate) as quantities of greenhouse gases in the atmosphere increase. There are three possibilities regarding the mean vertical temperature profile in the troposphere as greenhouse gases increase, and surface temperatures warm (Figure 12.7). If the surface warms more than the upper troposphere, the atmosphere will become more unstable (on average), and we would expect more convection and more thunderstorms. If the surface and the upper troposphere warm at the same rate, then no change will be seen in the lapse rate, and hence there will be no change in the mean stability (or instability) of the troposphere. Finally, if the upper troposphere warms more than the surface temperatures this will stabilize the atmosphere, with fewer thunderstorms likely to develop in a warmer world. Climate models of all complexity and sizes tend to support this third scenario, and show that as the climate warms at the surface, the tropical upper troposphere (exactly the location of most of the global thunderstorms) warms even more (Del Genio et al., 2007). The reason for this is that increased convection transports additional water vapour into the upper atmosphere, where it acts as a strong greenhouse gas, absorbing infrared radiation emitted from the surface of the Earth. This results in a greater warming in the upper troposphere than at the surface, and hence the average stabilization of the tropical atmosphere. However, within the thunderstorms themselves the instability, measured by the convective available potential energy (CAPE), tends to increase in a warmer climate (Futyan and Del Genio, 2007), especially for the most intense thunderstorms. And increases in CAPE in the present climate show clear increases in lightning activity (Williams et al., 1992; Pawar et al., 2011; Siingh et al., 2011). Therefore, when these same climate models are run under

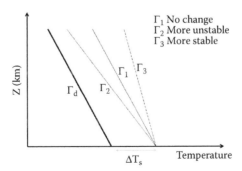

FIGURE 12.7   The atmospheric dry adiabatic lapse rate of 9.8C/km (Γ_d) is the equilibrium profile of the dry atmosphere. Γ_1, Γ_2 and Γ_3 give three scenarios of average future profiles due to global warming.

a scenario with a doubled-CO_2 atmosphere (a situation we are likely to reach by the middle of this century), the models show increases in lightning activity of approximately 10% for every 1°C of global warming (Price and Rind, 1994b; Shindell et al., 2006). Locally, that increase can be much larger. This appears to present a paradox of more lightning occurring even as the global mean atmosphere stabilizes. However, it appears that the CAPE in thunderstorms actually increases in a warmer climate, while in the fair-weather troposphere the overall instability decreases.

To understand this apparent paradox, we need to return to the link between rainfall and lightning in the present climate (Price, 2009). When we look at global maps of precipitation and lightning, we can rank the tropical continental regions according to rainfall: (1) Southeast Asia (most), (2) South America, and (3) Africa (least). The ranking according to lightning is opposite: (1) Africa (most), (2) South America and (3) Southeast Asia (least). In addition, if we look at the very rainy maritime regions along the intertropical convergence zone (ITCZ), there is very little lightning. Hence, on a spatial and temporal mean (climate scales), lightning is associated with drier regions of the globe. The wet monsoon regions and areas of maritime convection show little lightning, perhaps due to the warm rain processes (cloud temperatures above 0°C) being dominant in those regimes. An additional example of a short-term climate change is the El Niño phenomenon. The largest impact of El Niño occurs in the Pacific region, with the western Pacific (Indonesia, Borneo, northern Australia) experiencing major drought conditions during El Niño years. Satellite data from the Tropical Rain Measuring Mission (TRMM) have shown that during the severe drought period of 1997/1998 in Southeast Asia, the lightning activity increased by nearly 60% (Hamid et al., 2001; Yoshida et al., 2007). The explanation is that, while there were fewer thunderstorms during the El Niño dry period, those that did develop were much more explosive, producing much more lightning activity. Model-based studies suggest that this is likely to be true for the future as well, and that we may expect fewer thunderstorms in the future, but more explosive storms when they do occur (Price and Rind, 1994a; Del Genio et al., 2007; Siingh et al., 2011).

As the climate changes, and world population expands, we can also expect to see changes in aerosol loading of the atmosphere (air pollution, biomass burning, dust storms, etc.). This will also impact lightning activity. Some studies (Lyons et al., 1998; Steiger et al., 2002; Bell et al., 2009; Altaratz et al., 2010) show that changes in either polarity or amount of lightning occur as the aerosol loading changes. Every cloud drop nucleates on a hygroscopic particle called a cloud condensation nucleus (CCN). Nuclei, called ice nuclei (IN), are also needed for the initiation of ice crystal growth in clouds. Changing the concentration, size and chemistry of CCN and IN will change the microphysics of thunderstorms. In a polluted atmosphere, the CCN compete for moisture, and hence we get more, but smaller, drops in the cloud. These can be lifted to higher altitudes in the cloud by the updrafts, increasing the chances of entering the mixed phase region of the storm (electrification region), and therefore increasing the chances of charge transfer between hydrometeors (Rosenfeld et al., 2008). Furthermore, if the droplets reach the altitude of freezing, the additional release of latent heat due to freezing will add buoyancy to the cloud, further enhancing the development of the thunderstorm.

Altaratz et al. (2010) investigated the regional impact of increased aerosol loading in the Amazon region (due to biomass burning) on lightning activity. They used the MODIS aerosol optical depth (AOD) index as a measure of the number of aerosols in the atmosphere, together with World Wide Lightning Location Network (WWLLN) lightning flash data from a global VLF network (Abreu et al., 2010). They found that as the AOD increased from 0 to 0.25 (relatively clean atmosphere) there was an increase in the regional lightning activity. However, on days when the AOD increased beyond 0.25, up to values of 0.8, an inverse relationship was seen, with the lightning activity starting to decrease with increasing AOD, giving a 'boomerang' effect: at low AOD values lightning activity increases with increasing aerosols; while at high AOD values lightning activity decreases with increasing AOD values. The increasing lightning at low AOD values is explained above due to microphysical processes. However, on days when the AOD is very high the reduction in lightning activity is possibly related to the increased radiative effect of the aerosols. The aerosols in the lower troposphere absorb some of the incoming solar radiation, heating the mid troposphere while cooling the surface. This stabilizes the local atmosphere, choking the development of convection and thunderstorm cells. So, when the aerosol loading is too high, the radiative effect outweighs the microphysical effect, and lightning activity decreases. Hence, as the climate changes in the future, we will also need to understand the changes in aerosol loading in different regions of the globe to understand how thunderstorms and lightning will respond. Long-term aerosol changes have been linked to 'global dimming' before 1985, as the global pollution and aerosol loading increased, followed by a 'global brightening' since 1985 as air quality regulations have improved (Wild, 2012). Global increases in aerosol loading act to reduce global warming, cooling the surface temperatures either directly by the backscatter of solar radiation, or indirectly through changes in cloud albedo or lifetime (Altaratz et al., 2010). It is interesting to note that precipitation and thunder-day number also tend to follow this long-term trend in global aerosols (Chagnon, 1985; Gorbatenko and Dulzon, 2001; Lohmann and Feichter, 2005).

Future changes in climate will have important implications related to natural hazards linked to lightning and thunderstorms. Hence, when discussing the hazards resulting from lightning, we need to always consider how these hazards may change in the future due to global climate change.

12.5   CONCLUSION

Lightning is a major natural hazard that impacts many commercial and recreational sectors, while also often causing death or severe injury due to the high temperatures and high peak currents in the lightning channel. Global lightning and thunderstorm activity is driven first and foremost by the Earth's climate, driven by solar insolation that varies with latitude, longitude (land/ocean), season, and hour. The climate drives circulation patterns that promote thunderstorms in the tropics and mid-latitudes, and inhibit thunderstorms in the subtropics and polar regions. Locally, thunderstorm activity depends on surface temperature, water vapour, the tropospheric lapse rate, as well as aerosol loading. These parameters can impact the intensity and polarity of lightning in thunderstorms.

It has been shown by numerous studies that the types of severe weather phenomena produced by thunderstorms are linked to anomalous lightning signatures. Tornadoes, derechos and hail storms often show shifts in lightning polarity (more +CG), while sometimes jumps in total lightning activity act as precursors for tornadic activity. In recent years, hurricanes have also been shown to provide lightning signatures related to their development, with lightning activity peaking one day before the maximum intensity of tropical storms. Due to increasing concentrations of greenhouse gases in the atmosphere, we should expect changes in lightning and thunderstorm activity in the future.

REFERENCES

Abrahamson, J. and Dinniss, J. (2000). Ball lightning caused by oxidation of nanoparticle networks from normal lightning strikes on soil. *Nature*, 403, 519–521.

Abreu, D., Chandan, D., Holzworth, R.H. and Strong, K. (2010). A performance assessment of the WWLLN via comparison with the CLDN. *Atmos. Meas. Tech.*, 3, 1143–1153.

Altaratz, O., Koren, I., Yair, Y. and Price, C. (2010). Lightning response to smoke from Amazonian fires. *Geophys. Res. Lett.*, 37, L07801, doi:10.1029/2010GL042679.

Beard, K.V. and Ochs III, H.T. (1993). Warm-rain initiation: An overview of microphysical mechanisms. *J. Appl. Met.*, 32, 608–625.

Bell, T.L., Rosenfeld, D. and Kim, K.M. (2009). Weekly cycle of lightning: Evidence of storm invigoration by pollution. *Geophys. Res. Lett.*, 36, L23805, doi:10.1029/2009GL040915.

Black, R.A. and Hallett, J. (1999). Electrification of the hurricane. *J. Atmos. Sci.*, 56, 2004–2028.

Carey, L.D., Petersen, W.A. and Rutledge, S.A. (2003). Evolution of cloud-to-ground lightning and storm structure in the Spencer, South Dakota, tornadic supercell of 30 May 1998. *Mon. Wea. Rev.*, 131, 1811–1831.

Carey, L.D. and Rutledge, S.A. (1998). Electrical and multiparameter radar observations of a severe hailstorm. *J. Geophys. Res.*, 103(D12), 13,979–14,000, doi:10.1029/97JD02626.

Chagnon, S.A. (1985). Secular variations in thunder-day frequencies in the twentieth century. *J. Geophys. Res.*, 90, 6181–6194.

Chagnon, S.A. (1992). Temporal and spatial relations between hail and lightning. *J. Appl. Met.*, 31(6), 587–604.

Christian, H.J., Blakeslee, R.J., Boccippio, D.J. et al. (2003). Global frequency and distribution of lightning as observed from space by the Optical Transient Detector. *J. Geophys. Res.*, 108, 4005. doi:10.1029/2002JD002347.

Cooray, V., Cooray, C. and Andrews, C.J. (2007). Lightning caused injuries in humans, *J. of Electrostatics*, 65, 386–394.

Curran, E.B., Holle, R.L. and Lopez, R.E. (2000). Lightning casualties and damages in the United States from 1959–1994. *J. Climate*, 13, 3448–3464.

Deierling, W. and Petersen. W.A. (2008). Total lightning activity as an indicator of updraft characteristics. *J. Geophys. Res.*, 113, D16210, doi:10.1029/2007JD009598.

Del Genio, A.D., Mao-Sung, and Jonas, J. (2007). Will moist convection be stronger in a warmer climate? *Geophys. Res. Lett.*, 34, L16703. doi:10.1029/2007GL030525.

Dotzek, N. and Price, C. (2009). Lightning characteristics in severe weather. In Betz, H.D., Schumann, U. and Laroche, P. (Eds.), *Lightning: Principles, Instruments and Applications*. Springer Publications, 487–508.

Emersic, C., Heinselman, P.L., MacGorman, D.R. and Bruning, E.C. (2011). Lightning activity in a hail-producing storm observed with phased-array radar. *Mon. Wea. Rev.*, 139, 1809–1825.

Futyan, J.M. and Del Genio, A.D. (2007). Relationships between lightning and properties of convective cloud clusters. *Geophys. Res. Lett.*, 34, L15705. doi:10.1029/2007GL030227.

Gatlin, P.N. and Goodman, S.J. (2010). A total lightning trending algorithm to identify severe thunderstorms. *J. Atmos. Ocean Tech.*, 27, 3–22.

Glushakow, B. (2007). Effective lightning protection for wind turbine generators. *IEEE Trans. on Energy Conversion*, 22, 214–222.

Goodman, S.J., Blakeslee, R., Bovvippio, D. et al. (2010). The Geostationary Lightning Mapper (GLM) for GOES-R: A new operational capability to improve storm forecasts and warnings. In *Proceedings of 6th Annual Symposium on Future National Operational Environmental Satellite Systems—NPOESS and GOES-R*, AMS Annual meeting, 2010.

Gorbatenko, V. and Dulzon, A. (2001). Variations of thunderstorms, *KORUS '01 Proceedings. The Fifth Russian-Korean International Symposium on Science and Technology*, 2, 62–66.

Grenfell, J.L., Shindell, D.T. and Grewe, N. (2003). Sensitivity studies of oxidative changes in the troposphere in 2100 using the GISS GCM. *Atmos. Chem. Phys. Discuss.*, 3, 1805–1842.

Gungle, B. and Krider, E.P. (2006). Cloud-to-ground lightning and surface rainfall in warm-season Florida thunderstorms. *J. Geophys. Res.*, 111 (D19203), doi:10.1029/2005JD006802.

Hamid, E.Y., Kawasaki, Z., and Mardiana, R. (2001). Impact of the 1997–98 El Nino on lightning activity over Indonesia. *Geophys. Res. Lett.*, 28, 147–150.

Intergovernmental Panel on Climate Change (IPCC). (2013). *Climate change 2013: The physical science basis*. World Meteorological Organization (WMO) and UN Environment Programme (UNEP).

Johnson, D.B. (1981). The role of giant and ultragiant aerosol particles in warm rain initiation. *J. Atmos. Sci.*, 39, 448–460.

Kane, R.J. (1991). Correlating lightning to severe local storms in the Northern United States. *Wea. Forecasting*, 6(1), 3–12.

Kohn, M., Galanti, E., Price, C., Lagouvardos, K. and Kotroni, V. (2010). Now-casting thunderstorms in the Mediterranean region using lightning data. *Atmos. Res.*, 100, 489–502.

Kuhlman, K.M., MacGorman, D.R., Biggerstaff, M.I. and Krehbiel, P.R. (2009). Lightning initiation in the anvils of two supercell storms. *Geophys. Res. Lett.*, 36, L07802, doi:10.1029/2008GL036650.

Lagouvardos, K., Kotroni, V., Betz, H.D. and Schmidt, K. (2009). A comparison of lightning data provided by ZEUS and LINET networks over Western Europe. *Nat. Hazards Earth Syst. Sci.*, 9, 1713–1717.

Latham, D.J. and Schleiter, J.A. (1989). *Ignition probabilities of wildland fuels based on simulated lightning discharges*. USDA FS Report INT-411, Ogden, UT.

Liu, D., Feng, G. and Wu, S. (2009). The characteristics of cloud-to-ground lightning activity in hailstorms over northern China. *Atmos. Res.*, 91, 459–465.

Lohmann, U. and Feichter, J. (2005). Global indirect aerosol effects: A review. *Atmos. Chem. Phys.*, 5, 715–737.

Lyons, W.A., Nelson, T.E., Williams, E.R., Cramer, J.A. and Turner, T.R. (1998). Enhanced positive cloud-to-ground lightning in thunderstorms ingesting smoke from fires. *Science*, 282, 77–80.

MacGorman, D.R. and Burgess, D.W. (1994). Positive cloud-to-ground lightning in tornadic storms and hailstorms. *Mon. Wea. Rev.*, 122, 1671–1697.

Markson, R. (2007). The global circuit intensity: Its measurement and variation over the last 50 years. *Bull. Am. Meteorol. Soc.*, doi:10.1175/BAMS-88-2223, 223–241.

Markson, R. and Price, C. (1999). Ionospheric potential as a proxy index for global temperature. *Atmos. Res.*, 51, 309–314.

Molinari, J., Moore, P.K., Idone, V.P., Henderson, R.W. and Saljoughy, A.B. (1994). Cloud-to-ground lightning in Hurricane Andrew. *J. Geophys. Res.*, 99, 16665–16676.

Montanya, J., Soula, S., Pineda, N., van der Velde, O., Clapers, P., Sola, G., Bech, J. and Romero, D. (2009). Study of the total lightning activity in a hailstorm. *Atmos. Res.* 2009, 91, 430–437.

Pasko, V.P., Yair, Y. and Kuo, C.L. (2011). Lightning related transient luminous events at high altitude in the Earth's atmosphere: Phenomenology, mechanisms and effects. *Space Sci. Rev.* doi:10.1007/s11214-001-9813-9.

Pawar, S.D., Lal, D.M. and Murugavel, P. (2011). Lightning characteristics over central India during Indian summer monsoon. *Atmos. Res.*, 106, 44–49.

Perez, A.H., Wicker, L.J. and Orville, R.E. (1997). Characteristics of cloud-to-ground lightning associated with violent tornadoes. *Wea. Forecasting*, 12, 428–437.

Petersen, W.A. and Rutledge, S.A. (1998). On the relationship between cloud-to-ground lightning and convective rainfall. *J. Geophys. Res.*, 103(D12), 14,025–14,040, doi:10.1029/97JD02064.

Piepgrass, M.V., Krider, E.P. and Moore, C.B. (1982). Lightning and surface rainfall during Florida thunderstorms. *J. Geophys. Res.*, 87, 11,193–11,201.

Price, C. (1993). Global surface temperatures and the atmospheric electrical circuit. *Geophys. Res. Lett.*, 20, 1363–1366.

Price, C. (2006). Global thunderstorm activity. In M. Fullekrug et al. (Eds.) *Sprites, Elves and Intense Lightning Discharges*. Springer, Amsterdam, the Netherlands, 85–99.

Price, C. (2008). Lightning sensors for observing, tracking and nowcasting severe weather. *Sensors*, 8, 157–170.

Price, C. (2009). Will a drier climate result in more lightning? *Atmos. Res.*, 91, 479–484.

Price, C. and Asfur, M. (2006). Can lightning observations be used as an indicator of upper-tropospheric water vapor variability? *Bull. Amer. Meteor. Soc.*, 87, 291–298.

Price, C., Asfur, M. and Yair, Y. (2009). Maximum hurricane intensity preceded by increase in lightning frequency. *Nat Geosci.*, 2, 329–332. doi:10.1038/NGEO477.

Price, C. and Federmesser, B. (2006). Lightning-rainfall relationships in Mediterranean winter thunderstorms. *Geophys. Res. Lett.*, 33, L07813, doi:10.1029/2005GL024794.

Price, C. and Rind, D. (1994a). Modeling global lightning distributions in a General Circulation Model. *Mon. Wea. Rev.*, 122, 1930–1939.

Price, C. and Rind, D. (1994b). Possible implications of global climate change on global lightning distributions and frequencies. *J. Geophys. Res.*, 99, 10823–10831.

Price, C., Yair, Y., Mugnai, A. et al. (2011). Using lightning data to better understand and predict flash floods in the Mediterranean. *Sur. Geophys.*, 32, 733–751.

Price, C.G. and Murphy, B.P. (2002). Lightning activity during the 1999 Superior derecho. *Geophys. Res. Lett.*, 29(23), 2142, doi:10.1029/2002GL015488.

Reeve, N., and Toumi, R. (1999). Lightning activity as an indicator of climate change. *Q. J. Royal Meteorol. Soc.*, 125, 893–903.

Rosenfeld, D., Lohmann, U., Raga, G.B. et al. (2008). Flood or drought: How do aerosols affect precipitation? *Science*, 321, 1309, doi:10.1126/science.1160606.

Rudlosky, S.D. and Fuelberg, H.E. (2011). Seasonal, regional, and storm-scale variability of cloud-to-ground lightning characteristics in Florida. *Mon. Wea. Rev.*, 139, 1826–1843.

Samsury, C.E. and Orville, R.E. (1994). Cloud-to-ground lightning in tropical cyclones: A study of Hurricanes Hugo (1989) and Jerry (1989). *Mon. Wea. Rev.*, 122, 1887–1896.

Saunders, C.P.R., Bax-Norman, H., Emersic, C., Avila, E.E. and Castellano, E. (2006). Laboratory studies of the effect of cloud conditions on graupel/crystal charge transfer in thunderstorm electrification. *Q. J. R. Met. Soc.*, 132, 2653–2673.

Saunders, C.P.R., Keith, W.D. and Mitzeva, R.P. (1991). The effect of liquid water on thunderstorm charging. *J. Geophys. Res.*, 96, 11007–11017.

Schultz, C.J., Petersen, W.A. and Carey, L.D. (2009). Preliminary development and evaluation of lightning jump algorithms for the real-time detection of severe weather. *J. Appl. Met. Clim.*, 48, 2543–2563.

Shindell, D.T., Faluvegi, G., Unger, N. et al. (2006). Simulations of preindustrial, present-day, and 2100 conditions in the NASA GISS composition and climate model G-PUCCINI. *Atmos. Chem. Phys.*, 6, 4427–4459.

Siingh, D., Singh, R.P., Singh, A.K., Kulkarni, M.N., Gautam, A.S. and Singh, A.K. (2011). Solar activity, lightning and climate. *Surv. Geophys.*, 32, 659–703, doi:10.1007/s10712 -011-9127-1.

Steiger, S.M., Orville, R.E. and Huffines, G. (2002). Cloud-to-ground lightning characteristics over Houston, Texas: 1989–2000. *J. Geophys. Res.*, 107, 4117, doi:10.1029/2001JD001142.

Stocks, B.J., Mason, J.A., Todd, J.B. et al. (2002). Large forest fires in Canada, 1959–1997. *J. Geophys. Res.*, 107, 8149, doi:10.1029/2001JD000484.

Stuhlman, R. et al. (2005). Plans for EUMETSAT's third generation Meteosat geostationary satellite programme. *Adv. Space Res.*, 36, 975–981.

Takahashi, T. (1978). Riming electrification as a charge generation mechanism in thunderstorms, *J. Atmos. Sci.* 1978, 55, 1536–1548.

Takahashi, T. (2006). Precipitation mechanisms in East Asian monsoon: Videosonde study. *J. Geophys. Res.*, 111, D09202, doi:10.1029/2005JD006268.

Weber, M.E., Williams, E.R., Wolfson, M.M. and Goodman, M.M. (1998). *An assessment of the operational utility of a GOES lightning mapping sensor.* Project Report NOAA-18, MIT Lincoln Laboratory, Lexington, MA.

Wild, M. (2012). Enlightening global dimming and brightening. *Bull. Amer. Met. Soc.*, 93, 27–37.

Williams, E.R. (1992). The Schumann resonance: A global tropical thermometer. *Science*, 256, 1184–1187.

Williams, E.R. (2001). The electrification of severe storms. In Dowswell III, C.A. (Ed.) *Severe Convective Storms*, American Meteorological Society, AMS Monographs, Boston, 527– 561.

Williams, E.R. (2005). Lightning and climate: A review. *Atmos. Res.*, 76, 272–287.

Williams, E.R. (2009). The global electric circuit: A review. *Atmos. Rev.*, 91, 140–152.

Williams, E.R., Boldi, B., Marti, A. et al. (1999). The behavior of total lightning activity in severe Florida thunderstorms. *Atmos. Res.*, 51, 245–265.

Williams, E.R., Rutledge, S.A., Geotis, S.C. et al. (1992). A radar and electrical study of tropical hot tower. *J. Atmos. Sci.*, 49, 1386–1395.

Williams, E.R., Weber, M.E. and Orville, R.E. (1989). The relationship between lightning type and convective state of thunderstorms. *J. Geophys. Res.*, 94, 13,213–13,220.

Williams, E.R., Zhang, R. and Rydock, J. (1991). Mixed-phase microphysics and cloud electrification. *J. Atmos. Sci.*, 48, 2195–2203.

Wilson, C.T.R. (1921). Investigations on lightning discharges and on the electric field of thunderstorms. *Philosophical Transactions of the Royal Society of London. Series A*, 221, 73–115.

Wilson, C.T.R. (1924). The electric field of a thundercloud and some of its effects. *Proc. Phys. Soc. London*, 37, 32D–37D.

Yoshida, S., Morimoto, T., Ushio, T. and Kawasaki, Z. (2007). ENSO and convective activities in Southeast Asia and western Pacific. *Geophys. Res. Lett.*, 34, L21806, doi:10.1029/2007GL030758.

13 Natural Hazards in Poland

*Joanna Pociask-Karteczka, Zbigniew W. Kundzewicz,
Robert Twardosz and Agnieszka Rajwa-Kuligiewicz*

CONTENTS

13.1 INTRODUCTION

Poland is located in the mid-central part of Europe. Most of the country area belongs to the great postglacial plains of Europe, and a smaller southern part includes mountains (Carpathians and Sudetes) and uplands. The drainage basins of two large, international, rivers – the Vistula and the Oder – cover most of the country. These rivers flow northward from the mountains and uplands in the south, through the lowlands, and empty into the Baltic Sea. There are also several smaller coastal rivers within northern Poland, which also flow to the Baltic.

The weather in Poland is typically influenced by polar, maritime or continental air masses and for this reason the country's climate is known as transitional – between oceanic and continental. Violent weather phenomena occur mostly during the warm half of the year, when humid oceanic air meets dry and warm air from Eastern Europe or hot air from the south. The occurrence of extreme cold or hot conditions is driven by the air advection from outside of the temperate latitudes, that is, Tropical or Arctic air. The range of mean annual temperatures extends from 6°C in the northeast to 10°C in

317

the southwest part of the country. The altitudinal zonation in the mountainous regions in southern and south-western parts occurs, reaching the annual mean of −4°C at the highest elevations. The Baltic Sea coast has been influenced by west winds causing cooler summers and warmer winters. The average annual precipitation is 600 mm, while the most elevated mountain locations receive as much as 1700 mm per year. Approximately half of the precipitation in the country falls as snow. Precipitation in summer prevails providing a supply of water for vegetation (Woś, 1999).

Poland is located in a nonseismic area. It is generally free from significant earthquake risk and volcanic activity. Most natural hazards occurring in Poland have a meteorological nature or origin. In this chapter, an overview is given of the types of natural hazards in Poland, followed by a more detailed examination of floods, droughts, rapid air pressure changes, and extreme thermal conditions, landslides and radiological hazards of radon as specific classes of hazards.

13.2   TYPES OF NATURAL HAZARDS IN POLAND

Meteorological hazards (rapid air pressure changes, extreme thermal conditions, droughts), hydrological hazards (floods) and geological hazards (landslides, radiological hazards of radon) have the most negative effect on people and the environment in Poland. Most of these natural hazards are strongly related to atmospheric conditions, while some of them are dependent on the geology.

Some of the greatest meteorological threats in Poland include strong winds, violent maritime storms, extreme temperatures, exceptionally heavy precipitation or long-lasting lack of precipitation. Prolonged precipitation events cause dangerous floods and trigger landslides. Low precipitation or long dry spells cause droughts that can result in shortages of potable water. Dangerous atmospheric phenomena include thunderstorms and hailstorms, and even quasi-tornados. Freezing rain is a rare, but particularly painful phenomenon, especially for the transport sector. Many of such phenomena have caused natural disasters. At the turn of the millennium, there has clearly been an increase in the frequency and intensity of exceptionally hot months (Twardosz, Kossowska-Cezak, 2013; Kossowska-Cezak, Twardosz, 2017) and heat waves (Twardosz, Batko, 2012) and part of the change could be attributed to global warming. The increase in the number of extremely warm days in a year and the decrease in the number of extremely cold days have been observed (Graczyk, Kundzewicz, 2014). Reviews of historical sources reaching as far back as the Middle Ages demonstrate that Poland experienced natural hazards no less serious than today, as confirmed by these quotations: 'That year [1473] was memorable throughout Europe and the Kingdom of Poland with exceptional heat waves and persistent drought …, so much so that all sources ran dry and even the grandest rivers in Poland could be forded', or, mentioning the year of 1590: 'Rain did not fall for 38 weeks. Rivers ran dry' – as Jan Długosz wrote in his *Annales seu cronicae incliti Regni Poloniae* (Girguś, Strupczewski, 1965).

A drought is the result of deficiencies in precipitation during several rainless weeks. The rainless periods in Poland are connected with the persistence of a stationary east European high that joins with the Azores anticyclone via central Europe. In such conditions, with the prolonged lack or insufficiency of atmospheric precipitation (meteorological drought), a comprehensive drought begins to develop gradually, possibly amplified by high temperature and evapotranspiration (Łabędzki and Bąk, 2014). A soil drought (agricultural drought) appears, followed by hydrologic drought, during which the groundwater level decreases, resulting in the reduction of water flow into rivers. During such periods, drying of some springs and upper parts of river courses are observed (Somorowska, 2016).

FIGURE 13.1   Catastrophic floods in Poland in 1946–2010 and spatial distribution of potential natural hazards in Poland. (After Kozak, K., Kozłowska, B., Mamont-Cieśla, K., Mazur, J., Mnich, S., Olszewski, J., Przylibski, T.A. et al. Correction factors for determining of annual average radon concentration in dwelling of Poland resulting from seasonal variability of indoor radon. *Applied Radiation and Isotopes* 2011, 69(10), 1459–1465; Kundzewicz, Z.W. Ed. Changes in Flood Risk in Europe, Special Publication 2012, 10, IAHS Press, Wallingford, Oxfordshire, UK; Stachý, J. Ed. *Atlas hydrologiczny Polski*. Główny Geodeta Kraju, Warszawa 1986–1987.; changed and completed.) 1 – Sea storm floods (January – February), 2 – early snowmelt floods (February – March), 3 – spring snowmelt floods (March – April), 4 – rainfall floods (May – August), 5 – precipitation deficiency, 6 – landslides, 7 – radon radiation.

Floods are hydrological hazards that pose a major risk to people and cause significant economic impacts in Poland (Figure 13.1). Floods can be caused by several generating mechanisms, whose probability of occurrence depends on the season and location (Kundzewicz et al., 2012). Floods include such categories as river (fluvial) floods, flash floods, urban floods, rain-induced and snowmelt floods, ice jam floods, coastal floods, and these various classes of floods are generated by different mechanisms that either lead to increase of inflow to a river reach (as a consequence of intense and/or long-lasting rainfalls, snowmelt) or to a decrease of outflow from a river reach (usually flow obstructions resulting from ice jams or damming a stream by a landslide). Rain-induced floods have been the principal type of floods in Poland, responsible for 60% of all floods in a time interval of 60 years, and being prevalent as the cause of material damage (Ostrowski and Dobrowolski, 2000).

Relatively high differences in climatic conditions influence up to five runoff regime patterns in Poland (Dynowska and Pociask-Karteczka 1999; Wrzesiński 2016, 2017):

1. *Nival regime poorly developed*: Mean discharge in spring months does not exceed 130% of the mean annual discharge, and river runoff variability is low. This pattern concerns primarily the rivers of the eastern part of the coastal region, and the lakeland in the northern part of Poland (Figure 13.2a).
2. *Nival regime moderately well developed*: Represented by rivers with the mean discharge in spring months varying from 130–180% of the mean annual discharge, located mostly in the western part of the coastal region and in the central part of the country.
3. *Nival regime very well developed*: Represented by rivers with the mean discharge in spring months higher than 180% of the mean annual discharge and high variability in monthly runoff in the annual cycle; this pattern occurs in the central lowland part of Poland (Figure 13.2b).
4. *Nival-pluvial regime*: Characteristic of rivers with the mean discharge in spring months usually amounting to 130–180% of the mean annual discharge representing a marked increase in discharge in the summer months, amounting to at least 100% of the mean annual discharge; this pattern is represented by the Sudeten and the majority of Carpathian rivers as well as the transit Vistula River in its upper course (Figure 13.2c).
5. *Pluvio-nival regime*: Embraces rivers with the mean discharge in summer months higher or almost equal to the mean discharge in spring, in both cases usually amounting to 130–180% of the mean annual discharge and high runoff variability; this pattern is represented by rivers in the Sudetes and Carpathians.

According to the *Hydrological Atlas of Poland* (Stachý, 1986–1987), spring snowmelt floods from February to April are typical for the most of the country. Rainfall floods from May to August occur in the mountainous regions. Sea storm floods in January and February happen in the northern part of Poland (Figure 13.3).

Rain-induced floods can be subdivided into those generated by convective rains with high intensity, occurring locally over small areas, typically in the summer; and those generated by long-lasting frontal precipitation with lower intensity, that can

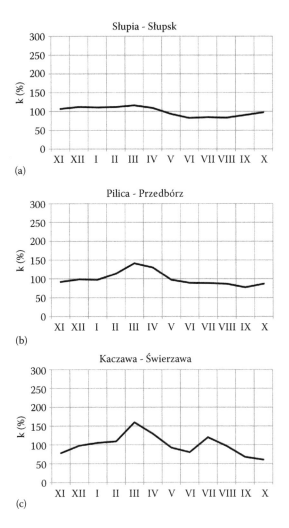

FIGURE 13.2 Examples of hydrological regime of Polish rivers: (a) Słupia (Słupsk) and (b) Pilica (Przedbórz), (c) Kaczawa (Świerzawa). k – Monthly discharge coefficient calculated as the ratio of the multiannual monthly discharge and multiannual yearly discharge. (After Wrzesiński D. Reżimy rzek Polski. In: P. Jokiel, W. Marszelewski, J. Pociask-Karteczka, Eds., *Hydrologia Polski*. Wyd. Nauk. PWN, Warszawa 2017, 215–221, modified by the authors.)

cover large areas. Local but intense flash floods associated with torrential rain of short duration, sometimes called a 'cloudburst', are generated by thunderstorm cells in an air mass or frontal convection. The frequency of such extreme events has likely increased since 1995 (Cebulak and Niedźwiedź, 2000).

Dramatic summer floods in southern Poland have been caused by prolonged orographic precipitation, duration of a few days, connected with cyclones following the Vb trajectory of van Bebber which causes an advection of tropical air into the higher latitudes (van Bebber, 1891; Kotarba, Pech 2002; Degirmendžić et al., 2014;

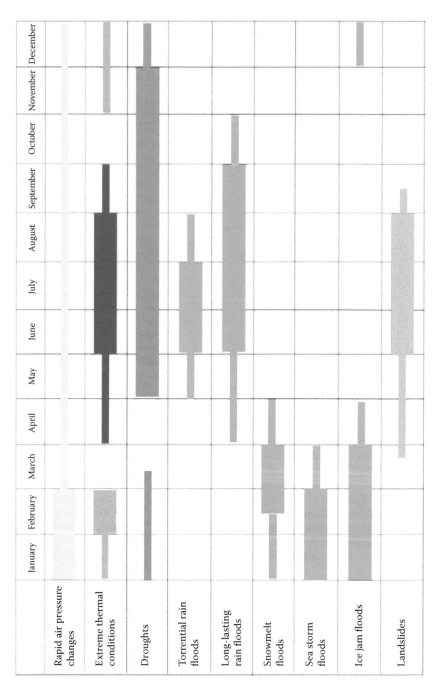

FIGURE 13.3   Typical time of occurrence of natural hazards in Poland.

Niedźwiedź and Łupikasza, 2016). When the cyclone is blocked by the Russian high, humid air masses are orographically lifted over the northern slopes of the Carpathians and the Sudetes Mountains and produce prolonged intense rainfalls (Figure 13.4). Typical cyclonic circulation types with an air advection from the northern sector, cyclonic centre, or cyclonic trough have been responsible for the most severe precipitation-induced flood events in mountainous regions of southern Poland (Niedźwiedź, 2003a, b).

Snowmelt-induced floods result from rapid melting of the snow pack, sometimes amplified by rainfall. Ice jam floods, related to the river freeze-up or break-up and thaw processes, occur in winter and spring. At the Polish coast, strong storm winds blowing landward impound the Baltic Sea waters thus obstructing the flow of river water and ice, causing sea storm floods at river mouths which propagate upstream. This may coincide with snowmelt, rainfall and ice jam flooding in the river basin, augmenting the effect of the storm surge (Bartnik and Jokiel 2012).

Landslides belong to *geological hazards*. They are a particularly complex natural and human-influenced hazard, resulting from an interaction of both internal factors such as topography, land cover, hydrology and slope-forming soil and rock materials, and external factors such as precipitation. Landslides are common morphogenetic processes in the contemporary slope system and valley development, and they are a major factor of landscape evolution in mountainous and upland

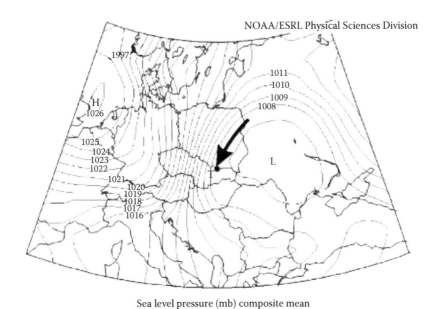

FIGURE 13.4 Sea level pressure distribution on days with the most severe flood in Poland on 8 July 1997 (NEc circulation type). (After Niedźwiedź, T., Łupikasza, E., Pińskwar, I., Kundzewicz, Z., Stoffel, M., Małarzewski, Ł., Variability of high rainfalls and related synoptic situations causing heavy floods at the northern foothills of the Tatra Mountains. *Theoretical and Applied Climatology* 2015, Springer, 119, 273–284.)

regions in Poland, especially in the Polish Flysch Carpathian (Oszczypko, 2004) and Sudetes Mountains (Figure 13.1). Landslides in the Carpathian Mountains favour the complex of shales, claystones, sandstones and the shale-derived waste mantle. Landslides are triggered by heavy or prolonged rainfall in spring and summer and snowmelt in early spring (Gorczyca, 2004; Figure 13.3). The regolith blanket in the form of coarse slope deposits is usually very thin, hardly exceeding a few metres, thus solid bedrock occurs very close to the surface in the Sudetes. The primary factor controlling the occurrence and effects of shallow landslides there appears to be a human impact – deforestation of mountain slopes that also accelerates water circulation (Gorczyca, 2004).

The potential of an extreme event to initiate landsliding processes depends on the balance between gravity (the driving force) and resistance (friction, the internal coherence of the slope materials). Factors that can affect this balance include angle of slope, the nature of the slope materials, the rock structure, the presence of built structures, triggers such as earthquakes or engineering works, tectonic forces of folding and faulting and, of particular significance in the Polish uplands, the availability of water to act as a lubricant. During extreme rainfall events, threshold values are exceeded for the occurrence of various processes including landslides. The potential of a single torrential rainfall to trigger a landslide is much greater when preceded by a long-duration rainfall than that of an isolated rainstorm of even a particularly great intensity. Other critical factors mostly involve the slope system, such as lithology, relative elevation and slope gradients (Gorczyca, 2008b).

Expansion of housing development into slopes increases the number of people who are threatened by landslides, especially in the southern part of Poland. On the other hand, landslides can also be triggered by such human activities as overloading of slopes by buildings and other objects and constructions, as well as land use changes, for example, deforestation. Although Poland is generally considered seismically inactive, one should not underestimate even earthquakes, which, although weak, also contribute to the activation of these processes (Guterch and Lewandowska-Marciniak, 2002). Epicentres of seismic events are distributed mainly in southern Poland along Sudetes and Carpathians, and in central and northwestern Poland, generally along the Teisseyre-Tornquist Zone. The best seismically recognized is the Orawa-Nowy Targ Basin in the Carpathians. A series of seismic events occurred in November 2004, with the main earthquake of magnitude M = 4.4 and epicentral intensity Io = 7 in the EMS-98 scale. Similar events were recorded there in years 1935 and 1717. More recently, earthquakes with epicentres in the Kaliningrad area (Russian Federation) were recorded (e.g., on 21 September 2004) that were felt by the population of Northeast Poland (Guterch 2015).

Radiological hazard of radon is another geological hazard. The Sudetes and Fore-Sudetic block (southwest Poland) are prone to the radiological hazard of radon, which is favoured by the occurrence of high radium concentrations in rocks, which associate with uranium and thorium. Radon migrates easily to the atmosphere from rocks and soils, where it is formed as a result of natural alpha transformation of Ra-226 in the uranium-radium series. Ra-222 is the most stable isotope of radon and it has a half-life of 3.8235 days. It is highly soluble in water. It is a colourless, odourless, tasteless, radioactive noble gas and is considered – particularly in Sudetes (especially its presence in the ground and groundwater) – as a threat to health due to the radioactivity of short-lived progenies of Rn-222 decay: Po-218, Pb-214, Bi-214 and Po-210. Radon is a very easily inhaled gas and causes the risk of developing lung cancer and lung tumours. Especially dangerous exposure to radon and its progeny particularly occurs in caves, mines and adits where touristic underground facilities occur (Fijałkowski-Lichwa and Przylibski, 2011a; Przylibski, 2011).

13.3 HYDROLOGICAL HAZARD

13.3.1 FLOODS

Floods have unleashed their destructive power a number of times in the history of the country (Girguś and Strupczewski, 1965; Kundzewicz, ed. 2012). According to Bielecka and Ciołkosz (2000) about 7% of Polish territory is threatened by floods every year. The vast majority of this area is concentrated along large river valleys. The first written mentions of floods on Polish territory come from different historical sources such as archives, annals and chronicles. These merely provide descriptive information. Though historical records are essential in terms of flood frequency analysis, they are limited to high-impact events and often suffer from the lack of homogeneity, unverified credibility and low accuracy (Nachlik and Kundzewicz, 2016).

One of the earliest documented examples was the flood of the Nysa Kłodzka in the Oder basin of Lower Silesia, in southwestern Poland, in July 1310. The flood covered the suburbs of Kłodzko town, killing more than 1500 people. Another memorable flood, in April 1813, affected a great part of central Europe including Poland (then under the administration of Russia, Prussia and Austria). The flood inundated Kraków with neighbouring villages and forced the river to change its course in many places.

Over the twentieth century, several serious floods have been recorded in the Oder River in 1902, 1903, 1977, 1985 and 1997 (Kundzewicz et al. 1999; Dubicki et al., 2005). After the flood of 1903, a flood control system was constructed in Wrocław (then Breslau), which was intended to protect the city against the maximum discharge of 2400 m^3/s. Unfortunately, the peak discharge during the flood of 1997 was about 30% higher than this threshold, and the system proved to be insufficient. Thus far, the flood of 1997 is considered to be the worst flood ever in the Oder basin (Kundzewicz et al., 1999).

During the twentieth century, the Vistula basin has experienced several great flood events in 1903, 1924, 1927, 1934, 1938, 1947, 1960, 1962, 1970, 1972, 1979, 1980, 1982, 1983 and 1997 (Figure 13.5). The flood of July 1934 was generated by a coincidence of flood waves on the Upper Vistula and the Dunajec, resulting from intensive rainfall in the mountain valleys of the Dunajec basin (with the two-day total exceeding 300 mm). The observed peak discharge of the Vistula upstream of the mouth of the Dunajec amounted to 3100 m^3 s^{-1}, while the Dunajec carried an additional 4500 m^3 s^{-1}. The flood inundated the area of 1260 km^2 and caused 55 fatalities (Cyberski et al., 2006). This event has accelerated further works on the construction of reservoirs in Porąbka on the Soła and in Rożnów on the Dunajec, that aim to reduce future peak flood waves (Kundzewicz, 2014).

FIGURE 13.5   The Vistula River in Cracow – flood in July 1997 – threatened historic centre of Cracow. (UNESCO's World Heritage List, photo by J. Sokołowski.)

A vast majority of floods in Poland were rain-induced (Figure 13.1). According to Bogdanowicz and Stachý (1998), almost 85% of floods in Poland are associated with low pressure systems moving along the Van Bebber track from the Mediterranean Sea toward Eastern Europe. The greatest flood risk is associated with advective and frontal precipitation characteristic for the upper course of the Vistula and the Oder.

Snowmelt floods predominantly occur in the middle and lower parts of the Vistula in March and April. They affect relatively large areas. Sometimes a flood can be induced by a simultaneous occurrence of snowmelt combined with intense or long-lasting rainfall. This usually happens on large lowland rivers such as Narew, Bug, Warta and Noteć (Kundzewicz, 2014). One of the largest recorded snowmelt floods occurred in 1979 in the Narew and Bug basins. The concentration time of the flood wave lasted 21 days, while its falling stage was more than one month. The total volume of runoff during the flood amounted to 18 km^3. The maximum discharge recorded at Dębe barrage reached Q = 3450 m^3 s^{-1} (Bednarczyk et al., 2006).

Ice jam floods largely occur during spring snowmelt and sudden winter thaw on the rivers: Vistula and its tributary Narew; the lower Oder and tributaries of the Oder; the Warta and the Noteć. Between 1946 and 2001, six cases of ice jam-related floods were recorded, two of which were catastrophic: in 1947 and 1982 (Dobrowolski et al., 2004). The latter was caused by adverse hydro-meteorological conditions (Majewski, 2007).

Storm surges constitute a major risk at the Polish coast. They are generated by the strong winds and passage of low pressure systems that induce the deformation of the sea surface (Majewski et al., 1983; Wiśniewski et al., 2009; Wiśniewski, Wolski, 2011). Storm surges cause a rapid rise of the sea level and inundation of coastal areas, which occurs most frequently during autumn and winter. Over the last 50 years, several storm surges have hit the Polish coast, that is, in 1983, 1988, 1993, 1995 and 2001. One of the biggest storm surge events occurred in January 1993 in the Puck Bay, raising the sea level in 1–2 hours by almost 2 m. Water rise usually occurs either in the eastern or western part of the coast, but rarely proceeds along the entire shore (Ciupak, 2010). According to Wiśniewski et al. (2009), the amplitude of extreme sea level fluctuation is higher in the western part of the Baltic coast, between Świnoujście and Kołobrzeg.

Flash floods at the local scale usually occur in mountain and upland areas. They are caused by convective rainstorms of high intensity lasting from one to a few hours (Pociask-Karteczka et al., 2017). The flooding occurs when the precipitation totals largely exceed the soil water capacity in the catchment. In Poland, the occurrence of rainstorms that trigger flash floods is conditioned by two synoptic situations: (1) a trough of low pressure and undulating cold fronts over central Europe, or (2) the area of low pressure over central Europe along with a quasi-stationary or occluded front. Local flash flood events are usually generated by rainfall lasting less than 2 hours with a mean intensity in the range of 20–80 mm h^{-1}, and may occur from April to October, with high frequencies in May, June and July. The floods usually affect mountain and upland areas differently than lowland and basin regions, and this might be explained by heavy rainfall distribution and topographical conditions (Bryndal, 2015). Flash floods cause not only high material losses, but also contribute to geomorphological changes of the river channels and floodplain terraces (Bryndal et al., 2017). As reported by Bryndal (2015), the maximum unit discharge of flash floods in Poland ranges between 2.6 and 11.8 m^3 s^{-1} km^{-2} in uplands and 1.6 and 5.8 m^3 s^{-1} km^{-2} in mountain regions. The magnitude and origins of flash floods in Poland are very similar to those observed in central Europe, but less acute than in Mediterranean countries.

13.3.2 Projections of Heavy Precipitation and Floods

The results obtained from climate models indicate a general increase in frequency and intensity of heavy precipitation in many areas of the world (IPCC, 2007). Hirabayashi et al. (2008) noted that some regions of the world including Eastern Europe show an increase in both drought frequency and annual precipitation. It is projected that over the twenty-first century the changes in mean annual precipitation will differ significantly among particular regions of Europe. In northern Europe, sums of annual precipitation are projected to increase, while southern Europe is expected to receive lower sums of annual precipitation. As reported by Romanowicz et al. (2016), the annual precipitation sums in Poland are projected to increase by less than 10%, while the mean annual temperature to rise by about 1°C in 2021–2050 in comparison with the reference period (1971–2000).

Changes in the potential occurrence of extreme rainfall events are likely to become more intense and more frequent in Poland (Christensen and Christensen, 2003; Kundzewicz et al., 2006; Spinoni et al., 2016). The projected increase in frequency and intensity of heavy precipitation poses a great risk of flood events, which cause the highest damage and mortality in urbanized areas. In Europe as a whole, an increase in both short- and long-duration extreme precipitation is projected; however, its magnitude is less certain (Fowler et al., 2007). The increase in heavy winter precipitation is more likely to occur in central and northern Europe. Amplitude of heavy summer precipitation events is predicted to increase in northeastern Europe, while the decrease in heavy precipitation is expected to occur during the whole year in southern Europe (Beniston et al., 2007).

Rojas et al. (2012) clarified that the increase in extreme precipitation in northern Europe is expected to occur mainly in winter, spring and autumn. A substantial reduction in extreme precipitation in south and southeastern Europe is projected to occur during summer and spring, and to a lesser extent during autumn. Accordingly, extreme river flows are projected to increase during winter in northern Europe, except northeastern Germany and Poland, where the winter maximum discharges are projected to decrease. Rojas et al. (2012) suggest that Q100 is supposed to decrease in these regions. These results are consistent with Roudier et al. (2016), who projected that flood magnitudes are more likely to decrease in the future.

The projections made by Dankers and Feyen (2008) showed a general decrease in extreme flows in northeastern Europe and several rivers in central and southern Europe, where warmer winters and shorter snow season restrain the volume of snowmelt runoff. In contrast, in many parts of western Europe, the simulations suggest that occurrence of an event reaching or exceeding current 100-year flood may become twice as frequent in the future. The projected decrease in flood hazard in northeastern Poland and simultaneous increases in the western and southern parts of Poland are in agreement with results obtained by Alfieri et al. (2015) and Osuch et al. (2016a). Although Dankers and Feyen (2008) confirmed the higher risk of extreme floods in many rivers in western and eastern Europe, they claim that the general trend is rather negative because floods will tend to be less extreme owing to decreasing snow cover.

Given the above, it should be remembered that projections of extreme flows (both magnitude and frequency) are burdened with large uncertainties associated with the selection of particular climate models (global climate models [GCMs], regional climate models [RCMs]) and assumed scenarios of atmospheric CO_2 concentration, statistical downscaling and bias correction methods. Therefore, the results obtained from particular models and scenarios should be considered with caution (Kundzewicz et al., 2010; Seneviratne et al., 2012; Meresa and Romanowicz, 2016; Osuch et al., 2016b).

13.3.3 FLOOD MANAGEMENT AND FLOOD RISK MANAGEMENT

In Poland, all issues related to water management and flood protection belong to the mandate of the National Water Management Authority (*Krajowy Zarząd Gospodarki Wodnej* [KZGW]), which is supervised by the Minister of Environment. The National

Water Management Authority consists of seven Regional Water Management Boards (*Regionalny Zarząd Gospodarki Wodnej* [RZGW]) that administer all hydrotechnical structures and are responsible for their periodic inspection. Meteorological and hydrological forecasts and warnings are prepared by the Institute of Meteorology and Water Management (*Instytut Meteorologii i Gospodarki Wodnej* [IMGW]).

Field services of RZGW and smaller operational teams continuously cooperate and provide information about the current hydrological and meteorological situations and actual conditions of hydro-technical facilities to crisis management centres of particular voivodeships, and to KZGW. At the same time, representatives of RZGW, KZGW and IMGW take part in meetings organized to counter flood risks that may occur in the country. Joint efforts undertaken by governmental and scientific institutions in recent years have contributed to the establishment of the IT system of the Country's Protection against Extreme Hazards (*Internetowy System Osłony Kraju* [ISOK]), which places ever-increasing emphasis on improving the effectiveness of flood risk management.

Flood defence in Poland consists of embankments with a total length of about 8500 km, 28 multipurpose water reservoirs, 11 dry reservoirs, relief channels and polders. It is estimated that the total area subjected to flood hazard in Poland covers 2 million hectares whereof only half is protected by embankments. In addition, existing reservoirs in Poland play a minor role in flood protection, allowing for a storage of only 6% of the mean annual runoff. By comparison, in other European countries this percentage is much higher, up to 30% (Kledyński, 2011). In this context, flood defence in Poland is still insufficient and requires further improvement and proper planning.

History shows that flood risk is related not only to the current efficiency of flood defence systems but also to growing urbanization, river regulation practices and the state of legislation (or lack of suitable regulations on spatial planning). The flood of 1997 also revealed shortcomings in Polish legislation. The flood coincided with the period of intensive systemic (economic, social, political) transformations that started in 1989. The legal system during 1997 was not yet prepared for emergency situations. The division of responsibilities was unclear. For instance, only high-level authorities (antiflood committees) were entitled to announce flood alerts, which led to poor communication and long-time delays. As a result, local authorities were forced to take joint actions without waiting for instructions from higher level authorities (Kundzewicz et al., 2005). In response to recent flood incidents in Europe, and future projections of a growing flood risk, the Floods Directive (CEC, 2007) was adopted in the European Union (EU) and implemented by its Member Countries (including Poland). The Directive obliges the EU countries to undertake a preliminary flood risk assessment, and to prepare flood hazard and risk maps and flood risk management plans to reduce the future flood risk.

Over the last few decades, considerable investments have been made in both a hydro-meteorological monitoring network and flood control systems. These include, for example, the development of a network of weather radars; automatic transmission of observational data; modernisation of the warning system; strengthening of international cooperation on sustainable flood management and increasing the role of emergency response. Considerable scientific efforts have also been undertaken, aimed at addressing various aspects of flood hazard and flood risk analysis

(Kiczko et al., 2013), flood mapping (Romanowicz et al., 2010; Romanowicz et al., 2013), improving modelling capabilities (Piotrowski et al., 2006) and flow routing tools (Kochanek et al., 2015).

13.4 METEOROLOGICAL HAZARD

13.4.1 Droughts

In Poland, atmospheric droughts, associated with high temperatures and precipitation deficit, are most likely to begin in spring–summer months (Figure 13.3) and end in winter–spring months (November–February). The size and spatial extent of atmospheric droughts depend in a large degree on the meteorological conditions prevailing at that time. Hydrologic droughts, associated with low river flows, are a direct consequence of atmospheric droughts. They usually start 2–3 months after a significant cumulative deficit of precipitation was reached, but their end usually converges with the month in which the precipitation returned to normal or higher level. Hydrologic droughts frequently occur in summer and autumn, but can also arise in winter because of the temporary retention of water in snow cover. A prolonged deficit of precipitation and ground overdrying may lead to the lowering of groundwater levels and recession of surface waters, preventing further replenishment of aquifers. These types of droughts, however, are relatively rare and mostly occur at the end of winter or beginning of spring (Farat et al., 1998).

The formation of severe droughts in Poland is most frequently associated with the occurrence of persistent long-lasting dry anticyclones, in particular the Azores and East European highs (Bąk and Łabędzki, 2002). Statistically, droughts in Poland occur every 4–5 years. Detailed analysis of all drought events shows that the phenomenon occurs most often within the latitudinal belt in central Poland (Figure 13.2). This part of the country is especially prone to droughts because it suffers from a permanent deficit of precipitation. The average annual precipitation in this region rarely exceeds 500 mm (in 1945–2004), and most of it (about 279 mm) is recorded in the growing season (Łabędzki, 2007). Considerably higher drought hazards occur also in areas of poor groundwater aquifers such as mountains and foothills or in regions of impaired water balance. The spatial distribution of droughts in Poland is indeed a matter of great concern, because the vast majority of Polish lowlands, recognized as the bread basket of Poland, are susceptible to droughts (Figure 13.1).

Droughts were given much less attention in the Polish chronicles than floods, owing to the fact that droughts develop relatively slowly, and because their effects (e.g. famine) are invisible at a regional or country scale, their description is very poor. Based on different sources, between 1900 and 2010 Poland experienced several severe droughts in 1904, 1913, 1920–1921, 1930–1931, 1943, 1950–1954, 1959–1960, 1963–1964, 1969, 1972–1974, 1976, 1982–1984, 1988–1990, 1992–1993, 2002–2003 and 2006 (Kaznowska, 2006; Sromowska, 2006; Limanówka et al., 2015; Piniewski et al., 2016), as well as in 2012 and 2015. Among them, atmospheric droughts constituted the most prevalent type.

In 2015, a prolonged period of high temperatures combined with rainfall deficit resulted in a reduction of energy production from power plants that suffered from the lack of cooling water (Van Lanen et al., 2016). Agricultural yields of major crops dropped to 50% in comparison with the preceding years. Water stages in many Polish rivers reached extremely low levels since the eighteenth century. In 2015, the stage in the Vistula in Warsaw dropped to 42 cm, that is, less than 20% of the mean water level at the gauging station (Kowalski et al., 2016).

Nowadays, droughts have become a subject of lively debate and scientific research worldwide. Drought indexes such as the Standardized Precipitation Index (SPI), Standardized Precipitation–Evapotranspiration Index (SPEI) and Standardized Runoff Index (SRI) are currently the most widespread methods to analyse drought severity and frequency retrospectively and in future projections. The analysis of droughts over the last 60 years has revealed that a relatively large part of Poland shows a significant drying trend. The trend expands approximately from the south-west toward the centre of the country and is especially visible in the growing season (Spinowska, 2009).

The response of particular catchments to droughts is, however, inconsistent and depends on location. For instance, Tokarczyk and Szalińska (2015) noted that the Nysa Kłodzka basin (southwestern Poland) exhibits much higher risk to meteo-rological droughts than the Prosna basin (central Poland). On the other hand, the Prosna basin seems to reflect greater inertia of the hydrological cycle because the transition from hydrologically dry conditions to normal or wet ones takes a longer time.

Future climate projections indicated that N-year drought in the control period is likely to occur more frequently in the future (Lehner et al., 2006). This might be associated with the projected increase of temperature in Poland (especially in northern and eastern Poland), driven to a large extent by atmospheric circulation (Piotrowski and Jędruszkiewicz, 2013). A more recent study of Meresa et al. (2016) shows, however, that future projections of drought frequency are much more compli-cated because results obtained from different projections may vary depending on the type of drought index. According to this research, the SPI and SRI indicate a reduc-tion in frequency of droughts in Poland in the near future, while the SPEI shows an opposite tendency.

13.4.2 RAPID AIR PRESSURE CHANGES

The territory of Poland is far away from the main pressure centres of the Northern Hemisphere, that is, the Azores high and the Icelandic low. While days with excep-tionally high pressure, above 1050 hPa, or with exceptionally low pressure, below 965 hPa, are sporadic, the travel of low pressure systems, with their extensive atmo-spheric fronts, across Poland can cause violent air pressure fluctuations that can be dangerous from a biometeorological point of view. The greatest such fluctuations span up to 36 hPa over a very short period of time, for example, 35 hours in Warsaw (an increase from 987.2 to 1023.1 hPa) or even 25 hours (a decrease from 1032.6 to 996.7 hPa). The greatest short-term pressure changes on record include 17 hPa in 3 hours and 12 hPa in 1 hour. This scale of change is bound to be felt by par-ticularly sensitive (meteoropathic) people. Such fluctuations are typical especially in winter seasons and are accompanied by strong winds that often exceed 100 km/h.

The average wind speeds in Poland, however, are low and strong winds are rare, except for the Baltic coast during sea storms and the southern mountain ranges that experience *foehn* type winds (known in Poland as *halny*; Ustrnul and Czekierda, 2009).

13.4.3 EXTREME THERMAL CONDITIONS

Even if Poland's climate is labelled as moderate, its thermal conditions are, in fact, less stable than would be required by this definition (Twardosz, 2009). Indeed, the range of this fluctuation is greater than 85°C between the record low of −45°C (10 February 1929 at Rabka in the Carpathian Mountains valley) and the record high of 40.2°C (29 July 1921 at Prószków near Opole in southwestern Poland). The greatest fluctuations occur in winter. In February, for example, the minimum temperature can drop to −40°C and below (e.g. in 1929 and 1940) while the maximum temperature can reach approximately 20°C (1990).

From the point of view of human health, wide ranging overnight changes of temperature can be particularly adverse. In Poland they can exceed 20°C, such as on 15–16 January 1940 when the temperature in Warsaw dropped by 21.6°C from 0.6 to −21.0°C. Very wide temperature fluctuations are characteristic of mountain basins in the south of Poland. There are two causes of this pattern: thermal inversions because of cold mountain air flowing down the slopes and the occurrence of the *foehn* winds causing the temperature to rise.

Extreme thermal conditions give rise to adverse biometeorological and economic effects. Cold weather is a more serious threat, especially to life, than hot weather in Poland. While heat waves do occur in this part of Europe, their persistence is relatively low. Indeed, during the last two decades, there were three exceptionally hot summers (1992, 2002 and 2006) when the average temperature exceeded the long-term average by more than two standard deviations (anomaly of approximately 2.5°C). In August 1992, the anomaly exceeded three standard deviations (4.0°C) in many locations of southern Poland. This scale of the anomaly is attributed not just to the atmospheric circulation, but also to a great forest fire about 120 km west of Krakow caused by a long drought. At the time, Krakow recorded a record number, that is, 13 days with the maximum temperature above 35°C (Twardosz and Kossowska-Cezak, 2013).

While in Western Europe global warming has caused a clear increase in the frequency of mild winters (Twardosz et al., 2016), some winters in Poland are still very cold. An example is the winter of 2011/2012, with a particularly large number of days with very cold weather ($t_{min} < -10$°C) and minimum temperatures of −40°C in some places. However, it is the winter of 1928/1929 that stands out, with the coldest February on record in Poland (the average temperatures between −13 and −16°C) and severe losses caused in gardens, fruit plantations and orchards (Gumiński, 1931). Also, thick and persistent snow cover can cause problems, such as in the winter of 2005/2006, when the roof of an exhibition hall collapsed in Katowice during a pigeon fair killing 65 and injuring more than 170 people. The weight of a thick layer of snow and ice on the roof was the direct cause of this disaster (Twardosz et al., 2011).

13.5   GEOLOGICAL HAZARD

13.5.1   Landslides

Landslides play an important role in the transformation of the morphology of mountain terrain in Poland, that is, in the Polish parts of the Flysch Carpathians and the Sudetes which are particularly prone to development of landslides (Figure 13.1). The average density is 1 landslide per 1 square kilometre in the Carpathian areas covered by landslides and mass movements often occupy 30–40% of some municipalities' area in that part of Poland. Almost every year mass movements are activated there, resulting in damage or destruction of housing, economic and communication infrastructure, as well as agricultural and forest areas. There are some years when these processes are very intense. Material damages caused by landslides in 2000–2001 were estimated to exceed 170 million PLN (over US$40 million) in Małopolskie voivodeship.

Landslides occur commonly after or during torrential or long-duration rainfall. Much less landslides are generated by snowmelt. Meteorological events triggering landslides may include situations where a torrential rainstorm coincides with long-duration rainfall, or where a spring thaw is accompanied by intensive rainfall in early spring (Figure 13.3). A series of events occurring in a short span of time (days, months or years) is known as 'clusters of events' (Starkel, 2003). If the time interval between the events is short enough, the system may have insufficient time to recover the balance and fails to return to its prior status.

Study on landslides in the flysch Carpathian Mountains, carried out by Rączkowski and Mrozek (2002), showed that the first threshold value of a landslide is a monthly precipitation total above 600 mm (as happened in years 1902, 1934, 1960, 1970, 1980, 1997 and 2000). The annual precipitation totals in these years were also above the average, and the precipitation totals in July amounted to one-third of the annual sum. Precipitation exceeding 200 mm, of duration of a few to a dozen days, favours mass movements of the ground of medium moisture; and precipitation exceeding 300 m and of the same duration triggers structural, rock-debris landslides on slopes built of shale and sandstone flysch. Precipitation exceeding 500 mm (monthly sums and sums of long-period precipitation) during several tens of days can trigger deep structural landslides on slopes built of sandstone flysch (Rączkowski 2007). The threshold values are usually reached locally, and they rarely occur simultaneously on the slopes. Deep, debris-weathered material landslides on shale-sandstone and shale rocks occur after 20–45 days of rainfall amounting to 250–300 mm. On the slopes with sandstone prevailing in the bedrock, landslides are likely to occur when precipitation amounts to 400–500 mm during 20–40 days and the total precipitation exceeding 250 mm is reached in 5–6 days. However, large landslides can be triggered by a 2-hour rainfall of at least 70 mm, following several days of steady rain amounting to almost 160 mm. The latter value is regarded as the landslide-activating threshold (Gil, 1996).

Over the last two decades, a series of extreme meteorological events caused considerable economic losses and transformation of slopes in various parts of the Carpathians and the Sudetes (Czerwiński and Żurawek, 1999). However, deep-seated slides were reported before. The largest one is located on the steep valley side of the Nysa Kłodzka River and dates to 24 August 1598. Its total length exceeded 350 m and the height of the slope affected was 170 m. The landslide is located at the structural boundary between Devonian shale and Carboniferous greywacke (Migoń et al., 2002).

In the period 1997–2010, a few rainfall events contributed to the activation of landsliding. After the disastrous rainfalls of July 1997, which caused flooding in the Vistula and Oder basins, intensified landslide activity was recorded in various parts of the Carpathians during subsequent years (1998, 2000, 2005, 2010). Geomorphological effects of extreme rainfall events were recorded in several areas including the Polish Tatra Mountains (debris flow), the Beskid Wyspowy Range and the Wielickie Foothills (Kotarba and Pech, 2002). During subsequent years, extreme events produced both new landslide forms and a further development of those that originated in 1997.

The rainfall in July 1997 was exceptionally severe; therefore, this event deserves special attention. On 9 July 1997, a torrential rainstorm ensued after several days of prolonged rain and landslide processes started in slope covers saturated with rainwater. Almost 20,000 landslides were activated. The subsequent event in April 2000 involved rapid snowmelt further accelerated by intense rain. As a result, the slope covers were highly saturated and landsliding was reactivated. In July 2005, the area in the Bieszczady Range experienced an extreme rainfall event that contributed to a large-scale transformation of the catchment. Closer investigation of this event by Gorczyca (2008a) showed that two factors influenced the potential of an extreme event to trigger landslide: complexity of the event (an isolated extreme rainfall is less likely to cause landslides than extreme rainfall preceded by long-duration rainfall); and the nature of the slope system (e.g. geologic structure, hypsometry, gradients).

After the intense and long-lasting rainfalls in 2010, many landslides were activated in the Carpathian Mountains (over 1300 cases in the Małopolskie voivodeship alone), causing material losses estimated to equal 2.9 billion euro. The horizontal displacements in the upper part of the landslides between June and July 2010 reached 85 m in some areas. In the same period, the vertical displacements varied from –20 m to +8 m in various parts of the landslides. In the autumn of 2010, the displacement velocities measured with SAR interferometry showed –0.76 mm per day to +0.57 mm per day (Borkowski et al., 2012). In the flysch Carpathian Mountains, large and deep rock and debris landslides tend to be transformed at a much slower rate than even very large shallow landslides developed just within the soil layer. Similar landslides – where compact and structurally unaffected packets have been displaced – are typified by a high degree of permanence in comparison with detritic landslides where the material has been mixed, crumbled and fragmented. Shallow debris slides prevail in the current climate and land use conditions. Trees play an important role in maintaining the salience of landslide features and forests tend to preserve the landslide relief by considerably restricting secondary movements. In contrast, the transformation rates of contemporary landslides on deforested slopes are incomparably greater. Moreover, almost half of the landslides occurring in 2010 involved anthropogenically created or modified landforms such as road escarpments

FIGURE 13.6   Cemetery destroyed by a landslide in Dobczyce, June 2012 (the Carpathian Foothill). (Photo by J. Sokołowski.)

or slopes on which many buildings were located, which indicates inappropriate land use in the area concerned (Figure 13.6).

The large scale of damage due to landslides should be a warning for decision makers. There is a need for recognition of landslide potential for proper regional planning at the level of the municipalities (smallest administrative units). The System of Protection against Landslide (SOPO) of the Polish Geological Institute is an important nationwide project whose primary purposes are identifying and mapping (at 1:10,000 scale) landslides and areas of different landslide potential over the whole Polish territory, based on standardized criteria and building up a national database, providing a tool for hazard and risk assessment and land use planning (http://geoportal.pgi.gov.pl/SOPO/). Mapping carried out by SOPO in 2008–2010 allowed the number of landslides in the Carpathian Mountains to be estimated as 50,000–60,000. They are a threat to the pipelines, high voltage electricity lines, road network and settlements. A small number of them (74 landslides) were mapped in detail and there is evidence that, apart from heavy precipitation, overloading of slopes by buildings or earthworks, or inappropriate construction on the slopes were the factors responsible for 52% of those investigated (Kaczmarczyk et al., 2012).

The results of the project are supposed to help in the management of landslide risk, and to reduce a large extent of damage caused by failure of road construction and housing within the active and periodically active landslides.

13.5.2 RADIOLOGICAL HAZARD OF RADON

Radon is a radioactive gas that has a long half-life and low chemical reactivity (Fijałkowska-Lichwa and Przylibski, 2011). It exists as a number of isotopes and, together with its radioactive decay products, is responsible for about 40–50% of the effective annual dose (1– 2.5 mSv year^{-1}) which a typical person receives from all natural and artificial sources of ionising radiation (Fijałkowska-Lichwa and Przylibski, 2011).

The Sudetes Mountains and Fore-Sudetic block are the Polish fragment of the European Variscan orogenic belt. The tectonic forces of orogeny gave rise to geological faults in the mountains and their northern foreland, leading to formation of horsts and basins. Rocks cut with numerous tectonic faults facilitate radon migration. Most of the rock material forming the region is Paleozoic, some part of the massif is built of the Archean rocks. Crystalline rocks predominate: metamorphic (mainly various kinds of gneiss and schist), and igneous (plutonic, such as granite and gabbro, or volcanic, such as basalt and porphyry). Crystalline rocks of the Sudetes are characterized by enriched quantities of the parent Ra-226 isotope. Granites and gneisses are especially U- and Ra-enriched (Przylibski et al., 2004, 2011a). For instance, the concentration of Ra-226 within gneisses (particularly orthogneisses) reaches up to 244 Bq kg^{-1}. Favourable conditions for radon exhalation occur especially in zones of intensive weathering, affecting fissure density and rock porosity, as well as zones disturbed by tectonic deformations. Numerous faults facilitate radon migration. Radon may be easily released and migrate even from very deep basement structures. Despite its short lifetime, radon can accumulate to concentrations that are far higher than normal, both underground (caves, tunnels) and on the ground surface, that is, in buildings, particularly in attics and basements. Dissolved radon may also be present in groundwater (springs, wells) and water supplies, from which it can outgas when the water is used in households (Przylibski et al., 2011b).

There are several tourist underground routes (i.e. caves, strategic-military spaces, adits) in the Sudetes Mountains, where guided tours are offered (Table 13.1). When visiting these locations, tourists are exposed to very high Rn-222 concentrations in

---

TABLE 13.1

Mean and Extreme Values of Mean Monthly Radon Ra-222 Concentration in the Air in Selected Underground Tourist Facilities in Poland

Name of Place	Rn-222 Concentrations in the Air [Bq m^{-3}]		
	Min.	Mean	Max.
Bear Cave	100	1260	4180
'Gold Mine' Museum	70	1880	18,500
Adit no. 9 in Kowary	340	580	690

Source: After Fijałkowska-Lichwa, L., Przylibski, T.A. Short-term ^{222}Rn activity concentration changes in underground spaces with limited air exchange with the atmosphere. *Nat. Hazards Earth Syst. Sci.* 2011, 11, 1179–1188.

the air, which can reach as much as about 23,000 Bq m^{-3}. Moreover, it is exceptionally dangerous for employees involved in underground work. Radon concentrations vary greatly with the season and atmospheric conditions. For instance, in the Jaskinia Niedzwiedzia (Bear Cave) nature reserve in Kletno, seasonal changes give high concentrations of Rn-222 from May to September and low values between November and March. In contrast, seasonal changes in Rn-222 activity at the Sztolnia Fluorytowa (Fluorite Adit), an abandoned uranium mine also at Kletno, is markedly different: high and variable values of radon activity concentration are observed from February to July, while low and stable values occur from August to October. During the rest of the year (late October to January), radon concentrations are also small but variable (Fijałkowska-Lichwa and Przylibski, 2011).

The risk of exposure to ionising radiation originating from radon and its decay products was examined in accordance to the Polish binding Act 'Atomic Law' (Ustawa Prawo atomowe, 2000) in three active underground tourist routes (Bear Cave and Old Mine of Uranium in Kletno, and Gold Mine in Zloty Stok). The effective doses received by the employees and visitors to the underground tourist routes were estimated in the years 2009–2011. Each of the tested sites posed exposure risks for the employees, and two of them significantly exceeded the permissible annual exposure limit of 20 mSv. Neither guides nor tourists are exposed to excessive concentrations of radon or its daughter products in the touristic routes (Rozporządzenie, 2005; Fijałkowska-Lichwa, 2012). However, the extension of the route to other workings will require the introduction of forced ventilation to lower the concentration (Przylibski 2001).

Because of the high concentration of Ra-226 in the rock one may note increased concentration of Rn-222 in groundwater. The highest values of Rn-222 activity (more than 100 Bq dm^{-3}) occur in groundwater flowing through strongly cracked, fissured reservoir rocks, where the radon concentrations are considered hazardous.

There is a wide spatial differentiation in the concentration of Rn-222 dissolved in groundwater in the Sudetes, which varies from 4900 to 1,517,000 Bq m^{-3} (Izera Massif – Świeradów Zdrój and Przylibski, 2011). Groundwater is classified into radon waters (100–999.9 Bq dm^{-3}), high-radon waters (1000–9999.9 Bq dm^{-3}) and extreme high-radon waters (>10,000 Bq dm^{-3}). A rise in Rn-222 activity concentration rises with the rise of emanation coefficient of reservoirs rocks (K) and the Ra-226 content (q) in these rocks, especially if $q > 30$ Bq kg^{-3} and $K > 0.1$. These three types of waters belong to shallow circulation systems, and are poorly mineralized. They occur in wells and springs, and are used in individual and centralized water supply systems. Such waters should not be used in households or at least they should be de-radoned before piping them to buildings. De-radoning of waters in an intake is relatively simple and does not require installing any complicated equipment, so the cost of it is also low (Przylibski, 2011).

The Rn-222 activity concentration level is monitored both in the air and in water, and it is a very important task regulated by law in many countries, as well as by the International Atomic Energy Agency (IAEA) and the International Commission on Radiation Protection (ICRP). The National Atomic Energy Agency (PAA), in cooperation with the Central Laboratory for Radiological Protection (CLOR), is responsible for the protection of the population and occupationally exposed persons against the hazards of ionizing radiation in Poland. The duties of PAA and CLOR include monitoring of radioactive contamination in food and environmental components; around-the-clock radiological emergency service assistance; monitoring of personal radiation doses; research on matters dealing with radiation; radiation protection; radiobiology and radioecology; professional training in radiation protection; and increasing the level of security of radioactive sources in Poland (http://www.clor.waw.pl).

Since Rn-222 concentration changes in time, diurnally and seasonally, it is very important to develop appropriate methods and procedures aimed at reducing the exposure of people to ionising radiation from radon. There should be correct usage of mechanical ventilation to protect unfavourable conditions in terms of radiation protection in underground locations, and the guidelines on management of radon in drinking-water supplies of the World Health Organization should be applied until Polish regulations with regard to radon concentration are introduced (Guidelines ..., 2011). Radon waters should not be used in households, or at least they should be re-radoned before piping them to buildings. Action for radiological protection of people living in the area of radon springs should be undertaken (Kozak et al., 2011).

13.6 CONCLUSIONS

Hydrometeorological extremes are the main cause of natural threats in Poland. They most prominently include intensive rainfall, which has been known to trigger floods, primarily in summer, and long spells of very low temperatures below the freezing point in winter.

The wealth of the nation has grown during the last 20 years, and except for the two first years after the systemic change in 1989, there has been steady annual economic growth, including the crisis year 2009. Also during the last 20 years, heat waves and accompanying droughts were observed to grow in intensity, and Graczyk et al. (2017) estimated that 1186 additional deaths in 10 large towns in Poland could be related to heat waves in the summer of 1994. The frequency of occurrence of torrential rainfalls has risen over recent decades as well, and consequently an increased number of local urban floods has been observed, also in large towns, with a corresponding significant increase in flood damage potential (Kundzewicz et al., 2017).

There have been ten individual years since the end of World War II, during which catastrophic river floods of regional extent have occurred in Poland, i.e. on average, one disastrous regional flood event every 6.5 years. In addition, there have been many more local floods, some of which were of disastrous severity, causing great human and economic loss.

The highest flood hazard can be attributed to the following situations of multiple-risk type (Kundzewicz, 2012):

- A flood wave on a tributary coincides with a flood wave on the main river. Especially dangerous locations are: the confluence of the River Nysa Kłodzka with the Oder, the confluence of the River Warta with the Oder, and the confluences of the Dunajec, the San and the Narew rivers with the Vistula.

- Intense rainfall during snowmelting.
- Intense rainfall in urban areas during passage of a flood wave on a river.

Most severe floods, in terms of flood fatalities and material damage, have occurred in the valleys of large rivers and their tributaries, and particularly in urban agglomerations and industrial areas protected by embankments. Since levees are designed based on probability theory, they do not guarantee a complete safety. When a very large flood comes, levees may fail to withstand the water mass and break. The degree of damage clearly depends on the duration of the flood wave.

Since the water resources of Poland are lower, per capita, than in most European countries, droughts also constitute an important hazard that leads to high economic damage, in particular in the agricultural sector.

Landslides occur randomly, irregularly, and only in some particularly prone mountainous areas due to specific geological settings. While extreme events normally occur very seldom, they tend to constitute a powerful factor in the development of the local relief. The frequency of landslide events is likely connected to recent climate changes, and intensified landsliding activity was recorded at the end of the twentieth and the beginning of the twenty-first centuries. Landslides triggered due to human activity occur more and more often in the Carpathian Mountains. Landslides often jeopardize people's safety and cause economic loss (e.g. in buildings, fields and roads). Compared to floods, the environmental and economic consequences of landslides are also substantial; however, the scale is not so widespread.

Projections for the future indicate that such extremes as heavy precipitation (triggering floods and landslides), drought, and heat waves are likely to become more frequent and more extreme in the future warmer climate.

ACKNOWLEDGEMENTS

Contribution on floods and droughts has been supported by the project PSPB No. 153/2010, Flood Risk on the Northern Foothills of the Tatra Mountains (FLORIST), supported by a grant from Switzerland through the Swiss Contribution to the enlarged European Union. The authors are very grateful to Dr. Elżbieta Gorczyca (Institute of Geography and Spatial Management, Jagiellonian University), Prof. Jerzy Wojciech Mietelski (The Henryk Niewodniczański Institute of Nuclear Physics, Polish Academy of Sciences) and Dr. Przemysław Wachniew (Faculty of Physics and Applied Computer Sciences, AGH University of Science and Technology, Krakow) for their comments on landslides and radiological hazards in Poland.

REFERENCES

Alfieri, L., Burek, P., Feyen, L., Forzieri, G. Global warming increases the frequency of river floods in Europe. *Hydrol Earth Syst. Sci.* 2015, 19, 2247–2260, DOI: 10.5194/hess-19-2247-2015.

Bąk, B., Łabędzki, L. Assessing drought severity with the relative precipitation index and the standardised precipitation index. *J. Water and Land Development* 2002, 6, 89–105.

Bartnik, A., Jokiel, P. *Geografia wezbrań i powodzi rzecznych.* Wyd. Uniwersytetu Łódzkiego, Łódź 2012.

Bednarczyk, S., Jarzębińska, T., Mackiewicz, S., Wołoszyn, E. Vademecum ochrony przeciwpowodziowej. Gdańsk 2006.

Bielecka, E., Ciołkosz, A. Flood susceptibility of the Odra Valley; its relation to land use changes. *Archiwum Fotogrametrii, Kartografii i Teledetekcji* 2000, 10, 26–1–26–8.

Bogdanowicz, E., Stachý, J. Maksymalne opady deszczu w Polsce. Charakterystyki projektowe. *Materiały Badawcze. Hydrologia i Oceanologia* 1998, 23, 85.

Borkowski, A., Perski, Z., Wojciechowski, T., Wójcik, A. LiDAR and SAR Data Application for Landslide Study in Carpathians Region (Southern Poland). *The XXII Congress of the International Society of Photogrammetry and Remote Sensing,* Melbourne, Australia, 25 August –1 September 2012.

Bryndal, T. Local flash floods in central Europe: A case study of Poland. *Norwegian J. Geography* 2015, 69(5), 288–298, DOI: 10.1080/00291951.2015.1072242.

Bryndal, T., Franczak, P., Kroczak, R., Cabaj, W., Kołodziej, A. The impact of extreme rainfall and flash floods on the flood risk management process and geomorphological changes in small Carpathian catchments: A case study of the Kasiniczanka river (Outer Carpathians, Poland). *Natural Hazards* 2017, DOI 10.1007/s11069-017-2858-7.

Cebulak, E., Niedźwiedź, T. Zagrożenie powodziowe dorzecza górnej Wisły przez wysokie opady atmosferyczne. *Monografie Komitetu Gospodarki Wodnej* PAN 2000, 17, 55–70.

CEC (Commission of European Communities) *Directive 2007/60/WE of the European Parliament and of the Council 23 October 2007on the Assessment and Management of Flood Risk.* http://eur-lex.europa.eu/LexUriServ/LexUriServ.do?uri=OJ:L:2007:288:00 27:01:en:htm, 2007.

Christensen, J.H., Christensen, O.B. Severe summertime flooding in Europe. *Nature* 2003, 421, 805.

Ciupak, M. Zagrożenia naturalne dla polskich miast portowych w świetle informacyjnego zabezpieczenia procesu zarządzania kryzysowego. *Roczniki Bezpieczeństwa Morskiego* 2010, 4, 157–199.

Cyberski, J., Grześ, M., Gutry-Korycka, M., Nachlik, E., Kundzewicz, Z. History of floods on the River Vistula. *Hydrological Sciences J.* 2006, 51, 5, 799–817.

Czerwiński, J., Żurawek, R. The geomorphological effects of heavy rainfalls and flooding in the Polish Sudetes in July 1997. *Studia Geomorphologica Carpatho-Balcanica* 1999, 33, 27–43.

Dankers, R., Feyen, L. Climate change impact on flood hazard in Europe: An assessment based on high resolution climate simulations. *J. Geophysical Research* 2008, doi:10.1029/2007JD009719.

Degirmendžić, J., Walisch, M., Szmidt, A., Pola opadów w Polsce związane z niżami Vb van Bebbera. *Acta Universitatis Lodziensis. Folia Geographica* 2014, 13, 3–15.

Dobrowolski, A., Czarnecka, H., Ostrowski, J., Zaniewska, M. Floods in Poland from 1946 to 2001 – Origin, territorial extent and frequency. *Polish Geological Institute Special Papers*, 2004, 15, 69–76.

Dubicki, A., Malinowska-Małek, J., Strońska, K. Flood hazards in the upper and middle Odra River basin – A short review over the last century. *Limnologica* 2005, 35, 123–131, DOI: 10.1016/j.limno.2005.05.002.

Dynowska, I., Pociask-Karteczka, J. Obieg wody, In: L. Starkel, Ed. *Geografia Polski. Środowisko przyrodnicze.* PWN, Waszawa 1999.

Farat, R., Kepińska-Kasprzak, M., Kowalczak, P., Mager, P. Droughts in Poland, 1951–1990. *Drought Network News (1994–2001), Paper 42,* University of Nebraska – Lincoln, 1998.

Fijałkowska-Lichwa L. Zagrożenie radonowe w wybranych podziemnych trasach turystycznych Sudetów, *Bezpieczeństwo Pracy i Ochrona Środowiska w Górnictwie* 2012, 8, 24–30.

Fijałkowska-Lichwa, L., Przylibski, T.A. Short-term [222]Rn activity concentration changes in underground spaces with limited air exchange with the atmosphere. *Nat. Hazards Earth Syst. Sci.* 2011, 11, 1179–1188.

Fowler, H.J., Ekstrom, M., Blenkinsop, S., Smith, A.P. Estimating change in extreme European precipitation using a multimodel ensemble. *J. Geophysical Research*, 2007, 112, D18104, DOI: 10.1029/2007JD008619.

Gil, E. Monitoring ruchów osuwiskowych. In *Zintegrowany monitoring środowiska przyrodniczego. Monitoring geoekosystemów górskich*. R., Soja, P., Prokop, Eds.; Biblioteka Monitoringu Środowiska, Szymbark, 1996, 120–130.

Girguś, R. Strupczewski, W. *Wyjątki ze źródeł historycznych o nadzwyczajnych zjawiskach hydrometeorologicznych na ziemiach polskich w wiekach od X do XVI* (Excerpts from historic sources on extraordinary hydrometeorological phenomena on Polish lands from the 10th to the 16th century) (in Polish). PIHM, Instrukcje i podręczniki, 87, WKiŁ, Warszawa, 1965.

Gorczyca, E. Przekształcanie stoków fliszowych przez procesy masowe podczas katastrofalnych opadów (dorzecze Łososiny). Jagiellonian University Press, Cracow, Poland, 2004.

Gorczyca, E. Rola płytkich ruchów osuwiskowych w kształtowaniu stoków fliszowych (na przykładzie Beskidu Wyspowego i Bieszczadów). *Przegląd Geograficzny* 2008a, 80, 1, 105–126.

Gorczyca, E. The geomorphological effectiveness of extreme meteorological phenomena on flysch slopes. *Landform Analysis* 2008b, 6, 15–27.

Graczyk, D., Kundzewicz, Z.W. Changes in thermal extremes in Poland, *Acta Geophysica* 2014, 62, 6, 1435.

Graczyk, D., Pińskwar, I., Choryński, A., Szwed, M., Kundzewicz, Z.W. Impacts of heat waves on health in large Polish towns. In: *Climate change and its impact on selected sectors in Poland*. Kundzewicz Z.W., Hov Ø., Okruszko T. (Eds.). Chapter 13, 2017b, 187–199.

Guidelines for Drinking-Water Quality, World Health Organization 2011, http://apps.who.int /iris/.

Gumiński, R. Zima 1928/29 w Polsce. *Przegląd Geograficzny* 1931, 11, 119–127.

Guterch B. Seismicity in Poland: Updated seismic catalog. In: *Studies of Historical Earthquakes in Southern Poland. Outer Western Carpathian Earthquake of December 3, 1786, and First Macroseismic Maps in 1858–1901*, Guterch B., Kozák J., Ed., GeoPlanet: Earth and Planetary Sciences 2015, Springer, 75–101.

Guterch, B., Lewandowska-Marciniak, H. Seismicity and seismic hazard in Poland. *Folia Quaternaria* 2002, 73, 85–99.

Hirabayashi, Y., Kanae, S., Emori, S., Oki, T., Kimoto, M. Global projections of changing risks of floods and droughts in a changing climate. *Hydrol. Sci. J.* 2008, 53(4), 754–773.

IPCC (Intergovernmental Panel on Climate Change), 2007: *Climate Change 2007: The Physical Science Basis – Contribution of Working Group I to the Fourth Assessment Report of the Intergovernmental Panel on Climate Change*. Solomon, S., D. Qin, M. Manning, Z. Chen, M. Marquis, K.B. Averyt, M. Tignor, H.L. Miller (Eds.). Cambridge University Press, Cambridge, United Kingdom and New York, NY.

Kaczmarczyk, P., Tchórzewska, S., Woźniak, H. Charakterystyka wybranych osuwisk z Polski południowej uaktywnionych po okresie intensywnych opadów w 2010 r. *Nowoczesne Budownictwo Inżynieryjne* 2012, 74–77.

Kaznowska, E. The characteristics of hydrological droughts based on the example of selected rivers in the northeastern part of Poland. *Infrastuktura i Ekologia Terenów Wiejskich* 2006, 4(2), 51–59.

Kiczko, A., Romanowicz, R.J., Osuch, M., Karamuz, E. Maximising the usefulness of flood risk assessment for the River Vistula in Warsaw. *Nat Hazards Earth Syst Sci* 2013, 13, 3443–3455.

Kledyński, Z. Ochrona przed powodzią i jej infrastuktura w Polsce. XXV Konferencja Naukowo-Techniczna, Międzyzdroje 2011.

Kochanek, K., Karamuz, E., Osuch, M. Distributed modelling of flow in the middle reach of the River Vistula. In: R.J. Romanowicz, M. Osuch M. (Eds.) *Stochastic flood forecasting system: The middle Vistula case study. GeoPlanet: Earth and Planetary Sciences* 2015, Springer, DOI: 10.5194/nhess-13-3443-2013.

Kossowska-Cezak, U., Twardosz, R. Anomalie termiczne w Europie (1951–2010). IGiGP UJ, Kraków 2017.

Kotarba, A., Pech, P. Geomorphic effect of the catastrophic summer flood of 1997 in the Polish Tatra Mountains. *Studia Geomorphologica Carpatho-Balcanica* 2002, 36, 69–76.

Kowalski, H., Magnuszewski, A., Romanowicz, R. Was the drought of 2015 on the River Vistula in Warsaw the lowest ever observed? *Geophysical Research Abstracts* 2016, 18, EGU2016-9716, EGU General Assembly.

Kozak, K., Kozłowska, B., Mamont-Cieśla, K., Mazur, J., Mnich, S., Olszewski, J., Przylibski, T.A. et al. Correction factors for determining of annual average radon concentration in dwelling of Poland resulting from seasonal variability of indoor radon. *Applied Radiation and Isotopes* 2011, 69(10), 1459–1465.

Kundzewicz, Z.W. Ed. Changes in Flood Risk in Europe, *Special Publication* 2012, 10, IAHS Press, Wallingford, Oxfordshire, UK.

Kundzewicz, Z.W. Lessons from river floods in central Europe, 1997–2010. In: S. Boulter, J. Palutikof, D. Karoly and D. Guitart (Eds) *Natural Disasters and Adaptation to Climate Change*, Cambridge University Press, 2013.

Kundzewicz, Z.W. Adapting flood preparedness tools to changing flood risk conditions: The situation in Poland. *Oceanologia* 2014, 56(2), 385–407, DOI: 10.5697/oc.56-2.385.

Kundzewicz, Z.W., Graczyk, D., Maurer, T., Pińskwar, I., Radziejewski, M., Svensson, C., Szwed, M. Trend detection in river flow series: 1. Annual maximum flow. *Hydrol. Sci. J.* 2005, 50(5), 797–810.

Kundzewicz, Z.W., Krysanova, V., Dankers, R., Hirabayashi, Y., Kanae, S., Hattermann, F.F., Huang, S. et al. Differences in flood hazard projections in Europe – Their causes and consequences for decision making. *Hydrological Sciences J.* 2017, 62, 1, 1–14, DOI: 10.1080/02626667.2016.1241398.

Kundzewicz, Z.W., Lugeri, N., Dankers, R., Hirabayashi, Y., Döll, P., Pińskwar, I., Dysarz, T. et al. Assessing river flood risk and adaptation in Europe – Review of projections for the future. *Mitig. Adapt. Strategies for Global Change* 2010, 15(7), 641–656, DOI: 10.1007/s11027-010-9213-6.

Kundzewicz, Z.W., Radziejewski, M., Pińskwar, I. Precipitation extremes in the changing climate of Europe. *Clim. Res.* 2006, 31, 51–58.

Kundzewicz, Z.W., Schellnhuber, H.-J. Floods in the IPCC TAR perspective. *Natural Hazards* 2004, 31, 111–128.

Kundzewicz, Z.W., Szamałek, K., Kowalczak, P. The Great Flood of 1997 in Poland. *Hydrol. Sci. J.* 1999, 44(6) 855–870.

Łabędzki, L. Estimation of local drought frequency in central Poland using the standardized precipitation index SPI. *Irrigation and drainage* 2007, 56, 67–77, DOI: 10.1002/ird.285.

Łabędzki, L., Bąk, B. Meteorological and agricultural drought indices used in drought monitoring in Poland: A review. *Meteorology, Hydrology and Water Management* 2014, 2, 2, 3–13.

Lehner, B. et al., Estimating the impact of global change on flood and drought risks in Europe: A continental, integrated analysis. *Climatic Change* 2006, 75(3), 273–299.

Limanowka, D., Cebulak, E., Pyrc, R., Doktor, R. Droughts in historical times in Polish territory. *Geophysical Research Abstracts* 2015, 17, EGU2015-14191.

Majewski, W. Flow in open channels under the influence of ice cover. *Acta Geophysica* 2007, 55(1), 11–22, DOI: 10.2478/s11600-006-0041-8.

Meresa, K., Romanowicz, R. The critical role of uncertainty in projections of hydrological Extremes. *Hydrol. Earth Syst. Sci. Discuss.* 2016, DOI: 10.5194/hess-2016-645.

Meresa, K.H., Osuch, M., Romanowicz, R. Hydro-meteorological drought projections into the 21-st century for selected Polish catchments. *Water* 2016, 8(5), 206, DOI: 10.3390/w8050206.

Migoń, P., Hrádek, M., Parzóch, K. Extreme events in the Sudetes Mountains. Their long-term geomorphic impact and possible controlling factors. *Studia Geomorphologica Carpatho-Balcanica* 2002, 36, 29–49.

Mrozek, T., Rączkowski, W., Limanówka, D. Recent landslides and triggering climatic conditions in Laskowa and Pleśna Regions, Polish Carpathians. *Studia Geomorphologica Carpatho-Balcanica* 2000, 34, 89–112.

Niedźwiedź, T. Extreme precipitation events on the northern side of the Tatra Mountains. *Geographia Polonica* 2003a, 76(2), 13–21.

Niedźwiedź, T. The extreme precipitation in central Europe and its synoptic background. *Papers on Global Change* IGBP 2003b, 10, 15–29.

Niedźwiedź, T., Łupikasza, E. Changes in circulation patterns. In: *Flood Risk in the Upper Vistula Basin*, Kundzewicz Z.W., Stoffel M., Niedźwiedź T., Wyżga B., Eds., *GeoPlanet: Earth and Planetary Sciences* 2016, Springer, 189–208.

Niedźwiedź, T., Łupikasza, E., Pińskwar, I., Kundzewicz, Z., Stoffel, M., Małarzewski, Ł. Variability of high rainfalls and related synoptic situations causing heavy floods at the northern foothills of the Tatra Mountains. *Theoretical and Applied Climatology* 2015, Springer, 119, 273–284.

Ostrowski, J., Dobrowolski, A. Eds. *Monografia katastrofalnych powodzi w Polsce w latach 1946–1998*. IMGW, Warszawa (CD-ROM), 2000.

Osuch, M., Lawrence, D., Meresa, H.K., Napiórkowski, J.J., Romanowicz, R.J. Projected changes in flood indices in selected catchments in Poland in the 21st century. *Stoch. Environ. Res. Risk Assess.* 2016a, DOI 10.1007/s00477-016-1296-5.

Osuch, M., Romanowicz, R.J., Lawrence, D., Wong, K. Trends in projections of standardized precipitation indices in a future climate in Poland. *Hydrol. Earth Syst. Sci.* 2016b, 20, 1947–1969, DOI: 10.5194/hess-20-1947-2016.

Oszczypko, N., The structural position and tectonosedimentary evolution of the Polish Outer Carpathians. *Przegląd Geologiczny* 2004, 52, 8/2, 780–791.

Piniewski, M., Szcześniak, M., Kundzewicz, Z.W., Mezghani, A., Hov, Ø. Changes in low and high flows in the Vistula and the Odra basins: Model projections in the European-scale context. *Hydrol. Process.* 2017, DOI: 10.1002/hyp.11176.

Piotrowski, A., Napiórkowski, J.J., Rowiński, P.M. Flash-flood forecasting by means of neural networks and nearest neighbour approach – A comparative study. *Nonlinear Processes Geophysics* 2006, 13, 443–448.

Pociask-Karteczka, J., Żychowski, J., Bryndal T. Zagrożenia związane z wodą – Powodzie błyskawiczne, *Gospodarka Wodna* 2017, 2, 37–42.

Przylibski, A.T. Shallow circulation groundwater – The main type of water containing hazardous radon concentration. *Nat. Hazards Earth Syst. Sci.* 2011, 11, 1695–1703.

Przylibski, A.T., Mamont-Cieśla, K., Kusyk, M., Dorda, J., Kozłowska, B. Radon concentrations in groundwaters of the Polish part of the Sudety Mountains (SW Poland). *J. Env. Radioactivity* 2004, 75(2), 139–209.

Przylibski, A.T., Stawarz, O., Wysocka, M., Zebrowski, A., Dorda, J., Grządziel, D., Kapała, J. et al. Mean annual Rn-222 concentration in homes located in different geological regions of Poland: First approach to whole country area. *J. Environmental Radioactivity* 2011a, 102(8), 735–741.

Przylibski, T.A. Radon and its daughter products behaviour in the air of an underground tourist route in the former arsenic and gold mine in Złoty Stok (Sudety Mountains, SW Poland). *J. Environmental Radioactivity* 2001, 57(2), 87–103.

Przylibski, T.A., Zagożdżon, K., Zagożdżon, P. Dependance of Rn-222 concentration changes in water of stream on the distance from the spring. *Geophysical Research Abstracts* 2011b, 13, 11774.

Rączkowski, W. Landslide hazard in the Polish Flysch Carpathians. *Studia Geomorphologica Carpatho-Balcanica* 2007, 61, 61–75.

Rączkowski, W., Mrozek, T. Activating of landsliding in the Polish Flysch Carpathians by the end of the 20th century. *Studia Geomorphologica Carpatho-Balcanica* 2002, 36, 91–111.

Romanowicz, R.J., Bogdanowicz, E., Debele, S.E., Doroszkiewicz, J., Hisdal, H., Lawrence, D., Meresa, H.K. et al. Climate change impact on hydrological extremes: Preliminary results from the Polish-Norwegian Project. *Acta Geophysica* 2016, 64(2), 477–509, DOI: 10.1515/acgeo-2016-0009.

Romanowicz, R.J., Kiczko, A., Karamuz, E., Osuch, M. Derivation of flood-risk maps for the Warsaw reach of the river Vistula. In: *Conference Proceedings 'Flood Estimation and Analysis in a variable and Changing Environment'* Volos 2012, 2013.

Romanowicz, R.J., Kiczko, A., Osuch, M. Determination of flood hazard maps, theory and practice, hydrology in engineering and water management. *Monografie Komitetu Inżynierii Środowiska Polskiej Akademii Nauk* 2010, 68, 337–346.

Roudier, P., Andersson, J.C.M., Donnelly, C., Feyen, L., Greuell, W., Ludwig, F. Projections of future floods and hydrological droughts in Europe under a +2°C global warming. *Climatic Change* 2016, 135, 341–355, DOI 10.1007/s10584-015-1570-4.

Rozporządzenie Rady Ministrów z dnia 18 stycznia 2005 r. w sprawie dawek granicznych promieniowania jonizującego (Dz. U. Nr 20, poz. 168).

Seneviratne, S.I. et al. Changes in climate extremes and their impacts on the natural physical environment. In: C.B. Field et al., Eds. *Managing the risks of extreme events and disasters to advance climate change adaptation*. A Special Report of Working Groups I and II of the Intergovernmental Panel on Climate Change (IPCC). Cambridge, UK, and New York, NY, Cambridge University Press 2012.

Somorowska, U. Wzrost zagrożenia suszą hydrologiczną w różnych regionach geograficznych Polski w XX wieku. *Prace i Studia Geograficzne* 2009, 43, 97–114.

Somorowska, U. Changes in drought conditions in Poland over the past 60 years evaluated by the standardized precipitation-evapotranspiration index. *Acta Geophysica* 2016, 64(6), 2530–2549, DOI: 10.1515/acgeo-2016-0110.

Spinoni, J., Naumann, G., Vogt, J., Dosio, A. High-resolution meteorological drought projections for Europe using a single combined indicator. *EGU General Assembly Conference Abstracts*, 2016.

Stachý, J. Ed. Atlas hydrologiczny Polski. Główny Geodeta Kraju, Warszawa 1986–1987.

Starkel, L. Extreme meteorological events and their role in environmental change, the economy and history. *Global Change* 2003, 10, 7–13.

Tokarczyk, T., Szalińska, W. Combined analysis of precipitation and water deficit for drought hazard assessment. *Hydrological Sciences J.* 2014, 59, 9, 1675–1689, DOI: 10.1080/02626667.2013.862335.

Twardosz, R., Fale niezwykłych upałów w Europie na początku XXI wieku. *Przegląd Geofizyczny* 2009, 54, 3–4, 193–204.

Twardosz, R., Batko, A. Heat waves in central Europe (1991–2006). *International J. Global Warming* 2012, 4, 3/4, 261–272.

Twardosz, R., Kossowska-Cezak, U. Exceptionally hot summers in Central and Eastern Europe (1951–2010). *Theoretical and Applied Climatology* 2013, 112, 617–628, DOI: 10.1007/s00704-012-0757-0.

Twardosz, R., Kossowska-Cezak, U., Pełech, S. Extremely cold winter months in Europe (1951–2010). *Acta Geophysica* 2016, 64, 6, 2609–2629, DOI: 10.1515/acgeo-2016-0083.

Twardosz, R., Niedźwiedź, T., Łupikasza, E. Zmienność i uwarunkowania cyrkulacyjne występowania postaci i typów opadów atmosferycznych na przykładzie Krakowa. Wyd. UJ, Kraków, 2011.

Ustawa Prawo atomowe z dnia 29 listopada 2000 r. (Dz.U. 2001 Nr 3 poz.18), http://isap.sejm
.gov.pl/.

Ustrnul, Z., Czekierda, D. Atlas of extreme meteorological phenomena and synoptic situations in Poland. Instytut Meteorologii i Gospodarki Wodnej, Warszawa 2009. (http://
geoportal.pgi.gov.pl/portal/page/portal/SOPO/) (accessed December 2012).

Van Bebber, W.J. Die Zugstrassen der barometrischer Minima. *Meteorologische Zeitschrift*
1891, 8, 361–366.

Van Lanen, H.A.J., Laaha, G., Kingston, D.G., Gauster, T., Ionita, M., Vidal, J.-P., Vlnas,
R. et al. Hydrology needed to manage droughts: The 2015 European case. *Hydrol.
Processes* 2016, 30, 3097–3104, DOI: 10.1002/hyp.10838.

Wiśniewski, B., Wolski, T. Physical aspects of extreme storm surges and falls on the Polish
coast. *Oceanologia*, 2011, 53, 373–390, DOI: 10.5697/oc.53-1-TI.373.

Wiśniewski, B., Wolski, T., Kowalewska-Kalkowska, H., Cyberski, J. Extreme water level
fluctuations along the Polish coast. *Geographia Polonica* 2009, 82(1), 99–107.

Woś, A. Klimat Polski. Wyd. Nauk. PWN, Warszawa 1999, 301.

Wrzesiński D. Use of entropy in the assessment of uncertainty of river runoff regime in
Poland. *Acta Geophysica* 2016, 64, 5, 1825–1839.

Wrzesiński, D. Reżimy rzek Polski. In: P. Jokiel, W. Marszelewski, J. Pociask-Karteczka,
Eds., *Hydrologia Polski*. Wyd. Nauk. PWN, Warszawa 2017, 215–221. http://geoportal
.pgi.gov.pl/SOPO/.

Index

Page numbers followed by f, t, b and n indicate figures, tables, boxes and notes, respectively.

T - #0103 - 111024 - C378 - 234/156/18 - PB - 9780367571924 - Gloss Lamination